British Expeditionary Warfare and the Defeat of Napoleon, 1793–1815

British Expeditionary Warfare and the Defeat of Napoleon, 1793–1815

Robert K. Sutcliffe

THE BOYDELL PRESS

First published 2016
The Boydell Press, Woodbridge

ISBN 978 1 84383 949 1

The Boydell Press is an imprint of Boydell & Brewer Ltd
PO Box 9, Woodbridge, Suffolk IP12 3DF, UK
and of Boydell & Brewer Inc.
668 Mt Hope Avenue, Rochester, NY 14620–2731, USA
website: www.boydellandbrewer.com

A catalogue record for this book is available from the British Library

Contents

Contents

Illustrations

Plates

Maps

Charts and Tables

Foreword

The workings of many parts of the British government 200 years ago are still unfamiliar to us today and this book has uncovered several mysteries. It is an original contribution to the study of the French Revolutionary and Napoleonic wars in several ways. Firstly, it looks not only at the navy, but right across the rapidly changing Whitehall government machine, as well as the City and the merchant shipping community. Secondly, it examines the continual process by which the British government solved the problem of transporting provisions and stores to warships on foreign stations, and troops, stores and provisions across the sea to confront the seemingly overwhelming land power of France and its conquered states.

The government was a customer in the shipping market and the Transport Office, headed by its Board, did not requisition merchant vessels, as happened in twentieth-century conflicts. Scholars have hitherto treated the Transport Board as part of the navy, but in fact it was a central procurement organisation, set up by William Pitt's administration in 1794. In the previous war, that of the American Revolution, the Navy Board, the army and the Board of Ordnance hired their own merchant ships, and frequently found themselves in competition when hiring merchant ships as government transports, much to the financial advantage of ship owners. After 1794 the three services were provided with transports by the Transport Board. Under the chairmanship of Sir Rupert George, who had the tricky job of taking orders from the three services, the Treasury and other government ministers, the Transport Board and department performed well, and the award of a baronetcy to George in the middle of a war was a strong indication of the success of his department.

However, in spite of the very large number of merchant ships registered in Britain, it was not easy to find suitable merchant ships at the right time. Setting the Transport Board's chartering rates, per ton per month, required constant and delicate decision-making. It was given the task of assembling thousands of tons of shipping for very large amphibious expeditions, often at short notice, which lay beyond the resources available. Every one of these operations against the near continent failed, none more so than the Walcheren expedition of 1809, perhaps the nearest parallel to the D-Day operation of June 1944. Late in sailing, badly led, Walcheren was a major disaster, and a sickly, decimated army struggled back to Britain. But where British transports really came into their own from 1808 was to the south of Europe in the long, year-round task of supplying Wellington's army in Portugal and Spain. Thousands of voyages, sailing in convoy from Britain to the Peninsula, ensured a supply train which the French armies could not match. Furthermore, the transports, and the warships which protected them, afforded tactical and military support for the British, Portuguese and Spanish armies which slowly forced the French northwards and over the Pyrenees.

Such was the length and difficulties of the 22-year war against Revolutionary and Napoleonic France that the British government had to improve its methods and administration and the Transport Office was one of the most efficient departments. In the end Britain survived, its economy and financial systems performing better and outlasting the French system of requisition and confiscation across Europe. But transports also provided the means by which Britain supplied specie, credit and munitions to her allies Russia, Prussia and Austria, countries which had the manpower to defeat the great armies which Napoleon put into the field. Dr Sutcliffe's book will ensure that the vital role of merchant ships hired by government is now better understood and appreciated.

Roger Knight
Institute of Historical Research
London University
May 2015

Preface

'All military campaigns begin with a conveyance by ship to the theatre of war.'[1]

Most military studies of British expeditionary warfare in the late eighteenth and early nineteenth centuries focus on the political context, military leadership, regiments, equipment, tactics and manoeuvres, the preparation and the battles themselves, generally without reference to the complex task of transporting the army to the foreign shores. Likewise most naval history books relating to the same period tend to focus on similar issues from the naval perspective including the sea battles, naval leadership, the political and administrative processes, ships and technology. There is a gap in the current literature that this book attempts to fill. The naval and maritime contribution to expeditionary warfare has not been fully appreciated, nor has the mode of conveying troops and sustaining the British army overseas been adequately addressed. The navy's role was essential: by achieving significant victories in famous fleet battles and less well known actions, together with the establishment of effective blockades, it achieved virtual, but not total, control of the oceans, where the main enemy was inclement weather. In addition it provided strong convoy protection for thousands of individual ship voyages, allowing trade to flourish and armies overseas to be sustained. This encouraged the Duke of Wellington to write in September 1813:

> If anyone wishes to know the history of this war, I will tell them that it is our maritime superiority that gives me the power to maintain my army while the enemy is unable to do so.[2]

Whereas the British launched more than fifty major seaborne expeditions during the Revolutionary and Napoleonic wars involving hundreds of thousands

[1] Philip J. Haythornthwaite, *The Armies of Wellington* (London, 1998), 194.

[2] Christopher D. Hall, *Wellington's Navy: Sea Power and the Peninsular War 1807–1814* (London, 2004), 111 and Michael Duffy, 'Festering the Spanish Ulcer: The Royal Navy and the Peninsula War 1806–1814', in Bruce A. Elleman and S.C.M Paine (eds), *Naval Power and Expeditionary Warfare: Peripheral Campaigns and New Theatres of Naval Warfare* (New York, 2012), 28, both quoting Sir R.V. Hamilton (ed.), *Letters and Papers of Admiral of the Fleet Sir T. Byam Martin*, vol. 2 (London, 1898), 409.

of troop voyages to Africa, India, the West Indies, America and the European mainland from the Baltic to the Mediterranean, the French army generally marched across national boundaries. In contrast to Britain, France launched few major seaborne expeditions involving large numbers of men – the two larger ones being the shipping of the Army of the Orient to Egypt in 1798, when 224 chartered merchantmen convoyed by thirteen ships of the line plus six frigates and other smaller vessels conveyed 24,000 infantry, 4,000 cavalry and 3,000 artillerymen to Aboukir Bay. There were smaller French colonial expeditions to the Caribbean; an unsuccessful attempted invasion of Ireland in 1798 in support of the Irish rebellion; and in the Mediterranean a more successful landing of 2,000 troops from sixty transports in Capri in 1808, followed by two failed attempts to take Sicily in early 1809 and then again in 1810.[3]

By 1811 the worldwide movement of British troops in the Napoleonic war achieved such success that France and its allies did not possess a single oversea territory.[4] This book is not about the battles at sea or on land; rather it is the study of this herculean task of moving large armies together with their essential support infrastructure by sea, to distant shores, and then of ensuring that such armies were adequately resupplied so that campaigns might be sustained.

The navy did not have the capacity for these tasks. Merchant ships were required. The government's *laissez-faire* approach meant that requisitioning the necessary vessels was not an option. Ships were procured through the market and managed by the Transport Board and the transport service was fulfilled by merchant seamen with great skill and courage, often in very challenging and dangerous conditions and frequently at a high personal cost.

Robert K. Sutcliffe
Little Chalfont, Buckinghamshire

[3] Philip J. Haythornthwaite, *Napoleon's Military Machine* (Staplehurst, Kent, 1995), 173.
[4] Roger Knight, *Britain Against Napoleon: The Organization of Victory 1793–1815* (London, 2013), 189.

Acknowledgements

It is impossible to complete a book such as this without a huge amount of encouragement and support from others. In this instance their generosity has been unbounded. In particular, I have to thank Dr Roger Knight for his guidance, skilled advice and encouragement, provided at a time when he was heavily committed to other major academic projects. He contributed significantly to making this research a very enjoyable project. Sim Comfort kindly allowed me to review documents relating to the proposed Brest expedition from his remarkable collection of naval papers. Joanna Ellis was particularly helpful in providing copies of correspondence between Spencer Perceval and Lord Mulgrave from the Mulgrave Papers and I am grateful to the Marquess of Normanby for allowing me to consult and quote from those papers. In addition I must thank the staff at the National Maritime Museum Caird Library and The National Archives for their assistance and support and for dealing with my enquiries in a timely manner.

I must also thank my editor Peter Sowden of Boydell & Brewer for his infinite patience and support throughout the process of preparing this book, Nick Bingham who patiently steered this book through the production stage, Cath D'Alton for her skillful preparation of the maps and finally Sarah Bryce, who painstakingly read every word, correcting and improving the text enormously. Of course any errors or omissions remain the responsibility of the author alone.

However, those who carry the heaviest burden throughout the preparation of a book such as this are those closest to us, and I will be eternally grateful to my wife Zena for her unwavering support throughout but in particular during the more testing periods. To her I dedicate this book.

Abbreviations

BL	British Library
Castlereagh, *Correspondence*	Second Marquess of Londonderry, *Memoirs and Correspondence of Viscount Castlereagh*, vols I–VIII (London, 1851)
Commission on Fees (5th)	Fifth Report of the Commissioners appointed to Inquire into the Fees, Gratuities, Perquisites and Emoluments which have lately been received in the Several Public Offices 1802–03 (111), 249
Commission on Fees (6th)	Sixth Report of the Commissioners appointed to Inquire into the Fees, Gratuities, Perquisites and Emoluments which have lately been received in the Several Public Offices 1803–04 (11), 1
Commission on Fees (8th)	Eighth Report of the Commissioners appointed to Inquire into the Fees, Gratuities, Perquisites and Emoluments which have lately been received in the Several Public Offices 1803–04 (11), 637
Commission for Revision (9th)	Ninth Report of the Commissioners for Revising and Digesting the Civil Affairs of His Majesty's Navy 1807, 73e of 1809, XXV, 354
Commission for Revision (13th)	Thirteenth Report of the Commissioners for Revising and Digesting the Civil Affairs of His Majesty's Navy 1809, 73e of 1809, XXV, 530
Committee on Finance (18th)	Eighteenth Report from the Select Committee on Finance 1798 (119), 191, 206
Committee on Finance (31st)	Thirty-First Report from the Select Committee on Finance 1798 (113), 03, 74
Committee on Finance (18th+31st) Further Proceddings	Further Proceedings on the 18th and 31st Report from the Select Committee on Finance 1715–1800 (114), 24

Condon, 'Transport Service'	Mary Ellen Condon, 'The Administration of the Transport Service during the War against Revolutionary France 1793–1802' (unpublished University of London Ph.D. thesis, 1968)
Dalley Evidence	House of Lords Sessional Papers 1802–03, 174, IV.327, Fifth Report of the Commissioners of Naval Enquiry into the Sixpenny Office, Appendix 7 – Examination of John Dalley
Gurwood, *Dispatches*	John Gurwood, *Dispatches of Field Marshal the Duke of Wellington* (reprinted, Cambridge, 2010)
HL	Huntington Library
HM	His Majesty King George III
HoCPP	House of Commons Parliamentary Papers (found on parlipapers.proquest.com)
IHR	Institute of Historical Research
Melville *Speech*	*Substance of the speech of Viscount Lord Melville in the House of Peers, Monday 21 May 1810 on the subject of Troop Ships* (London, 1810), BL, 8807.c.6
NMM	National Maritime Museum, Caird Library
Scheldt Inquiry	HoCPP, 1810 (12) VIII.1, Minutes of Evidence taken before the Committee of the Whole House, appointed to consider the policy and conduct of the late expedition to the Scheldt (Session 23, January–June 1810)
TB	The Transport Board
The Board	The Transport Board
TNA	The National Archives
VB	Victualling Board
Wars	The Revolutionary War of 1793 to 1801 and the Napoleonic War 1803 to 1815

Notes and Conventions

Arthur Wellesley, 1st Duke of Wellington (1769–1852) has been referred to as the Duke of Wellington or Wellington throughout this book although the title was not bestowed upon him until 11 May 1814. However, he was elevated to viscount following the battle of Talavera in 1809 after the fall of Napoleon. From 1808 his military ranks were lieutenant-general 1808, general 1811, field marshal 1813, commander-in-chief of the army in occupied France, until 1818.

Sir Rupert George Bart (1749–1823), Chairman of the Transport Board. Born in Ireland, son of Dennis George. He became a naval lieutenant in September 1770 serving on the *Rose*, then, in 1775, on the *Enterprise* as second then first lieutenant. In 1779 he joined the *Robust* as first lieutenant. He became commander of the *Charleston* in 1781. In this ship he served on the North American station in the American war, possibly under Samuel, Viscount Hood, and whilst there he met and married Margaret Cochran in June 1782. She was also of Irish origin, daughter of an influential family from Halifax, Nova Scotia.[1] He commanded the sloop *Vulture.* He made captain in November 1781. In 1790 he joined the *Thistle* and in 1792 the *Hussar*, in which he was again in North American waters in 1793. He succeeded Admiral Hugh Christian as Chairman of the Transport Board in 1795. However it is not clear why an apparently undistinguished naval captain came to the attention of the Commissioners of the Treasury. It is noteworthy that his first son, born in 1789, was named Samuel Hood George; he may have been named after Samuel, Viscount Hood. George may have benefited from the patronage of Hood with whom he had served on the American station in the latter stages of the American war and who was a Lord of the Admiralty from 1788 to 1795. Hood was renowned for looking after his following and his position at the Admiralty made him well placed to promote George's interests.[2]

It proved to be an auspicious appointment. George was to serve twenty-two years, throughout seven administrations, surviving the consequent changes in policies, personalities and interdepartmental relationships and the rivalries of

1 Rev. A.W.H. Eaton, *The Cochran-Inglis Family of Halifax Nova Scotia* (Halifax, N.S. 1899), 8. George and his wife had two sons and six daughters.

2 Michael Duffy, 'Samuel Hood, First Viscount Hood', in Peter LeFevre and Richard Harding (eds), *Precursors of Nelson: British Admirals of the Eighteenth Century* (London, 2000), 249–77.

other departments until the Board's dissolution in 1817.[3] This must rank amongst the longer civil service tenures of the period. By comparison Evan Nepean served as first secretary to the Admiralty for almost nine years.[4] Captain Sir Andrew Snape Hamond was Comptroller of the Navy for almost twelve years, followed, in 1806, by Captain Sir Thomas Boulden Thompson who was in post for just less than ten years. To have retained his position for such an extensive period, George must have possessed some likeable and remarkable qualities. He was obviously very competent, a very able administrator and a shrewd political operator.[5] In 1800 he declined promotion to flag rank, so that he could remain with the Board.[6] He was knighted in 1803 and created baronet in August 1809. George died in 1823 and was buried in the family crypt at St Mary's Church, Battersea, London. His title became extinct on the death of his son Rupert Denis George in 1856.

King's German Legion (KGL). This was effectively the Hanoverian Army in exile, founded in England in 1803 when thousands of Hanoverians arrived following the French invasion of Hanover. The regiments became British army units paid by the British government. They included artillery, infantry, cavalry and engineers. They served with great distinction in the Peninsula.

Austrian Netherlands were the provinces located in the southern part of the Low Countries, roughly comprising present Belgium and Luxembourg. On 1 October 1795 the Austrian Netherlands were annexed to France. In 1815 after the Napoleonic war, the Congress of Vienna merged the area with the Dutch provinces to become the Kingdom of the Netherlands. An independent Belgium was established in 1831.

Monetary values quoted are the original historic values; those over ten pounds have been rounded to the nearest pound rather than quoting the shillings and pence.

Relating the historic value of money to modern values continues to be the subject of much academic research because of its complexity.[7] However, based on

3 TNA, ADM, 1/ 3770, Sir Rupert George was still employed at the Transport Office on 3 April 1817 where he was preparing to hand over to Commissioner Boyle who was to complete the settlement of all outstanding accounts assisted by the secretary, Mr McLeay, and Mr Harding, the general accountant. On 25 March George was notified that the Commissioners of the Admiralty to the Navy and Victualling Boards would begin 'to execute duties now taken up within the TB'.

4 Evan Nepean, Secretary at the Admiralty, 3 March 1795 to January 1804.

5 Knight, *Britain Against Napoleon*, 178: 'witness his advice to Thomas Grenville, first Lord of the Admiralty in the cost-conscious Ministry of All the Talents, when stating the estimate for transports in 1806 in Grenville's first Naval Estimates: the sum could be reduced, "but perhaps it would be more convenient to Government, to have money in hand, than to run the risk of a deficiency, which might require a pre-mature calling of Parliament next year. I hope that you will excuse the liberty I take in making these observations"'.

6 A. Aspinall (ed.), *The Later Correspondence of George III* (Cambridge, 1966–67), Earl Spencer to HM King George, 31 December 1800.

7 N.A.M. Rodger, *The Command of the Ocean: A Naval History of Britain 1649–1815* (London, 2004), xxv and E. Victor Morgan, *The Study of Prices and the Value of Money* (London, 1950).

research by Jim O'Donoghue, Louise Goulding and Graham Allen, 'Consumer Price Inflation Since 1750' (ISSN 0013-0400, Economic Trends No. 604), 38–46, a Historical UK Inflation and Price Conversion calculator can be found on www.safalra.com which indicates the following approximate values of £100 in 2013:

Year	Approximate value of £100
1793	£13,000
1800	£9,700
1815	£6,800

Introduction

> 'Expeditionary warfare entails the deployment of forces far from their normal base of operations. Execution requires enormous logistical capabilities to transport, land and sustain forces, often at great distances.'[1]

Although Britain emerged from Trafalgar as the dominant naval power, that alone was not going to defeat Napoleonic France. Napoleon had to be defeated on land and in particular on the European mainland; this was only finally accomplished when Britain operated in conjunction with its principal allies. Over fifty amphibious expeditions were launched during the wars. Despite this, the British army did not achieve meaningful success in the mainland European theatre until 1807 at Copenhagen, 1808 to 1814 in the Peninsula and 1815 at Waterloo. In addition there were numerous modest successes but there were also some monumental failures.

To accomplish these adventures large numbers of British troops and an immense support system had to be transported overseas, often at relatively short notice. This was generally followed by a constant relief and reinforcement programme to convey the seriously injured and prisoners of war back to England, to ship out fresh recruits to replace those killed and injured, and to replenish supplies. In 1793 the British army abroad numbered only 18,194 men, but between 1793 and 1801 at least 89,000 rank-and-file troops were shipped to the West Indies alone, and thousands more were shipped to the Netherlands, France, the Mediterranean and Egypt.[2] In January 1805 there 144,500 rank and file overseas, rising to 180,991 in December 1813. These numbers increase by about 12½ per cent when sergeants and officers are included and possibly up to 30 per cent when officers' staff, drummers, artillerymen, artificers, engineers, medical and commissariat staff and women followers are included.[3] Of course horses, wagons, materiel, including ordnance

[1] Bruce A. Elleman and S.C.M. Paine (eds), *Naval Power and Expeditionary Warfare: Peripheral Campaigns and New Theatres of Naval Warfare* (New York, 2012), 1.

[2] Knight, *Britain Against Napoleon*, 76–7.

[3] Haythornthwaite, *Armies of Wellington*, 21, suggests an increase of one eighth to include sergeants and officers. Army strength numbers from: HoCPP, 1814–15 (195) IX.321, 14 Mar. 1815, which is a summary of reports issued in June and December each year by the Adjutant-General's Office. These totals include British cavalry, foot guards, British

and camping equipment, and provisions also had to be shipped to support the troops. Then when the theatre of war changed or evacuation was required the troops and equipment had to be repatriated.

These tasks were mainly accomplished by merchant ships. Thousands of them were essential to support the government's military operations. However, there were rarely enough transports readily available for government service, particularly at very short notice; rather than resorting to requisitioning ships, the government competed with the demands of trade by chartering ships on the open market. There the availability of shipping was already restricted by a shortage of seamen: throughout the wars there were never enough experienced sailors to meet the demands of both the navy and the rapidly expanding merchant fleet.[4] Despite this the government spent more than £42 million on hiring transports between 1794 and 1815. Morriss suggests that 'Britain's overseas achievements [during this period] reflected the marriage between key maritime resources and the state's bureaucracy.'[5] This sums up the achievement of the Transport Board, which was established by Prime Minister William Pitt in 1794 to manage this service.

Sir Charles Oman refers to naval operations in his *History of the Peninsular War* but otherwise, with few exceptions, military historians have not acknowledged the significance of the naval and maritime contribution in these conflicts.[6] Rather they have naturally tended to focus on events ashore. However, Christopher Hall's *Wellington's Navy* provides a comprehensive review of the role of sea power during the Peninsular war. It demonstrates the way that control of the sea influenced the outcome of events on land, as does Michael Duffy's *Soldiers, Sugar and Seapower: The British Expeditions to the West Indies and the War against Revolutionary France* and 'Festering the Spanish Ulcer: The Royal Navy and the Peninsular War, 1808–1814'.[7] David Syrett's *Shipping and the American War* and *Shipping and Military Power in the Seven Years War: The Sails of Victory* are considered to be the seminal works on the transport service, providing a detailed account of the successes and failures of the service during the earlier

infantry, foreign cavalry and foreign infantry. The split between those troops at home and those abroad is not shown in these reports until 1808. Generally these numbers are for rank-and-file men and do not include officers. Neither do they include the numbers of the Royal Regiment of Artillery, the artillery of the King's German Legion or foreign artillery. See appendix 3. See also chapter 1, planning for Brest expedition (p. 13) where the proposed force of 60,000 rank and file expanded to 83,620 individuals, an increase of 30 per cent.

4 For further consideration of the shortage of seamen, see Rodger, *Command of the Ocean*, 443–53; Roger Morriss, *The Foundations of British Maritime Ascendancy: Resources: Logistics and the State, 1755 to 1815* (Cambridge, 2011), 321, and J. Ross Dancy, *The Myth of the Press Gang: Volunteers, Impressment and the Naval Manpower Problem in the Late Eighteenth Century* (Woodbridge, 2015), 223–70.

5 Morriss, *British Maritime Ascendancy*, 321.

6 Sir Charles Oman, *A History of the Peninsula War*, 7 vols (Oxford, 1902–30).

7 Hall, *Wellington's Navy*, 15–28; Michael Duffy, *Soldiers, Sugar and Seapower: The British Expeditions to the West Indies and the War against Revolutionary France* (Oxford, 1987); Duffy, 'Festering the Spanish Ulcer'.

American war (1775–1783).[8] In addition, in 1968 Mary Ellen Condon produced an unpublished Ph.D. on the Transport Board in the Revolutionary war.[9]

It is only recently that the role of the transport service has been recognised by naval historians, particularly Knight in *Britain Against Napoleon*; Davey in *The Transformation of British Naval Strategy*; Mackesy in *War in the Mediterranean 1803–1810*; Morriss in three publications – *Naval Power and British Culture 1760–1850: Public Trust and Government Ideology*, 'Colonization, Conquest and the Supply of Food and Transport: The Reorganization of Logistics Management 1780–1795' and *The Foundation of British Maritime Ascendency: Resources, Logistics and the State, 1755–1815*;[10] Hall in *British Strategy in the Napoleonic War* and *Wellington's Navy*;[11] and Cole in *Arming the Royal Navy, 1793–1815: The Office of Ordnance and the State*.[12] In contrast there are volumes of publications on the navy and the army, on sea and land battles, on the political and economic context, and on specific subjects referred to in this work such as the various trade protection measures: the Navigation Acts, the Continental System and the Licence Trade.[13] Unfortunately the history of the merchant fleet in the period 1790 to 1820 has received limited attention. This is possibly due to the destruction of considerable volumes of records in the various Custom House fires, most notably that of 1814. This loss is exacerbated by the absence of surviving shipping company records, with the notable exception of the archive of Michael Henley & Son, London-based ship-owners and providers of transports, which survives at the National Maritime Museum. The history of the company has been well documented by Simon Ville in *English Shipowning during the Industrial Revolution: Michael Henley & Son, London Shipowners 1770–1830*.[14] Ralph Davis's important work *The Rise of the English Shipping Industry in the 17th and 18th Centuries* ends just before the start of the period, while Hope devotes only one chapter to the whole period in *A New*

8 David Syrett, *Shipping and the American War 1775–83: A Study of British Transport Organization* (London, 1970) and *Shipping and Military Power in the Seven Years War: The Sails of Victory* (Exeter, 2008).

9 Mary Ellen Condon, 'The Administration of the Transport Service during the War against Revolutionary France, 1793–1802' (University of London Ph.D. thesis, 1968).

10 Duffy, *Soldiers, Sugar and Seapower*; James Davey, *The Transformation of British Naval Strategy: Seapower and Supply in Northern Europe 1808–1812* (Woodbridge, 2012); Piers Mackesy, *The War in the Mediterranean 1803–1810* (Cambridge, MA, 1957); Roger Morriss, *Naval Power and British Culture 1760–1850: Public Trust and Government Ideology* (Aldershot, 2004); Morriss, 'Colonization, Conquest and the Supply of Food and Transport: The Reorganization of Logistics Management 1780–1795', *War in History 14* (2007), 310–24; Morriss, *British Maritime Ascendancy*.

11 Christopher D. Hall, *British Strategy in the Napoleonic War 1803–1815* (Manchester, 1992) and *Wellington's Navy*.

12 Gareth Cole, *Arming the Royal Navy, 1793–1815: The Office of Ordnance and the State* (London, 2012).

13 Sarah Palmer, *Politics, Shipping and the Repeal of the Navigation Laws* (Manchester, 1990).

14 Simon P. Ville, *English Shipowning during the Industrial Revolution: Michael Henley & Son, London Shipowners 1770–1830* (Manchester, 1987).

History of British Shipping.[15] Probably one of the more comprehensive studies of the period was *The Trade Winds*, edited by Northcote Parkinson in 1948.[16] All of these historians, including Creswell in his paper 'British Shipping at the End of the Eighteenth Century', explain the difficulty of defining the number and tonnage of British shipping involved in the overseas trade during the period.[17] These ships formed the pool from which transports were generally hired. More recently Woodman has published a comprehensive five-volume history of the merchant shipping fleet; the second volume relates to this period.[18]

Nevertheless, a comprehensive history of the logistics of expeditionary warfare, and the central role that the Transport Board played in most military expeditions launched between 1794 and 1815, has not yet been published. This book will seek to address this by demonstrating how the Board engaged with the shipping market to secure transports and by considering the impact of their diversion from trade. It will examine the role of the transport service, particularly between 1805 and 1815 when a significant number of large military forces were transported overseas to mount critical expeditions. Chapter 1 will review the various elements of expeditionary warfare during the period, particularly the planning of a campaign and the landing of military forces. Chapter 2 will consider how transports were procured and, by interpreting the output of a database of over 2,000 charter contracts, determine when, in what quantities and under what terms transports were brought forward. Chapter 3 will show that the transport demands were far more considerable than has previously been perceived and will assess the availability of shipping to support the combined demands of both government and trade. It will demonstrate that the transport service supported the British shipping industry during the earlier part of the wars when its activity was restricted because British shipping was prohibited from entering enemy-controlled ports. That trade was lost to licensed foreign ships. It will also explain the intensely competitive demand for the limited number of ocean-going ships. Chapter 4 will consider the impact on operational efficiency of the government's determination to contain the expense of war by attempting to minimise the costs of the transport service. This chapter also demonstrates that the preparation for major expeditions required co-ordinated activity from many departments of state and in consequence delays invariably occurred. The causes of these delays will be illustrated but the research has demonstrated that, contrary to current thinking, the preparation of transports rarely consumed time beyond that which might reasonably have been contemplated. This raises the question of whether there were practical limitations on the size of expeditions that the various services

15 Ralph Davis, *The Rise of the English Shipping Industry in the 17th and 18th Centuries* (Newton Abbott, 1962); Ronald Hope, *A New History of British Shipping* (London, 1990), 235–63.

16 Northcote Parkinson (ed.), *The Trade Winds: A Study of British Overseas Trade during the French Wars 1793–1815* (Leicester, 1948).

17 J. Creswell, 'British Shipping at the End of the Eighteenth Century', *Mariner's Mirror* 25 (2 Apr. 1939), 197–207.

18 Richard Woodman, *A History of the British Merchant Navy: Vol. 2 Britannia's Realm: In Support of the State, 1763–1815* (Stroud, 2009).

could deal with effectively, particularly when considering the provision of horse transports for cavalry regiments, of horses and artillery equipment, and of horses, mules and wagons for commissariat transport facilities. The shortage of heavy artillery, horses and wagons created constant dilemmas for military commanders, particularly the Duke of Wellington in the Peninsula and at Waterloo. Chapter 5 will consider the important role of the navy in supporting expeditionary warfare, demonstrating that achieving dominance of the seas was vital to ensuring that trading and military maritime activity could progress with limited interference from enemy navies and privateers. It will also illustrate the transformation of naval involvement in the war and the evolution of amphibious warfare, particularly during the Peninsular war and the war against America (1812–14). Chapter 6 will review the Admiralty's decision to resist the use of naval vessels as troop transports instead of smaller merchant ships. Chapter 7 will demonstrate the effectiveness of the military shipping involved in Castlereagh's European expeditions between 1805 and 1808. Chapter 8 will consider the Transport Board's performance during the most difficult year of the war, 1809, described by N.A.M. Rodger as 'a year of military disappointments', including the evacuation from Corunna and the Walcheren campaign.[19] Finally chapter 9 will review the transport service during the later years of the Napoleonic war, which witnessed huge transfers of manpower as Britain and her allies paved the way to ultimate victory over Napoleon in 1815.

By 1792 Britain had the largest merchant fleet in the world.[20] Its growth had been supported by the protectionist policies of succeeding governments who were anxious to protect international trade, which was critical to the economic survival of the nation. During the wars this trade provided markets for the output of the burgeoning industrial revolution and funded the increasing national debt which expanded from £273 million in 1792 to £792 million in 1816.[21] It also secured the country's survival when the corn harvests failed, as in 1795, between 1799 and 1801 and between 1805 and 1813, and generated funds to provide substantial subsidies to the nation's allies. Equally importantly, it supported the operational effectiveness of the navy, which was heavily dependent on imported supplies of iron, timber, flax and hemp.

During the period 1792 to 1815 imports rose by 83 per cent and exports grew by 172 per cent. This expansion in international trade combined with the shipping demands of the government fuelled the growth of the merchant shipping fleet. The number of registered ships increased during the period from 16,079 of 1,540,145 tons in 1792 to 25,864 of 2,783,000 tons in 1815, representing an 80 per cent growth in tonnage.[22] This rate of growth was unsustainable; there was a serious

[19] Rodger, *Command of the Ocean*, 556.

[20] B.R. Mitchell and Phyllis Deane, *British Historical Statistics* (Cambridge, 1962), 217; Duffy *Soldiers, Sugar and Seapower*, 387.

[21] Glyn Davies, *A History of Money* (Cardiff, 2002), 172.

[22] Inclusive of ships registered in England, Scotland, Ireland, plantations in America and the West Indies, the Channel Islands and the Isle of Man. Only ships over 15 tons were registered. For a table of the number and tonnage of ships registered between 1799 and 1817, see appendix 1.

overcapacity when peace was finally declared in 1815 and the government no longer needed to charter significant amounts of shipping.

It has been said that 'By any standards, the achievements of the transport service during the American war of 1776 to 1783 rank among the greatest military and administrative feats of the eighteenth century.'[23] However, there remained some real weaknesses in the co-ordination of chartering and utilising transports. Charles Middleton (later Lord Barham), Comptroller of the Navy Board 1778–90, was very critical of the way the government brought forward merchant shipping during the American war. The crux of the matter was that during the American war, merchant ships were hired by three boards – the Navy Board, the Victualling Board and the Ordnance Board – acting independently and often in competition. This competition impacted the freight rates but more importantly the availability and effective use of transports. Middleton wanted the Navy Board to assume responsibility for controlling the chartering of all shipping because he thought that this was the only way to have 'a rational policy for the procurement of shipping for government service'.[24] He was also, no doubt, mindful of the opportunity to expand both his and the Navy Board's influence. He continued to voice his opinions on this matter and his influence on the 1788 Parliamentary Commission appointed to 'Inquire into the Fees, Gratuities, Perquisites and Emoluments which have been lately received in the several Public Offices' became obvious when his views were reflected in the published recommendations.

The commissioners commented on issues relating to transports in their fifth report on the Commissioners of the Navy, in their sixth report on dockyards and in their eighth report on the Victualling Office. In the sixth report they determined that 'the practice of purchasing or hiring ships and vessels, when required for public service, by different Boards, has been found by experience very expensive, inconvenient and detrimental to the other services carried out in the dockyards'.[25]

One of the commissioners' criticisms was that, despite the immense care taken to ensure that the ships taken up were fit for service, lack of seafaring skills on the various boards, combined with the competition between the boards for tonnage, had often led to the hiring of some vessels that were unfit for service. They agreed with Middleton's proposal that the Navy Board should manage the process.[26] They also highlighted that some dockyard officers spent over 200 days a year on transport activities, which severely impeded the performance of their main tasks.[27] These officials were involved in surveying, measuring, valuing and reporting upon all ships tendered as transports. It was a very complex process which generally took at least a week when the tides were favourable and often much longer if they were not.[28]

23 Syrett, *Shipping and the American War*, 248.
24 Syrett, *Shipping and the American War*, 23, citing The Shelburne Papers, Middleton to Shelburne, 28 Jun. 1782.
25 Commission on Fees (6th), 305.
26 Commission on Fees (5th), 103.
27 Commission on Fees (6th), 139.
28 Commission for Revision (9th), 14.

A survey of a potential transport was instigated by an order from the Navy Board to the agent and dockyard officials; usually this occurred on the river Thames. A small armada of boats was involved: the Board's agent at Deptford accompanied the dockyard's master attendant and clerk of survey in one boat, the master shipwright's assistant was in another and the foreman afloat was with the ship's agent in the third boat. If a ship was found to be fit for the service, the master was directed to put her into dock or on the ways so that her bottom might be inspected and her dimensions taken. When she was ready for the second examination, the same officers and the agent were again involved. If she was approved, the master was ordered to take provisions and stores on board and proceed to Deptford. There the same officers – with the addition of the master mast-maker, foreman of the riggers, clerk of the surveys, clerk of the master sail-maker and some of his people, the master joiner for marking out the cabins and the clerk of the cheque for mustering the crew – were involved in the third inspection. After the calculation of the value of the ship and stores had been made in the clerk of the survey's office, the Navy Board was advised that she was ready to enter into pay and the agent commenced the fitting out to hasten her departure for the service to which she was appointed.[29]

Finally, in the eighth report on the Victualling Office, the commissioners exposed abuses promulgated by the hoy taker, the official who supervised all the shipping for the Victualling Board. He had received payment to favour some owners with charters and had also had an interest in some of the ships hired. They recommended that the duties of the hoy taker should be restricted to 'the hiring, superintendence and employment of lighters, barges and small craft on the River Thames and to loading or unloading vessels employed in the conveyance of provisions or victualling stores'.[30] The commissioners also referred, in the eighth report, to practices which in their view led to the government overpaying for transport services. Firstly, the Victualling Office had generally hired 'on freight', which tended to be more expensive than ships hired by the month that could usually carry a greater tonnage of supplies.[31] They noted that the Commissioners for Stating the Public Accounts, in their twelfth report dated 1784, had made similar comments in respect of the Ordnance Board.[32] Secondly, ships had been overvalued, causing the government to be defrauded when these ships were subsequently captured by the enemy and the owners reimbursed at the inflated rate. They did suggest that it might be more economical to pay an increased hire rate if the owner agreed to bear the responsibility for any loss but there is no indication that this suggestion was considered further. They also restated the view that competition between boards had caused the greatest detriment

[29] Commission for Revision (9th), 14.

[30] Commission on Fees (8th), 210.

[31] Hire 'on freight': by this method payment was based on the weight of the supplies to be shipped rather than on the tonnage of the vessel.

[32] Report of the Commissioners Appointed to Examine, Take and State the Public Accounts of the Kingdom. Twelfth Report Relative to Passing the Accounts of the Treasurer of Ordnance, in the Office of the Auditor of the Impress, (1784), 11–12, 43.

and loss by inflating prices and by diverting tonnage urgently needed for an important service by one board onto 'trifling' tasks elsewhere.[33] In summary, these commissioners concluded, as had those of earlier reports, that if the Navy Board was responsible for the service these abuses would not arise.[34]

Despite much lobbying by Middleton, supported by the Treasury, it was not until two years into the Revolutionary war that Prime Minister William Pitt finally ordered the consolidation of the transport service. However, it was not to be under the Navy Board, as the various commissions had recommended.[35] Instead, by an Order in Council dated 4 July 1794, he established a Transport Board as a subsidiary board of the Treasury.[36] There had previously been such a board between 1690 and 1724, initially established after delays in raising, fitting and provisioning shipping to convey troops to Ireland.[37] It is not clear why the establishment of the Board was delayed beyond the outbreak of hostilities, or what precipitated its introduction. It might have been motivated by the difficulty of procuring transports to convey troops to the West Indies which the Navy Board experienced in May 1794. By that time the West Indies campaign was well underway and the various boards had already chartered a large number of ships to convey troops and supplies. When the Navy Board received yet another order for transports it was forced to advise Henry Dundas, Secretary of State for War:

> The trade to the Baltic has carried away all those [ships] which might have been had for the service in the previous month, and very few are expected from the East Country for more than five or six weeks. The speculation to the newly acquired islands in the West Indies has employed a considerable number of ships, which is another cause of the present scarcity.[38]

This, together with strong support within the Treasury where this was viewed this as an important cost-saving opportunity, may have brought the matter to a head.[39] Then in September 1795 the Transport Office was instructed to take over the duties of care and custody of prisoners of war 'in health', which had previously been the responsibility of the Sick and Hurt Board.[40] In February 1806, all of the remaining business of the Sick and Hurt Board was transferred to the Transport Board. After the end of the war, in early 1817 the responsibilities of the Transport Board, excluding the medical services which were transferred to the Victualling Board, were devolved to the Navy Board to reduce costs.

33 Commission on Fees (8th), 210.
34 Commission on Fees (8th), 210.
35 See Condon, 'Transport Service', 32–40, 48, for the background to Middleton's lobbying.
36 Commission for Revision (9th), 11.
37 Commission for Revision (9th), 11.
38 TNA, HO, 28/63, Captain Andrew Snape Hamond, Navy Office to the Admiralty, 12 May 1794.
39 George Rose, *Observations respecting the Public Expenditure and the Influence of the Crown* (London, 1810), 32.
40 TNA, ADM, 1/3730/301, Patent dated 30 Sep. 1795, signed by the Duke of York.

1

British Expeditionary Warfare, 1793–1815

Campaigns were ultimately decided not only by fighting qualities but also by the effectiveness of pre-planning, the efficiency of supply and logistics, and the extent of support from indigenous population and allies.

The Principal Theatres of War, 1793–1815

Between 1793 and 1795 the British Cabinet launched attacks on two fronts. In northern Europe it attempted to establish footholds on the European mainland and in the West Indies it resolved to destroy France's financial stability by attacking the islands which provided the source of its wealth. The European expeditions launched in 1793, to Flanders, Dunkirk, Brittany and Holland, failed, as did subsequent expeditions to the Vendée in 1795, Ostend in 1798, den Helder in 1799 and Ferrol in 1800. Following these setbacks Henry Dundas, Secretary of State for War (1794–1801) admitted that he was 'impressed with a deep conviction that we cannot take a direct part in the military operations on the Continent, we can only act indirectly and collaterally with our Continental allies'.[1] After Ferrol there were no more attempts to land forces in northern Europe until 1805 when a force was despatched to Hanover in northern Germany; but this campaign was abandoned and the troops withdrew without engaging the enemy. However, in 1807, there were two successful campaigns against Heligoland and Copenhagen. An expedition in 1808 to the Baltic to support the Swedes was less so: it returned to England within two months. This was followed, in 1809, by the Scheldt (also known as Walcheren) expedition, which was to be the most monumental failure of the wars. This disaster was followed by partial success in Holland in 1813–14 and finally Anglo-Prussian success at Waterloo in June 1815.

Despite operational successes, the early expeditions to the West Indies were contentious: the excessive death rate, which accounted for 50 per cent of the army, was attributable to yellow fever, rather than conflict fatalities. More recently, the significance of the economic legacy of these expeditions has been exposed, rebalancing the negative perceptions. Duffy says that 'Overall, the Caribbean war had served its purpose to establish a decisive British maritime

1 John Hattendorf, Roger Knight, A.W.H. Pearsall, N.A.M. Rodger, Geoffrey Till (eds.), *British Naval Documents 1204 to 1960*, Navy Records Society 131 (Aldershot,1993), 345, Dundas's Memorandum to the Cabinet, 31 Mar. 1800.

and commercial supremacy over France.'[2] Possession of the Caribbean Islands was strongly contested: Britain captured Martinique in March 1794 and in April that year, St Lucia, which it lost in June 1795 and regained in May 1796; then in 1797 the Spanish surrendered Trinidad after little resistance. In 1798 Britain determined that its four-year occupation of St Domingo exceeded 'all the bounds of expense that we had resolved upon, and what the country can or will bear': the decision was taken to make a tactical withdrawal.[3] In accordance with the terms of the 1802 Treaty of Amiens, all conquests in the region were relinquished, except Trinidad.

Shortly after the restart of the war, in 1803, the French recaptured St Domingo and the British occupied Tobago and St Lucia. With the exception of Sir John Duckworth's naval victory off St Domingo in 1806, this theatre then disappeared from the headlines until 1809 when the British took Martinique. In 1810, Guadeloupe, which was France's last remaining foothold in the West Indies, was captured. Meanwhile the Spanish colonies in the region, Cuba and Puerto Rico, had become friendly with Britain following the 1808 revolt in Spain. The capture of Guadeloupe marked the end of West Indies campaigning.[4]

The other theatres of war during this period were the East Indies, the Mediterranean, the Peninsula and North America. Britain's war in the East Indies was principally against the Dutch and it secured the Cape of Good Hope and Trincomalee in Ceylon in 1795; regular garrisons were maintained until the peace in 1802, when the Cape was returned to the Dutch but Ceylon was retained. In 1806 3,000 troops under Sir David Baird recaptured the Cape. Mauritius was taken in 1810, Réunion and Java in 1811, as well as Amboyna in the Spice Islands.

India itself was an important stronghold of Britain. It provided a huge market for British manufactures and a rich source of supply of spices, cotton products, and saltpetre for gunpowder. Successive administrations worried almost as much about France extending her influence in India as they did about her invading Britain. Several campaigns were launched specifically to avoid this, including the invasion of Egypt in 1801 and Mauritius in 1810. Britain's control of India was effected through The British East India Company, which was much more than just a dominant trading company. It had acquired quasi-governmental status; it had its own army to protect its interests and, in effect, its own navy of armed East Indiamen. Its army, supported by several British regiments garrisoned in India, fought numerous campaigns to defend and extend British interests.[5]

[2] Michael Duffy, in his excellent narrative of the West Indies expeditions in *Soldiers, Sugar and Seapower*, 385–93, sets out a powerful case demonstrating the significant impact of these expeditions, not only in denying French merchant's huge profits and the French government substantial revenue, but also in enabling Britain to use the produce of the conquered islands for a tremendous commercial profit.

[3] Duffy, *Soldiers, Sugar and Seapower*, 304, quoting Henry Dundas.

[4] Hall, *British Strategy*, 185.

[5] Haythornthwaite, *Armies of Wellington*, 219. On 25 May 1809 there were 18,843 British Regular troops in India plus 5,115 in Ceylon compared to the East India Company force comprising 4,051 Europeans and 128,418 native sepoys.

The Mediterranean was never a major British theatre of war although strategically it was a potential gateway to India. The early activity focused on Toulon, which Britain occupied for almost four months in 1793, and then Corsica between 1794 and 1796. In 1800 the Cabinet decided to send a force of 14,000, led by Lieutenant-General Sir Ralph Abercromby, to drive the French out of Egypt. This was a high risk enterprise with two goals: to dislocate possible French progress into India and to increase the negotiating power in the anticipated peace talks. Abercromby's force was triumphant, although he died in the process. However, the success helped to salvage the army's battered reputation. The French had occupied Malta between 1798 and 1800 but, following a Maltese rebellion, the French surrendered to British forces in 1800. The island became the subject of dispute when Britain refused to hand it back to the French following the Peace of Amiens, thus contributing to the resumption of war. Late in 1805 an Anglo-Russian force under Sir James Craig invaded Naples for a short period, ultimately withdrawing to establish a garrison at Messina, Sicily, where a regular presence was maintained until 1815.[6] In 1807 there was another British landing in Egypt under General Alexander Mackenzie Fraser to capture Alexandria. The city surrendered to the British but lack of supplies and inconclusive operations against the Egyptian forces persuaded Fraser to re-embark. The war in the Peninsula did not commence until after the Spanish rebellion against the French invasion in 1808, and although there were significant setbacks, the British army led by the Duke of Wellington finally drove the French back into France.

The Americas were briefly targeted in 1806–07 when attempts were made to capture the South American towns of Buenos Aires and Montevideo, neither of which was held for very long. Following the declaration of war on Britain by the USA on 18 June 1812 and its subsequent attempts to annex Canada, this region became a significant focus of British military and naval intervention for the next eighteen months until peace was agreed by the Treaty of Ghent in 24 December 1814.[7]

The Scale of Expeditionary Warfare, 1793–1815

Of the fifty or so British military expeditions launched during the twenty-two years of war, very few involved more than 30,000 troops in the initial landing force.[8] Those that did were invariably unsuccessful. The army's demand for large

6 Comprehensive reviews of the war in the Mediterranean can be found in William Henry Flayhart, *Counterpoint to Trafalgar: The Anglo Russian Invasion of Naples 1805–1806* (Columbia, SC, 1992) and Mackesy, *War in the Mediterranean*.

7 Comprehensive reviews of the war against the USA in Brian Arthur, *How Britain won the War of 1812: The Royal Navy's Blockade of the United States 1812–1815* (Woodbridge, 2011) and Andrew Lambert, *The Challenge: Britain Against America in the Naval War of 1812* (London, 2012).

8 Brian Lavery, *Nelson's Navy: The Ships, Men and Organisation* (London, 1989), 310. He suggests that there were at least sixty-eight major landing operations and dozens of small ones.

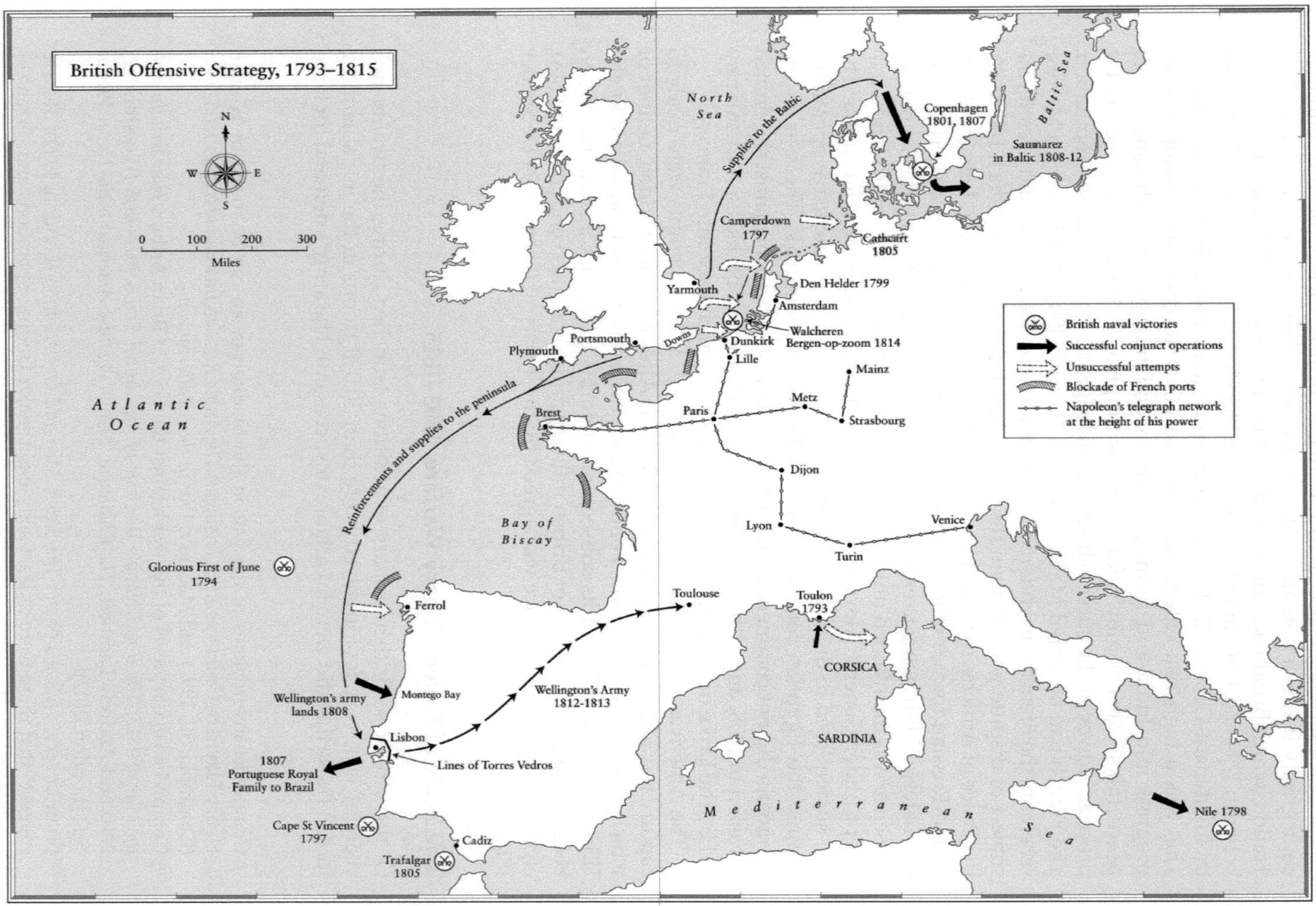

Map 1 British offensive strategy against Europe, 1793–1815

expeditions to invade the continent was a logical one but expeditions of such scale were simply too complex and unmanageable in the late eighteenth and early nineteenth centuries.[9] It became obvious to the government that attempts to establish a foothold in mainland Europe by direct attack on France and its north European conquests were unrealistic without significant support from major allies because Britain was simply unable to raise enough troops to defeat the French army unilaterally. However, Britain lived in the fear of an attack from across the North Sea, from the French-occupied Austrian Netherlands and Holland. This explains why it was not difficult for Henry Dundas and later Viscount Castlereagh (Secretary of State for War, July 1805–January 1806 and again March 1807–September 1809) to persuade the Cabinet to launch a number of expeditions in that direction, even though previous attempts had failed miserably. In reality Britain was only able to inflict real damage on the French forces when it attacked them on the fringes of the European mainland, such as in Spain and Portugal.[10]

The more successful British expeditions, prior to the Peninsular battles of 1808–14, were the smaller ones where the navy was able to support disembarkation, implement blockades and protect the rear-guard from attacks from the sea, and participate in coastal assault. These included the successes in the West Indies both in the early stages of the Revolutionary war and again in the Napoleonic war, in Egypt in 1801 and Copenhagen in 1807. Later such tactics supported Wellington's success in Spain, Portugal and France 1808–14 and the army in North America from 1812–14.

It is difficult to envisage the tonnage of shipping required to transport an army. Shipping requirements went far beyond what was needed for the rank-and-file troops. This was clearly demonstrated in 1799 by William Huskisson, Undersecretary in the Office of the Secretary of State for War, during the planning for a proposed expedition to capture the naval base at Brest.[11] It was envisaged that a force of 60,000 rank and file would be essential to be confident of success. This became 71,040 (an increase of 18 per cent) after accounting for the associated officers, non-commissioned officers, drummers, women followers and servants. In addition the force required the support of cavalry regiments of 4,000 rank and file, expanding to 5,420 individuals on a similar basis. It also required battalions of artillery of 2,000 rank and file rising to 2,400 when all the associated personnel were included, as well as 600 artificers, 100 engineers, 1,500 medical and commissariat staff and 500 general officers and their staff. This increased the total force to 83,620 (an increase of 30 per cent over the original rank-and-file strength) who would require 12,000 horses. In addition transports would also be needed for ammunition, camping equipment, provisions and

[9] Robert K. Sutcliffe, 'Bringing Forward Shipping for Government Service: The Indispensable Role of the Transport Service, 1793 to 1815' (University of Greenwich Ph.D. thesis, 2013), 305.

[10] Knight, *Britain Against Napoleon*, 460.

[11] William Huskisson, MP, Superintendent, Alien Office Jan. 1793 to Jul. 1794, Undersecretary to the Secretary of State for War 1 Mar. 1795 to 18 Mar. 1801.

forage for horses. Huskisson estimated that 350,125 tons of transports would be required as illustrated in table 1.1.[12]

Table 1.1 General idea of the transport tonnage required for an army of 60,000 rank and file and their support, including cavalry and officers, supplied for three months in a temperate summer climate.

	Transport tonnage required
83,500 individuals and stores for 3 months at 1¾ tons per man	146,125
12,000 horses at 8½ tons per horse	102,000
Artillery stores, pontoons and wagons	20,000
Engineer's stores	15,000
Victuallers, Wagon Train	18,000
Fuel for whole	13,000
Forage for horses	20,000
Hospital ship and medical stores	9,000
Quartermaster-General's department	7,000
Total tonnage required	350,125

Source: Dundas Letter Book, Sim Comfort Collection, Huskisson to Dundas, 30 Nov. 1799.

Given an average transport tonnage was 200 to 300 tons, this would represent 1,400 vessels, far more than were ever chartered at any one time throughout the wars. On the basis of Huskisson's assessment, Dundas concluded that the expedition was utterly beyond England's resources in manpower and shipping. The plan was abandoned and the invasion of France from the sea was not contemplated again. However, in 1805 and again in 1809 large forces were shipped to Hanover (67,000) and Walcheren (40,000). In both cases Castlereagh proposed that, given the relatively short voyage across the North Sea, each ship would make several trips. This meant that transports had to be unloaded quickly and returned to be prepared for a second load rather than being retained as floating barracks and store ships, which was certainly preferred by the army in the belief that it led to a more orderly dis-embarkation.

12 Henry Dundas Letter Book, The Sim Comfort Collection, Huskisson to Dundas, 30 Nov. 1799.

The Strength of the British Army Overseas

The effective rank-and-file strength of the regular army can be determined from reports prepared in the Adjutant-General's Office, but although these reports included the foreign corps such as the King's German Legion, they excluded the artillery, the Wagon Train and the Royal Wagon Train because these were attached to the Ordnance Department. Overall, the strength of the rank and file of the regular army increased more than six-fold during the period, from 38,945 on 1 January 1793 to peak in December 1813 at 239,431 (including foreign corps).[13] During the Revolutionary war the numbers increased to 124,262 in 1795, falling back to 102,563 in 1798 then increasing again to 149,865 in 1801.[14]

No formal reports relating to the number of troops based overseas during the Revolutionary war have been located but the figure is likely to have been significant. As already highlighted, there were probably about 20,000 in regular garrisons in overseas possessions. We do know that almost 89,000 officers and troops were transported from Europe to the West Indies between July 1793 and June 1797.[15] Later 46,000 participated in den Helder expedition in 1799 and 14,000 in Egypt in 1801. In total, during the Revolutionary war at least 135,000 troops and their equipment were transported to five theatres of war.[16]

Between 1804 and 1808 the rank-and-file strength, including infantry, cavalry, artillery and engineers, but excluding foreign corps, increased from 150,593 to 226,427. The proportion based overseas is not known except for January 1805 when there were 161,956 rank and file comprising 124,531 infantry, 20,316 cavalry and 17,109 artillery and engineers. They were stationed as follows: 71,456 in England, 35,000 in Ireland, 22,500 in the East Indies and Ceylon, 15,500 in the West Indies and Jamaica, 6,500 in Malta, 4,500 in Gibraltar, 4,500 in Canada , and 2,000 spread over numerous locations in smaller numbers. In total 55,500 (34 per cent) were abroad and a further 35,000 (20 per cent) in Ireland.[17]

Chart 1.2 and table 1.3 demonstrate the strength of the regular (rank-and-file) army at six-monthly intervals between 1808 and 1815 including British and foreign cavalry, foot guards and British and foreign infantry but not including militia and artillery. They also show the proportion that was serving abroad at that time. This increased between June and December 1808 as more troops were sent to Portugal. The numbers reduced shortly after the retreat from Corunna. The Walcheren campaign was launched in July 1809 but most of the troops were either home again or had returned to Portugal by December that year. The army

[13] Compiled from various HoCPPs: 1810–11 (30) X.169; 1812–13 (34) XIII.1; 1814–15 (195) IX.321; and 1816 (99) XII.419. 1806 (173) Returns of the Effective Strength of the Regular Army, Adjutant General's Office.

[14] HoCPP, 1806 (173), Returns of the Effective Men in the British Army, as far as can be determined from documents in this office, Adjutant General's Office, 17 May 1806. These numbers appear to exclude foreign corps.

[15] Duffy, *Soldiers, Sugar and Seapower*, 184–92.

[16] Condon, 'Transport Service', 311.

[17] HoCPP, 1813–14 (16) Adjutant General's Office, 13 Nov. 1813.

of the Peninsula increased throughout 1810 –13, by which time troops were also being sent to Canada, the North American coast and Holland, and in 1815 when they were sent via Ostend and Antwerp to Brussels in preparation for the battle of Waterloo.

There were 23,266 rank and file and officers and 8,916 horses serving in the artillery in 1808, rising modestly to 25,245 and 9,255 in 1810.[18] By September 1813 there were 27,014 artillery troops and engineers.[19]

Not included in any of these statistics was the militia which was raised for home defence and, with the exception of Ireland, did not serve abroad. It was, however, used as a source of recruits for the regular army. It was 86,000 strong in 1808, reducing to 71,000 in 1813 as a result of transfers into the regular army and an easing of recruitment.[20]

The French army, which relied upon conscription, was significantly larger: in 1789 it was 170,000 strong, it marched 300,000 troops into Russia in 1813 and at its peak it expanded to over 600,000.[21]

Despite the limited size of many operations, vast numbers of troops were constantly shipped to and from overseas locations throughout the wars, not only at the early stages of expeditions but also to sustain the army's progress in battle and to maintain garrisons in occupied territories. Brigades were withdrawn from the battle to rest and to train new recruits to replace losses from death, desertion and discharge. Deaths in battle were comparatively few compared to the number of deaths from disease. Between 1803 and 1813 approximately 225,000 troops died, averaging 20,500 each year.[22] In addition, there were losses due to discharges and desertions. In both 1811 and 1812 the number of discharges was 2,300 and there were about 2,400 desertions, rising to 6,546 and 5,424 respectively in 1813.[23]

[18] TNA, WO, 55/1315, Return of the Effective Strength of the Royal Regiment of Artillery, Artillery of the King's German Legion and Royal Foreign Artillery 1 Jan. 1808 to 1 Jan. 1810, Woolwich, 5 Feb. 1810.

[19] HoCPP, 1806 (173) and 1813–14 (16) (Army). Appendix 3 shows in more detail the effective strength of the regular and militia forces.

[20] Compiled from various HoCPP: 1810–11 (30) X.169; 1812–13 (34) XIII.1; 1814–15 (195) IX.321; and 1816 (99) XII.419; 1806 (173), Returns of the Effective Strength of the Regular Army, Adjutant General's Office.

[21] Knight, 'British Strategy at Sea in the Great Wars with France', Oxford Naval Historians' Conference, April 2014 and Michael Duffy, 'Science and Labour: The Naval Contribution to Operations Ashore in the Great Wars with France, 1793 to 1815', in Peter Hore (ed.), *Seapower Ashore: 200 Years of Royal Navy Operations on Land* (London, 2001), 39.

[22] Haythornthwaite, *Armies of Wellington*, 43. In 1811 fatalities and deaths from wounds were 2,000 in the Peninsular war but the total deaths in the army were 19,019 British and 3,441 foreign troops.

[23] HoCPP, 1812 (52) IX.95 and 1813–14 (73) XI.227. Return of Deaths, Discharges and Desertions which have occurred in the Regular Army, 25 Dec. 1810 to 24 Dec. 1813. The death rate was between 5,000 and 6,000 each year, somewhat below the average of 20,500. However, the number increased in 1813 to 13,855.

Chart 1.2 and Table 1.3. The effective strength of the regular army, 1808–15, showing the number and the percentage of those serving abroad.

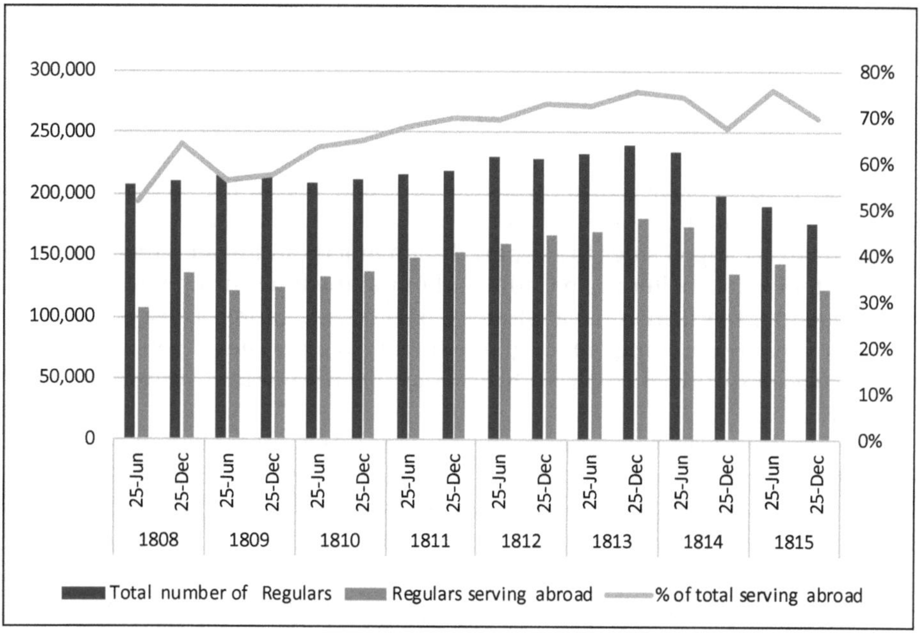

Year	Date	Total number of Regulars	Regulars serving abroad	% of total serving abroad
1808	25-Jun	207,216	107,569	52%
	25-Dec	210,614	135,335	64%
1809	25-Jun	215,171	121,359	56%
	25-Dec	216,159	123,865	57%
1810	25-Jun	208,668	132,572	64%
	25-Dec	211,159	137,207	65%
1811	25-Jun	216,757	147,613	68%
	25-Dec	218,949	153,040	70%
1812	25-Jun	229,666	159,662	70%
	25-Dec	229,149	167,131	73%
1813	25-Jun	233,316	169,516	73%
	25-Dec	239,431	180,991	76%
1814	25-Jun	234,108	173,841	74%
	25-Dec	199,457	134,956	68%
1815	25-Jun	190,573	144,288	76%
	25-Dec	176,417	123,298	70%

The Royal Navy and Expeditionary Warfare

The early military successes in the West Indies had allowed the burgeoning British mercantile sector to flourish and consequently generate essential tax revenue. However, though there were several significant naval victories during the Revolutionary war, the British navy could not claim dominance of the oceans until after the defeat of the combined French and Spanish fleet at Trafalgar in 1805. This was further secured by the subsequent capture of the Danish fleet in 1807, the removal of the Portuguese fleet to Brazil also in 1807 and the capture of a Russian squadron of seven ships of the line in the Tagus in 1808.[24] The Spanish fleet changed sides to support Britain in 1808 and seven French ships of the line that were sheltering in Cadiz harbour at the time surrendered to Spain. Subsequently, the British naval strategy of blockade more or less cleared the oceans of enemy shipping, although the threat from privateers persisted and in fact increased during the war with America when American privateers joined the fray.

The naval contribution to major expeditions was multifaceted. It ceaselessly provided armed protection to the vast number of convoys of merchant shipping and transports carrying troops, materials and supplies. Often, in the early stages of an expedition, the navy was called upon to provide essential artillery fire to support contested landings, such as in Egypt in 1801 and at Copenhagen in 1807, where it provided heavy firepower to support the attack on the town. Naval marines often joined landing parties ashore alongside the army to create havoc upon coastal targets.[25]

Much of the manhandling of heavy artillery during disembarkation, and occasionally ashore, was done by the navy. During some landings the navy demonstrated its expertise in manhandling heavy guns by hoisting them from the decks of ships anchored close inshore and then up cliff faces to achieve a commanding position over the surrounding countryside. Where it was practical, the navy continued to support the movement of troops and materials around coastlines as the army progressed to new locations. In Spain it supported guerrilla operations to attack coastal fortresses and other military targets. It blockaded ports to prevent the enemy receiving further troops, artillery, supplies and communications by sea. In 1813 Wellington told Earl Bathurst, the Secretary of State for War, that he could not undertake an attack on San Sebastian until he knew 'whether we are secure at sea'.[26] In 1814 Wellington relied upon the

[24] These include: in 1794 Admiral Hood at the glorious battle of the First of June; in 1797 Admiral Sir John Jervis (later Earl St Vincent) at the Battle of Cape St Vincent and Admiral Adam Duncan at Camperdown; and in 1798, Nelson's Battle of the Nile at Aboukir Bay.

[25] Marines served on board all warships over ten guns under naval discipline. Besides guard duty they worked alongside the naval gun crews. They were granted the title Royal marines on 29 Apr. 1802. In 1801 roughly one sixth of a ship's company would be marines, increasing to a fifth in 1808.

[26] Gurwood, *Dispatches* I, 583: Wellington to Earl Bathurst, 3 Jul. 1813.

navy to enter the river Gironde and attack the forts protecting Bordeaux after his advance force had taken the town, to allow navigation of the river to supply his forces.[27] Finally, the navy's ability to support evacuations saved the day on more than one occasion when events demanded the removal of large numbers of troops at short notice. The more successful of these evacuations were from Toulon in 1793, from Puerto Rico in 1797 and from Corunna in 1809. Probably the most unsuccessful, resulting in the greatest loss of life, was that of the émigré force from Quiberon Bay following the ill-fated Vendée expedition in 1795.[28]

The evacuation at Corunna in January 1809 demonstrated to the Cabinet that a large number of troops could be removed from a theatre of war quickly if necessary but that to achieve such a removal sufficient tonnage of transports had to be retained in the vicinity until the contingency was no longer required. Up to this time consideration of emergency evacuation had rarely been particularly explicit in expedition planning. That lesson was learned. Large numbers of transports were consolidated in Portugal's river Tagus in 1810 in anticipation of an emergency withdrawal, which subsequently proved to be unnecessary.

Although he was a regular critic of the navy, even Wellington recognised the importance of its protective role. In October 1812 he wrote about the security of navigation to and from Britain and then around the northern and western coasts of Spain and Portugal, describing this as 'very important for our hospitals, our provisions, our stores, our money, the stores and clothing for the Spanish and Portuguese army … the whole of which must be transported by sea'.[29] In 1813 he spoke of 'our maritime superiority [which] gives me the power of maintaining my army while the enemy is unable to do so'.[30]

There were generally more naval ships than crews to man them and many commissioned ships were undermanned. The Admiralty was adamant that it did not have enough ships to enable vessels to be spared to convey troops and materials, so, as in earlier wars, the government relied comprehensively upon privately owned merchant shipping to accomplish this. Hundreds of such ships were needed, and rather than the government commandeering these vessels they were chartered on the open market, in competition with merchants who wanted to hire them for trade.

The Transport Service

Merchant transports supported all major campaigns and at the same time conducted the regular business of maintaining the supply lines to support the army and navy abroad. An army overseas needed constant replenishment of officers, men, horses, heavy artillery, light weapons, ammunition, camping equipment, clothing, shoes, food, forage, and specie to pay the troops and

[27] Hall, *Wellington's Navy*, 231–2.
[28] Duffy, 'Science and Labour', 51.
[29] Gurwood, *Dispatches* II, 439–40.
[30] Hall, *Wellington's Navy*, 111 quoting from Sir R.V. Hamilton (ed.), *Letters and Papers of Admiral of the Fleet Sir T. Byam Martin*, vol. 2 (London, 1898), 409.

purchase local supplies. Once overseas, transports were often retained to be used as mobile magazines (warehouses) and to support the numerous tactical movements by sea as the focus of battle shifted. Wellington acknowledged this important contribution during the Peninsular campaign. He advocated this philosophy to his subordinate commanders:

> I kept the sea always on my flank; the Transports attended the movements of the army as a magazine; and I had at all times, and every day, a short and easy communication with them. The army, therefore, could never be distressed for provisions or stores, however limited its means of land transport; and in case of necessity it might have embarked at any point of the coast.[31]

Later, in August 1811, Wellington confirmed his army's dependence on the transport service when he wrote that his force subsisted almost entirely upon supplies brought from the coast whether imported from England or elsewhere.[32] The availability of transports which, under naval protection, provided the capacity for Britain to move battalions of troops, horses, artillery and supplies and then to keep them resupplied meant that by 1811 France and its allies did not possess a single overseas territory.[33]

The Commissariat

The Commissariat was responsible for establishing magazines (warehouses) of stores for all the materials and clothing, provisions and supplies that the army needed to sustain battle readiness. Where possible it procured food and forage locally and it hired transport carts for moving the supplies. The Commissariat service was headed by the commissary-in-chief. Like the Transport Board it was a subsidiary board of the Treasury.[34] Although it fulfilled a critical role and clearly achieved some degree of success, its reputation was not high. Wellington made his views clear in a letter to Castlereagh: 'This department deserves your serious attention. The existence of the army depends upon it, and yet the people who manage it are incapable of managing anything out of a counting house.'[35] The Commissariat was the subject of various government committee reports including the Eighteenth Report of the Office of the Commissioners of Military Enquiry, which closes with the following comment:

> various salutary and effective regulations have been latterly introduced into it; and that from the experience which must be gained by the active campaigns in Portugal, it may be expected that the Department will daily advance towards

31 Duffy, 'Festering the Spanish Ulcer', 16, and Hall, *Wellington's Navy*, 32.

32 Gurwood, *Dispatches* VII: Wellington to the Earl of Liverpool, 1 Aug. 1811.

33 Knight, *Britain Against Napoleon*, 189.

34 The commissary-in-chiefs were Sir Brooke-Watson 1798–1806, Thomas A. Coffin 1808–10, Lieutenant Colonel Willoughby Gordon 1810–11, John Charles Herries 1811–16.

35 Gurwood, *Dispatches* III, 62: letter to Viscount Castlereagh, Lavos, 8 Aug. 1808.

> a higher degree of perfection. Yet we think that considerable attention is still requisite to give … all the efficiency they are capable of.[36]

In May 1812 the Commissariat in Portugal had a complement of 552 made up of 3 commissaries general, 9 deputy commissaries general, 2 acting deputy commissaries general, 10 acting assistant commissaries general, 39 deputy assistant commissaries general, 2 acting deputy assistant commissaries general, 183 clerks, 61 storekeepers, 20 interpreters, 28 issuers, 114 conductors, 22 coopers, 3 bricklayers, 2 wheelwrights, 3 carpenters, 2 blacksmiths, 1 farrier and 37 bakers.[37]

The main task of the Commissariat was to ensure that the forward forces were constantly supplied with equipment and provisions. This was a very significant task as is illustrated by the following instruction issued by Wellington to Colonel Framingham, Colonel Fletcher, his chief engineer, and Mr Kennedy, Commissary, in July 1811. A total of 1,600 barrels of gunpowder needed to be transported from Lisbon to Oporto. In addition heavy ordnance, siege artillery and stores and engineers' stores were embarked in eleven transports moored in the Tagus in March and then two months later were shipped via Oporto as far as was navigable up the river Douro. The supplies were needed at Ciudad Rodrigo in preparation for the forthcoming battle, which occurred the following January, and were also used at Badajoz in April 1812.[38] They had to be transhipped into local boats at Oporto and ferried up the Douro to Lamego where they were landed and then hauled overland to Ciudad Rodrigo via Trancoso. This arduous task required 384 pairs of bullocks for the ordnance, 892 carts for the heavy guns and mortars, 200 carts for the engineering equipment and hundreds of men. It was estimated that this operation would take 12 days for the journey to Oporto, 6 days to remove stores and ordnance to boats, 6 days from Oporto to Lamego, 4 days from Lamego to Trancoso, 4 days for carts to return and load, 4 days for two trips to Trancoso, 6 days to Ciudad Rodrigo, 6 days to return to Trancoso, and 6 days to bring a second quantity of stores. In total the operation would take 62 days.[39]

The Administration of War Policy

Successive British Cabinets were well aware that they needed partnerships with other nation states to enable them to defeat France.[40] Many of the British military campaigns were initiated to support or encourage the uprising of indigenous populations or co-ordinated actions against France by other states. There were

[36] HoCPP, 1812 (119), Eighteenth Report of the Commissioners of Military Enquiry, Office of the Commissariat, 20 Mar. 1812, 308.

[37] HoCPP, 1812 (232), Return of the Number of Persons Employed in the Commissary General's Department in Portugal, J.C. Herries, Commissary in Chief's Office, 5 May 1812.

[38] Hall, *Wellington's Navy*, 114.

[39] Gurwood, *Dispatches* IV, 9 Jul. 1811: Wellington to Cols Framingham and Fletcher and Mr Kennedy.

[40] See p. 9.

seven coalitions of states opposed to France during the wars. The first was established in 1792 by Prussia and Austria; Britain and Holland joined in 1793 followed by Spain, Portugal, the Two Kingdoms of Sicily and the Kingdom of Sardinia. The general pattern of these coalitions was that the partner states were successively forced to sign peace treaties with France, leaving Britain to stand alone. The extent of support that Britain received from foreign powers was generally based on the level of subsidy it was prepared to pay. During the twenty-two years 1793–1815, £65,830,228 was paid out in subsidies.[41] Nevertheless, Britain frequently found former allies became adversaries after defeat by the French, who then used their armies and navies against her. Spain declared war on Britain on 8 October 1796 and was a French ally until 1802. She was again at war with Britain from December 1804 until 1808 when Britain supported the Spanish uprising against France. Russia declared war on Britain on 2 December 1807 and Sweden declared war on Britain on 17 November 1810 but both joined Britain in the sixth coalition in 1813.

For most of the period the initiative lay mainly with the enemy and Britain could do no more than react to events as they occurred.[42] Amphibious operations were the most complex and costly operations attempted by the British state. As such they were overseen and co-ordinated by Cabinet ministers and sanctioned by the prime minister because the decision to send an expedition, to commit so many resources and so much money, was likely to have a direct effect on any plans for other theatres of war and on foreign policy. Small Cabinet cabals generally made the ultimate decision to embark upon a campaign, based upon available intelligence and limited pre-planning. They had to wrestle with the complex issues of monitoring the progress of the war, maintaining diplomatic relations, encouraging and supporting the suspicions and discordant ambitions of allies, the limitations of sea power, the conflict between military and political objectives, and the difficulty of controlling British military efforts from London. Invariably there were divergent views within the Cabinet on how to proceed. During the Revolutionary war, when Pitt was prime minister, war policy was largely created by Pitt and his two senior ministers, Henry Dundas, the Secretary of State for War, and Lord Grenville, the Foreign Secretary, occasionally joined by Lord Spencer, the First Lord of the Admiralty. Dundas preferring to attack the enemy trade centres in the West Indies whilst Grenville favoured European intervention.[43] When a policy was decided, the Cabinet, through the secretary of state for war and the first lord of the admiralty, issued orders to the War Office and the Admiralty. These departments made their own plans and issued their own instructions.

The secretary of state for war was also able to give orders direct to more junior boards including the Transport Board, the Ordnance Board, the

[41] Knight, *Britain Against Napoleon*, 388, quoting John Sherwig, *Guineas and Gunpowder: British Foreign Aid in the Wars Against France 1793–1815* (Cambridge, Mass, 1969), 345, 367–8.

[42] Mackesy, *War in the Mediterranean*, viii.

[43] Knight, *Britain Against Napoleon*, 64.

Commissariat and the Victualling Board. These boards caused ships to be chartered, ordnance to be made available, regiments to prepare their battalions to march to designated embarkation points and provisions to be procured. Once the embarkation location had been determined it was the responsibility of the transport agents to co-ordinate the proceedings by ensuring that sufficient transports were prepared, that the troops, horses and supplies were embarked safely and that sufficient stores and provisions were available at the points of embarkation. The navy was responsible for ensuring that the convoy was delivered speedily and safely to the disembarkation point and assisted with the landing. Once the army, its equipment and provisions were ashore the responsibility for success fell to the military command supported by the Commissariat, whilst the naval station commander-in-chief generally played an integral role by defending the supply lines.

Military Planning

Systematic strategic planning as we understand it today was unknown in the early nineteenth century. Indeed, the principal military policies of the various British administrations were driven mainly by the actions of others, both allies and enemy; they were essentially reactive and opportunistic. The major handicap to military planning was the lack of reliable, up-to-date intelligence. Throughout the wars considerable volumes of intelligence were available to the government, from a host of sources. The need to organise a formal process of integrating and analysing the reliability and importance of the information and communicating it to the relevant departments was recognised, but plans for establishing a 'Depot for Military Knowledge' failed simply because there appears to have been insufficient accommodation to house the department.[44] Thus the Cabinet received a plethora of raw data, often conflicting. In particular, the foreign secretary received regular commentaries from diplomats based overseas, each of whom had a large number of contacts and spies. There were numerous other sources of intelligence, including naval officers, merchants and merchant seamen.

Detailed local knowledge was important when planning campaigns. Whilst there was often a senior naval officer with some experience of the relevant coast and there were political representatives who had experience of the country in question, there was usually a dearth of data on the terrain because there were few reliable maps. Insufficient knowledge of the enemy strength was also a significant weakness. When planning the Egyptian expedition in January 1801 Sir Ralph Abercromby was advised by Sir Sydney Smith's secretary, who had visited Cairo, that the French had 20,000 troops. At the same meeting Consul

[44] Knight. *Britain Against Napoleon*, 287.

Baldwin said it was 12,000, which roughly matched the advice that Abercromby had received from London. The reality was much closer to 30,000.[45]

Governments were less stringent with security than is the case today. Intelligence that was very helpful to the enemy was regularly published in newspapers, not only in Britain but also in France. Both governments devoured foreign newspapers for intelligence. On Wednesday 27 July 1808 *The Morning Chronicle* published a 'Return of the arrangements of the forces destined for Spain and Portugal'. This identified each battalion that had sailed under Wellington and Major-General Spencer, detailing its strength and commanding officer, and reported that a second expedition was being prepared at Deal and Chatham.[46] It was quite usual for all preparations for expeditions to be fully reported and it can be assumed that as such they were well known in France, thus losing any benefit of surprise. Similarly, foreign newspapers were important sources of intelligence, if allowances were made for their propaganda content and accuracy.[47]

Politicians were not military or logistical experts and yet they often resented negative commentary on their plans, even from those with relevant experience. Pitt was culpable of this during the preparations for the 1799 den Helder campaign when, despite 'admitting his own ignorance of military organisation', he was irritated by objections raised by Sir Ralph Abercromby to plans conceived by the Foreign Secretary Lord Grenville for political reasons, with no regard for military problems'.[48] In what has been described as 'a final triumph of political over military considerations', Pitt finally assured Grenville that 'all military difficulties are over-ruled'.[49] This interference was contrary to the position adopted by Henry Dundas, Secretary of State for War. He had refrained from giving Abercromby prescriptive instructions:

> I have avoided entering into any detail of the military operations which this service may require, wishing to leave this entirely to your own judgement to direct the movements of the Army according to the intelligence you may procure on the spot, of the strength of the enemy, the disposition of the inhabitants and other circumstances which must govern your discretion.[50]

Surprisingly, the designated commander-in-chief was often excluded from the early stages of planning. When the Duke of York commanded the army in Flanders between 1793 and 1795, and again in 1799, he was not involved in the

45 Piers Mackesy, *British Victory in Egypt: The End of Napoleon's Conquests* (London, 1995), 45.

46 *The Morning Chronicle*, Wednesday 27 Jul. 1808.

47 Gurwood, *Dispatches* VII, 210: Wellington to Lieutenant General Sir T. Graham, 19 Dec. 1813.

48 Mackesy, *Egypt*, 11.

49 Mackesy, *Egypt*, 10.

50 TNA, WO, 6/20, Instructions to Expedition Commanders-in-Chief. Correspondence relating to the planning of den Helder campaign, Dundas to Lieutenant-General Sir Ralph Abercromby, 10 Aug. 1799.

planning of the operations prior to embarkation. Chatham was not appointed to lead the Walcheren campaign until May 1809; by that time Castlereagh had already taken advice from others and determined the overall strategy. Such planning as occurred was generally restricted to the military campaign; it does not appear to have considered the wider consequences of the operation. For instance the launch of major expeditions in 1794–95 and 1801 to the West Indies, the Mediterranean and the Baltic seriously exacerbated the already serious food shortages being experienced in Britain. Similarly the government's demand for feed for horses required for den Helder in July 1799 amounted to 10 million pounds (4,464 tons) 'of good sound sweet heavy oats … three million pounds of good fresh wheat, three hundred thousand pounds of rye meal' and had to be procured as secretly as possible. Nevertheless this had a significant temporary impact on availability and price.[51] Neither was the impact anticipated of withdrawing large numbers of collier ships from their normal trade to be used as horse ships. Coal shortages occurred and prices rose.

Whilst den Helder campaign was in progress, Dundas was planning another major offensive, an ambitious operation to capture the French naval base at Brest. On 22 August 1799 Dundas asked Lord Spencer, the First Lord at the Admiralty, to recommend the most favourable landing site for the assault, which would involve 30,000 to 40,000 men, on the French coast between St Malo and L'Orient. He sought information on the coastal defences and the state of the roads, to support inland transportation of artillery and supplies. In addition he wanted to know the enemy strength in the region, an estimate of the size of force that might be required to capture Brest and how long it would take to bring them forward for service. He asked what level of support the navy might supply to effect the task. Spencer, being a political appointee rather than an experienced naval officer, turned to Admirals Sir Edward Pellew, Sir Richard Strachan and Sir John Warren for advice. The admirals identified Douarnenez Bay, south of Brest, on the Brittany coast as the preferred landing site. The bay would afford a good anchorage for a large fleet, even in south-westerlies, the prevailing winds in the latter part of the year. However, though this location was advantageous for the navy, it would create challenges for the army. Firstly, it was some 35 to 40 miles from Brest. The army would have to cross two or three rivers to reach Brest, and the roads had been neglected and were not in a good state.[52]

When General Sir Charles Grey was advised of the plan he explained to Dundas that it 'appears so full of difficulties', referring to the distance from Brest and the four rivers (not two or three as had been assumed) to be crossed. He was also concerned that attempting a landing so late in the year risked the ships of war and transports being driven from their anchorages during the disembarkation. He was inclined to think that the expedition should be deferred until the following spring. He also suggested that a larger force of no fewer than 70,000 men would need to be committed to this expedition, allowing for two

[51] Knight, *Britain Against Napoleon*, 104 and 159.

[52] Henry Dundas Letter Book, The Sim Comfort Collection, Spencer to Pitt, 2 Sept. 1799.

Plate 1. Invasion of Holland: landing of British troops near Callantskoog, 27 August 1799

forces of 6,000 to operate as feint attacks elsewhere to distract the enemy from the real landing site and for sufficient forces to be stationed to protect the rear-guard and means of escape. He also pointed out that Brest straddled a river and the operation would require a force capable of occupying both banks. A previous study had indicated that Brest might be defended by some 37,000 French troops, but because it was only 276 miles from Paris many more troops could be rushed forward very quickly. It was Grey's belief that more up-to-date intelligence was required and that above all secrecy was essential for the success of such an operation.[53]

Only in November did Dundas invite William Huskisson, his under-secretary, to review the plans. He quickly identified the magnitude of the proposed operation and its 'bearings and influence on all subordinate plans you have in contemplation'. He highlighted the practical difficulties of bringing such a large body of troops together quickly and also of chartering at short notice the large number of troop transports, ordnance ships, victuallers and store ships that would be required. He estimated that the planned force would take up all the transports then on hire (approximately 100,000 tons) and would require a further 200,000 tons, still leaving 40,000 men to be transported in converted naval ships.[54] Despite winter being the time of greatest availability

[53] Henry Dundas Letter Book, The Sim Comfort Collection, Sir Charles Grey to Henry Dundas, 15 Sept. 1799.

[54] Henry Dundas Letter Book, The Sim Comfort Collection, Huskisson to Dundas, 30 Nov 1799.

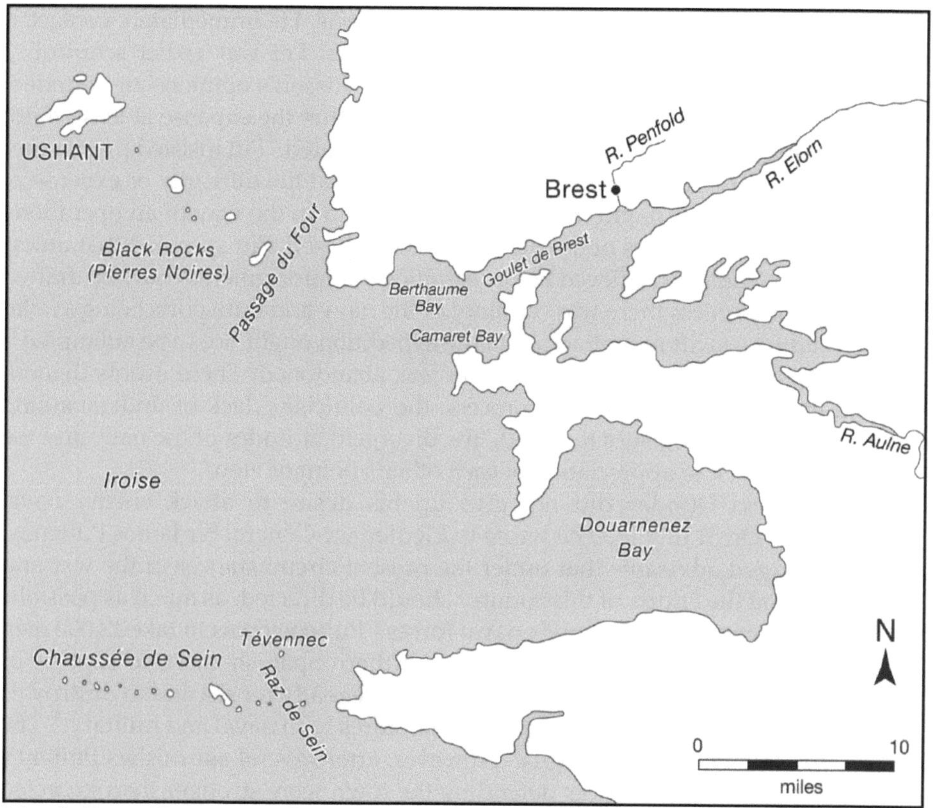

Map 2. Brest, Douarnenez Bay

of merchant ships, this volume of shipping could just not be readily chartered without creating an immense adverse impact on trade, particularly the Baltic trade which was so important for grain and shipbuilding materials. Huskisson's assumption about capacity stands up to careful scrutiny. The peaks in 1805, 1808, 1809 and 1814 of tonnage on hire were 261,000, 250,000, 278,000 and 231,000 and these transports were taken, at a high cost and with great difficulty, from a significantly larger merchant shipping fleet than was available in 1799.[55] Huskisson was firm in his belief that this expedition could not be prepared before early March in the following year: 'I entreat you to bear in recollection the experience you have of the time requisite for fitting, preparing and collecting the tonnage required by the different departments for great expeditions, and the time they are in getting ready even after tonnage is allotted to them.'[56]

[55] Sutcliffe, 'Bringing Forward Shipping for Government Service', 58.
[56] Henry Dundas Letter Book, The Sim Comfort Collection, Huskisson to Dundas, 30 Nov. 1799.

Dundas was shocked by Huskisson's conclusions. He immediately contacted Pitt to explain the potential impact on their plans. Pitt was rather scornful: 'I should like to know the particulars on which Huskisson's opinions are founded, but from all that I recollect I cannot help thinking that the expense at least, if not the difficulty, of providing transports, is much overrated.' Pitt insisted that if other concerns could be addressed then 'I cannot think that the difficulty or expense of procuring Transports ought, for a moment, to stand in the way of an operation.' However, the weaknesses of the plan had been exposed. Pitt specified that unless the objectives could be achieved before significant reinforcements could be drafted from Paris and unless there was certainty of the navy and transports being available to conduct a swift re-embarkation, the expedition ought not to be attempted.[57] After some further deliberation the plan was abandoned. These events demonstrated the gaps in the planning process: the politicians' lack of understanding of the extent of the logistics required, 'the divergent attitudes of the navy and the army and the lack of appreciation of each other's point of view'.[58]

Nevertheless Dundas did not give up his desire to attack enemy naval strongholds. On 31 July 1800 he wrote to Lieutenant-General Sir James Pulteney, 'It being judged advisable that under the present circumstances of the war and of Europe that the efforts of this country should be directed, as much as possible, on the destruction of the enemy's naval forces.' Pulteney was to take 12,000 men to Ferrol, a significant naval base on the northern Spanish coast, to capture or destroy the six Spanish ships of the line that were ready for sea and to destroy its arsenal, dockyards, fortifications and public stores both naval and military.[59] The attack was launched on 25 August. However, after several skirmishes Pulteney decided that the enemy forces defending the town were stronger than expected and so re-embarked his troops without achieving his objectives.

Egypt, 1801

Following his appointment as commander-in-chief of the Mediterranean forces in 1800, Abercromby sailed with 5,000 troops southwards to join Admiral Lord Keith, Commander-in-Chief of the Mediterranean station. He was ordered to launch an attack upon Cadiz but the attempted landing was a fiasco. It was aborted when Abercromby realised that the size of the initial landing force would not be strong enough to launch the attack. Keith bore the responsibility for the shambles; his apparent incompetence was described by General Moore to his father as 'bad management and confusion which attended the assembly of the boats' for landing the army.[60] Another army officer described the attempt as 'most sadly bitched'. The convoy then fell foul of a strong gale.

[57] Henry Dundas Letter Book, The Sim Comfort Collection, Pitt to Dundas, 2 Dec. 1799.

[58] Knight, *Britain Against Napoleon*, 193.

[59] WO 6/21, Dundas to Lieutenant-General Sir James Pulteney regarding plans to attack Ferrol, 31 Jul. 1800.

[60] Mackesy, *Egypt*, 38–9. Keith is described as 'a selfish careerist who held the record for accumulating prizes … and who would never share the responsibility of a risky

Whilst regrouping and recovering from the recent setbacks Abercromby received instructions from Dundas to proceed to Egypt to dislodge the French and restore Turkish rule. Napoleon had invaded Egypt in 1798 and routed the Turks but Nelson's destruction of his fleet at Aboukir Bay had stranded his army. Napoleon himself escaped, but for some time it suited the British government to have this substantial French force removed from the European theatre. However, after the French defeated the Austrians at Marengo there were indications that a negotiated peace might be inevitable. The British government was fearful that France would attempt to retain possession of Egypt in such negotiations, a situation that it could not contemplate, envisaging that this foothold would prepare the way for a French invasion of India which was a vital trading resource for Britain.

Dundas had spent many weeks trying to persuade his Cabinet colleagues that it was essential to drive the French from Egypt, to protect Britain's Indian flank, and that Abercromby's force was ideally positioned to achieve this. He was to take an army of 14,000 to Egypt and to send 8,000 others to Portugal to assist against a threatened attack from Spain. In October 1800 Dundas wrote a remarkably frank letter to Abercromby, to update him on the various discussions which had taken place with regard to the expedition.[61] He was convinced that the loss of Egypt would have a significant impact on the French but recognised that 'it would be unfair and uncandid in me not to acknowledge that if the country meets with any serious disappointment in the event or consequences of this enterprise, the responsibility does and ought to rest upon me'. He pointed out that many doubts had been expressed against this enterprise by 'persons of great respectability' and that they had made a serious impression on him. They all agreed that whilst the period between December and March was perceived to be the best for any operation on land, it was absolutely the wrong season of the year to conduct such an operation from a naval perspective.

Dundas had sought further advice from Rear Admiral Sir Thomas Troubridge (Naval Commissioner of the Admiralty, 19 February 1801–15 May 1804), whom he believed to be the most competent authority on this subject. Troubridge concluded that 'this is not the season which one could adopt as a matter of choice'. However, he did not advise against a naval operation at that time of year. The earliest time reckoned to be best for both sea and land operations was not until the following May. Abercromby was invited to use his own discretion. Dundas confirmed that if he decided that the attack was not feasible then he should postpone the operation. Dundas confided that 'in your hands I feel that the interests of the country are safe … no false point of honour or feeling will induce you to attempt what is too hazardous to be attempted'. However, there is a hint of pressure being applied by his further comment: 'when great objects are at stake, and great services are in contemplation, you are not to be startled by

decision if he could avoid it'.

61 Mackesy, *Egypt*, 14.

those difficulties to which characters of less energy or less standing than yours [Abercromby's] may allow themselves to yield'.[62]

However, after months at sea and after sustaining damage during recent storms the fleet needed repairs and replenishment of stores. Horses were also required for the cavalry and more artillery. Lord Keith intended to take the fleet to Rhodes but a south-east gale prevented this and so he settled on the Bay of Marmaris, on the Turkish coast, for his temporary anchorage whilst the expedition was prepared.

Preparing an Expedition

A major aspect of any expedition planning process was the selection of the commanders-in-chief. The secretary of state for war nominated the military commander and the Admiralty appointed the senior naval officer, but sometimes they do not appear to have selected compatible individuals, and as neither was appointed supreme commander this led to some very strained relationships. The success of an operation was more assured if there was mutual respect between the senior commanders. Dundas's appointment of Sir Charles Grey and the Admiralty's appointment of Vice-Admiral Sir John Jervis guaranteed compatibility for an operation dependent upon inter-service co-operation because Jervis and Grey were old friends.[63] The success at Copenhagen in 1807 could partially be attributed to the good working relationship between Admiral Gambier and Lord Cathcart. Indeed Wellington and Vice-Admiral Berkeley in Portugal 1808–12 also had considerable mutual respect.[64] Less successful was the relationship between Abercromby and the 'prickly and defensive' Admiral Lord Keith, particularly following the aborted Cadiz landing. Nevertheless after receiving stern words from the secretary of state for war via the Admiralty, Keith modified his relations with Abercromby to the extent that they subsequently achieved significant success at Aboukir Bay on 8 March 1801. Later, the ill-advised combination of Rear Admiral Sir Richard Strachan and Lord Chatham to command the Walcheren expedition in 1809 inevitably led to weak decision-making and poor communications.[65] Hot-tempered Strachan was described as 'an irregular impetuous fellow, possessing very quick parts', while Chatham, who was then Master-General of the Ordnance Department, was 'without experience as a leader, or with the qualities necessary to the success of an enterprise which demanded decision of character and activity of mind and body'. Indeed, Chatham attempted to blame Strachan and the Admiralty for the ultimate failure of the expedition.[66] The problems arose partly because of the personalities involved but were also due to the significant gap in understanding of the challenges that each

[62] TNA, WO, 6/21, Dundas to General Sir Ralph Abercromby, Downing Street, 6 Oct. 1800.
[63] Duffy, *Soldiers, Sugar and Seapower*, 43.
[64] Rodger, *Command of the Ocean*, 428.
[65] Knight, *Britain Against Napoleon*, 197 and 208.
[66] Hall, *British Strategy*, 68.

commander faced. It has been observed that even Wellington 'was and remained, despite numerous explanations, woefully ignorant of what was nautically viable'.[67] Despite these illustrations it seems that in general the army and navy commanders on the ground found ways of working together amicably to achieve the common goal, even in Wellington's case, despite his frequent criticism of the Admiralty.[68]

Bringing together a large body of troops, horses, wagons, ordnance, victuals and shipping, preparing the ships, embarking the troops, arranging the stowage of supplies and provisions and finally assembling the fleet and arranging convoy protection constituted a huge operation. It required good communications and co-operation across the wide spectrum of government boards, ship-owners and contractor suppliers.

For large expeditions numerous ports were used not only to facilitate loading and avoid complete congestion in the harbours, but also to ease the strain on local accommodation and food supplies. Local food and oats prices generally increased when large numbers of troops and horses were being moved through a region. Marches were carefully scheduled in the Quartermaster-General's Office, in Horse Guards, London, to avoid too many troops arriving in the vicinity of the towns en route at the same time. During the preparations for Corunna in late 1808 six regiments and two troops of horse artillery, including 4,000 horses, had to be marched from their various barracks to their port of embarkation. It took two weeks for a whole brigade of the 15th Hussars to march from their barracks in Woodbridge (Suffolk) to Portsmouth via Romford (Essex), Kingston and Guildford (Surrey), and Chichester (Sussex). Although the troops were divided up into squadrons to minimise disruption at the places en route, some towns on the main routes would often experience weeks of battalions marching through every day.[69]

Vast influxes of army personnel who were to be embarked put enormous strain on the main departure points such as Ramsgate, Deal, Chatham, Falmouth, Portsmouth, Weymouth and Southampton. In April 1800 it was reported that an encampment of 30,000 men, mainly cavalry, was to be established at Southampton. They were expected to remain there for six weeks prior to embarkation.[70] It was not unusual for troops having marched to a designated embarkation point to have to march to an alternative location to meet transports that had been diverted by adverse weather conditions. The 63rd Regiment was at Deal in May 1800 awaiting embarkation but a north-east wind with a heavy on-shore surf prevented this and consequently the troops had to be marched to Dover to take ship.[71]

[67] Hall, *Wellington's Navy*, 223.

[68] Hall, *Wellington's Navy*, 224.

[69] Knight, *Britain Against Napoleon*, 185, quoting Brian Austin, *English Provincial Posts 1633 to 1840* (Chichester, 1978), 81.

[70] Burney Collection Newspapers, *Morning Post and Gazetteer*, Wednesday 30 Apr. 1800.

[71] Burney Collection Newspapers, *True Briton*, Friday 16 May 1800.

It was not an exercise that could be conducted quickly or clandestinely. Preparations for den Helder expedition at Ramsgate in August 1799 were described in one newspaper as follows:

> Last night and early this morning about fifty transports with troops on board sailed from this port to the Downs; each transport had about 300 men on board. The wind is contrary and they are still in sight (at one o'clock) beating against it. More troops are constantly marching in for embarkation. The 85th arrived here this morning. A considerable number of transports are lying in the harbour prepared for cavalry, some of which are expected to embark this night. This place is all bustle and confusion; troops marching in and embarking, the parting of friends present altogether a very interesting spectacle.[72]

The press regularly speculated upon the destinations of the fleets in preparation, as in 1804: 'We hold from unquestionable authority that the transports for which the government are so largely contracting in London are going to Gibraltar, as the season does not admit the Baltic nor other northern destinations, probably sending an army to aid in Portugal.'[73] Such press reporting and speculation inevitably alerted the enemy to the preparations which were in hand.[74]

Often expedition plans failed to include sufficient contingency for transporting enough cavalry horses, artillery horses and wagons in the initial wave of transports. Indeed one of the major complaints of the military commanders was that inevitably too few horses were sent with the force or sufficient horses were not sent in the early shipments. During the den Helder preparations Abercromby told Dundas that 'no army can march without horses or wagons'.[75]

Sailing vessels had restricted manoeuvrability so even in the largest harbours there was insufficient space for safe mobility and too few moorings for all the necessary troop ships, victuallers, store ships and naval vessels. Once loaded, transports had to be moved on to allow others into the harbour. They gathered in huge numbers at Spithead and in the Downs to await the embarkation of the whole force, the loading of the supply ships and finally the arrival of the naval escort. It was not unusual for those ships loaded early in the process to have to be at anchor for four to six weeks before commencing their voyage, with the obvious impact on the morale and fitness of the troops and the well-being of the horses. At such times Spithead and the Downs were exceptionally busy stations: ships were constantly coming and going; supply ships and smaller troop ships were arriving daily to offload additional troops or further supplies.

72 Burnley Collection Newspapers, *Courier and Evening Gazette*, Saturday 10 Aug. 1799.

73 Burnley Collection Newspapers, *Hampshire Telegraph*, 10 Dec. 1804.

74 Richard Glover, *Peninsular Preparation: The Reform of the British Army 1795–1809* (Cambridge, 1963), 34, regarding the speed of French spies in obtaining intelligence of British operations.

75 Mackesy, *Egypt*, 10.

Disembarkation

The success of the dis-embarkation played a significant role in the immediate fortunes of military campaigns. The British army did not have a good track record of successful, co-ordinated dis-embarkation, despite having significant experience.[76] At den Helder in 1799 landings of the Anglo-Russian force 'were far too slow, unloading was chaotic thus losing any strategic advantage of surprise'.[77] The troops were landed on the beach in no particular order, mingling battalions and brigades; men had to resort to searching the beaches for their units. Later, the attempted landing at Cadiz was abandoned when it deteriorated into a fiasco.[78] An ideal landing required reasonable weather to ensure that the early assault wave could be reinforced and that supplies including heavy artillery could be landed speedily. The selection of a suitable landing site was important. The ability to offload along riverbanks was preferred, particularly for horses and heavy equipment. However, the draught of larger vessels frequently prevented them from crossing sandbanks, and other obstacles generally associated with river entrances meant that such items had to be offloaded into smaller vessels to facilitate landings. Most frequently landings were onto beaches in a bay as close as possible to the military target, yet sometimes this was not possible, forcing the army to march some distance, hauling equipment and supplies, before engaging the enemy.

Landings were directed by the use of predetermined flag signals by day and lights during hours of darkness. Instructions issued in May 1808 illustrate this process. There was a specific triangular flag to denote the brigade which was combined with further flags to identify the appropriate regiment. For instance the 1st brigade of the 28th Regiment was identified by a plain red flag over a white pendant but when the red flag was over a red pendant it related to the 1st brigade of the 4th Regiment. A red flag with a central white stripe identified the 2nd brigade of the 79th Regiment when over a red pendant and the 2nd brigade of the 92nd Regiment when over a white pendant. In this manner the 'controller' on the beach was able to direct instructions to specific brigades.[79] When anchoring at the dis-embarkation point, ships with officers commanding brigades on board were to hoist the triangular flag of that brigade. This would enable transports with troops from the same brigade to anchor close by. Further instructions were relayed via a series of numerical flags, often between 50 and 100. Each flag denoted a specific predetermined instruction relating to the disembarkation process.

[76] Mackesy, *Egypt*, 40, Mackesy suggests that forty years earlier in the Seven Years war Commodore Hotham had made a reputation for smooth and successful landings on the American coast and in the Caribbean but that these experiences had not been systematised.

[77] Knight, *Britain Against Napoleon*, 192.

[78] Mackesy, *Egypt*, 38.

[79] TNA, WO, 37/10/7.

Assault force landings could be particularly hazardous in anything but a relatively calm sea condition. This is demonstrated by the loss of troops during the landing at Aboukir Bay in 1801 and of forty-one soldiers from the 93rd Highland Regiment during the landing at the Cape of Good Hope in January 1806.[80] The experience of a beach landing is illustrated by this description of a landing at Maceira Bay, just north-west of Lisbon, Portugal, on 28 August 1808, where the raging breakers were 'raising their heads houses high as they rolled in from the Atlantic':

> the flags were flying directing the landing, informing the fleet of the ebb and flow of the tide and the state of the breakers so that the disembarkation of the troops could be properly timed. ... At the first line of surf 20 to 30 British seamen, all quite naked, made several attempts to caste a rope to us and then ran at top speed through the advancing breakers up the beach, dragging us until the boat stuck fast. ... When the retreating wave had withdrawn sufficiently each of them took a soldier onto his back and carried him to the dry shore.[81]

The Landing in Aboukir Bay, 1801

Following the Cadiz debacle, Abercromby was determined that the landing in Egypt would be better planned. He was convinced that the arrangements could not be left purely in the hands of the navy: there needed to be army involvement in planning the order of precedence of the landings to ensure that the troops landed in the correct order of battle. Lord Keith and Abercromby waited in Marmaris Bay whilst negotiations with the Turks were finalised and further horses, transports, landing craft and supplies procured. The seemingly endless delays at Marmaris Bay provided the time to develop the plans for the disembarkation, to better prepare the troops for landing and to build the equipment that would be necessary to move heavy artillery across the soft sand of the targeted landing beaches in Aboukir Bay, which was some 12 miles from Alexandria, the first objective. The challenges faced by the two commanders-in-chief in the winter of 1800 illustrate how the difficulties of the normal process of preparing a force at home were exacerbated when that force was thousands of miles from Britain. The team that Abercromby had sent ahead of the fleet to secure provisions, horses and transports from Rhodes had met with little success and even the Turks had refused to provide supplies, so alternative time-consuming arrangements had to be made.

Meanwhile Keith appointed Captain Alexander Cochrane and Abercromby appointed his deputy quartermaster-general, Colonel Anstruther, to plan and rehearse the landing. The term 'beach-master' developed during this exercise. The strength of the initial assault force was important, but it was restricted by

[80] Knight, *Britain Against Napoleon*, 231.

[81] August Ludolf Friedrich Shaumann, *On the Road with Wellington* (London 1999), 1.

the available number of ships' boats, even though there were 180 boats for the infantry and twenty-eight for the artillery. The shallow waters of Aboukir Bay forced the larger warships and troop ships to lie well out to sea. Nevertheless Cochrane and Anstruther instituted timed practice landings and constantly revised the instructions, managing the process by flag signals from Cochrane's cutter which were repeated to the landing craft by captains of the boat divisions. The final plans, which included provision for the first wave of troops and artillery to be offloaded from the larger troop ships onto the smaller ships that could anchor much further into the bay, impressed even the battle-worn officers who had witnessed the confusion at den Helder and Cadiz.[82]

The fleet arrived at Aboukir Bay on the morning of 2 March. The force included 5 ships of the line, 2 frigates, 3 sloops, 3 bomb vessels, 3 armed cutters, 4 gunboats, 40 naval vessels converted to troop ships, 60 transports with troops and supplies and 57 Turkish vessels converted into supply ships.[83] Due to the shallow water, the ships of the line could not approach more than 7 miles from the shore, the transports 5 miles. Because of the distances involved Abercromby estimated that he needed ten hours of calm weather to get his army ashore, but the strong winds were creating a large swell. Whilst the selected landing site provided good anchorage for the warships in those conditions it was not just not suitable for landing troops. The heavily laden flatboats would not be able to make the long approach through the swell. This highlighted the gap between naval and military requirements, which even the careful planning for this expedition had failed to resolve.[84] This sea-state continued until the evening of 7 March when preparations were finalised to commence the disembarkation before daybreak on the following morning. At 2.00 am signals were made for the boats to begin receiving the troops in accordance with the previously rehearsed plan.

Map 3 shows the proposed anchored dispersement of the fleet, the furthest from the shore being the victuallers, ordnance ships, hospital ships and others not immediately wanted (FK), then the line-of-battle ships (AB), then in front of them the larger troop ships (CD), then the small troop ships and transports gathered around their respective transport agents (EF). The troop ships were protected by small armed craft and gunships (GH), with more gunboats as near as possible to cover the landing (LM).[85] Cochrane had planned that the first division of troops would be landed in five sub-divisions – the first to land to the right of the beach and the fifth to the left. The first was to consist of 751 men of the 28th Regiment and 111 of the 50th to be landed from HM's ships *Winchelsea*, *Druid* and *Blonde* with 203 rangers from the *Argo* transport: a total of 1,065 men to be landed in 21 flatboats each containing 50 to 55 troops and a crew of oarsmen.

[82] Mackesy, *Egypt*, 44.
[83] Christopher Lloyd (ed.), *The Keith Papers*, vol. 2 (London, 1950) 267–9.
[84] Mackesy, *Egypt*, 52.
[85] TNA, MFQ, 1/133/1, Egypt, Aboukir Bay, Plan of the coast showing the invading force in boats.

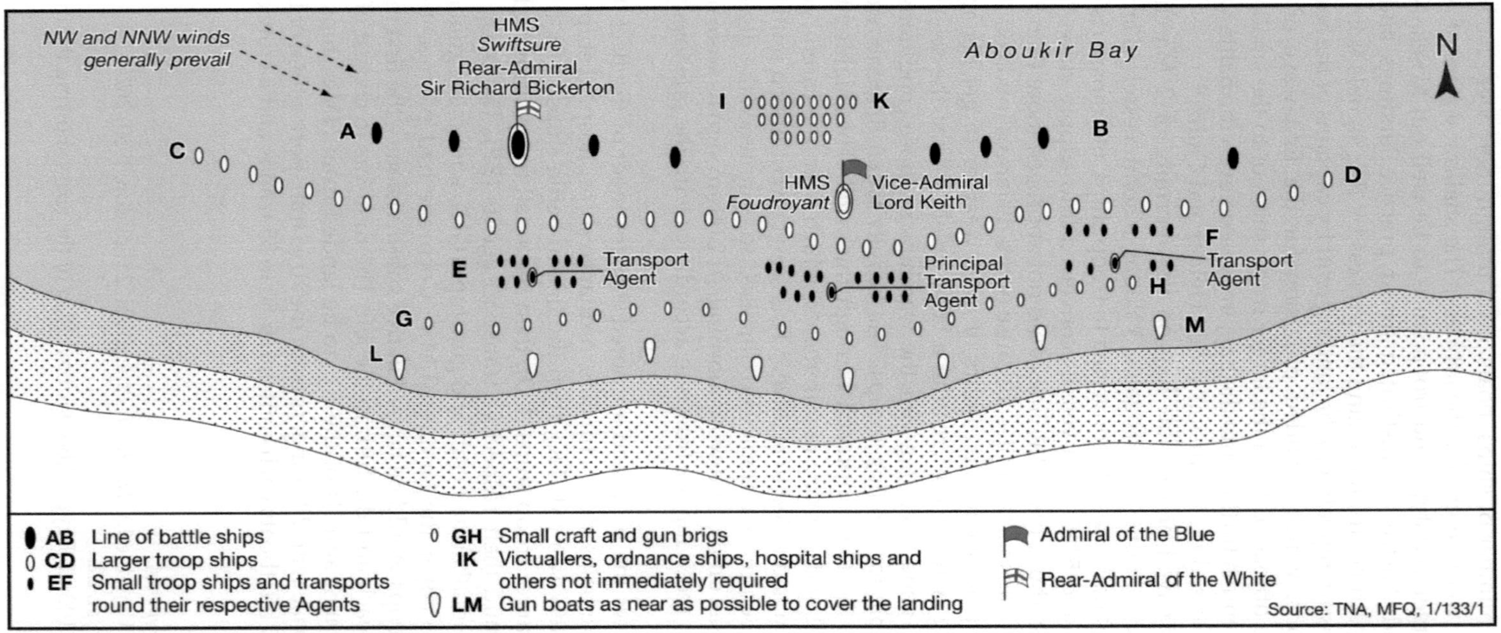

Map 3. Plan of landing at Aboukir Bay in March 1801.

The second sub-division consisted of 500 men of the 50th and 521 men of the 58th Regiment on board *Thistle, Alligator* and *Ulysses* in 20 flatboats, the third 1,050 men of the 42nd Regiment, the fourth 996 guards and the fifth 1,006 guards from *Dictation* and *Trusty*. In total this division was to consist of 5,132 troops. It would require 102 flatboat transfers, resulting in the boats making several trips.[86] Those ships with troops intended for the first division were to hoist an ensign on their main top gallant mast head which was to be hauled down when all the troops had transferred to the boats.[87] Cochrane's instructions for the landing, which had been drawn up at Marmaris Bay, were very comprehensive. He insisted that the troops were assembled on the ships by regiment, in the order of landing. The first boat for each regiment would be distinguished by flying the army colours of that regiment; the other boats with troops from that regiment were to keep to the left of that boat. The flatboat commander was to be instructed on which part of the beach he was to land his troops and was to maintain a distance of no less than 50 feet from the nearest flatboat. He was also to ensure that none of the troops stood up during the transit to avoid endangering the stability of the boat. On no account were the flatboats to cross each other; nor were they to break the line by getting too far ahead or astern. Also in the first line would be the launches with the artillery, which had to be towed by cutters. The second line was to be cutters only, which were to afford immediate assistance should any flatboat need it. The landing points were to be established by signals to ensure that the troops had sufficient room to form up.

When approaching the shore the flatboats were to throw their grapnel from the stern at the proper distance so that they could haul off the moment the troops had landed. If it became necessary for the flatboats to turn around quickly to effect a retreat then it was important that they did this together and in one direction. To achieve this, strict attention to the signals was essential so that all the boats could react at the same time. However, if no signal was made they should always pull to starboard, if the appropriate circumstances arose. The captains commanding each division were to repeat all the signals made by the beach-master and each captain was to have a roving boat attending him, to carry his orders to the boats of his division. Flag signals were to be reinforced by a musket shot from Cochrane's boat which was to be repeated by the captains of each division. To make sure that the flag signals would be seen should the weather be calm, each boat with signal flags on board was to be provided with flag stretchers. Finally each boat was to be provided with four or five small casks of water so that immediate assistance could be given to troops upon landing, should they require it. When the first landing was completed the boats were to proceed to those ships that still had ensigns at their top gallant mast head and

[86] Mackesy, *Egypt*, 43. Fifty-eight of the assault craft were flatboats, which drew only nine inches of water and could be run high into the shallows carrying fifty infantrymen, crewed by a naval officer, a gunner and twenty oarsmen. There were also thirty-seven launches that carried twenty-five men each and eighty-four that carried ten men.

[87] TNA, HRS/J/1, Captain Cochrane's instructions for landing troops at Aboukir Bay in 1801.

then afterwards to those having ensigns at the mizzen until all the troops were ashore.[88]

Whilst at Marmaris, Abercromby had identified that he would need naval support to supply his troops regularly with water, using naval personnel ashore to unload heavy artillery, ammunition and engineering stores and move them to forward bases. This put enormous pressure on Keith, for even though his total force was 3,339, the troop ships carried a smaller crew than that size of vessel would normally require to handle the ships at their moorings and for manhandling anchors and cables. His commitments to the army for boat crews and reliefs would absorb 820 men, so by the time he had allocated another 545 men to support the army ashore more than 40 per cent of his strength would be serving the army requirements. However, he was not in a position to resist the requests.

The first division was landed in the manner pre-planned at Marmaris. The beach-head was secured despite fierce resistance from the French, who had the good fortune of time to prepare for the assault whilst the British waited for fair weather. The defences were manned by 2,000 infantry and cavalry and included three iron cannon and a dozen field guns. Only three of the landing boats were sunk but 97 officers and men were killed and wounded during the landing. The rehearsals at Marmaris had been time well spent. Colonel Henry Bunbury described the scene:

> So closely were our soldiers packed into the boats that they could not move, and indeed the strictest orders had been given that they should sit perfectly still. The seamen pulled steadily onward, the pace of each boat being regulated by the boat on the extreme right. In this calm order on they came, till they were within reach of grapeshot, and then the fire became terribly severe and destructive. Some boats were sunk, and many of our men were killed or wounded as they sat motionless and helpless.[89]

For the next ten days all the boats of the fleet were occupied in landing the remainder of the troops, the marines and the stores most necessary for supplying their immediate needs.[90] The subsequent land campaign was successful but at a cost, including the death of Sir Ralph Abercromby. The expedition earned the epitaphs 'our service [the army] regained its ancient standing in the estimation of the British people' and 'after Egypt the soldiers could look the navy in the face again'.[91]

88 TNA, HRS/J/1, Captain Cochrane's instructions for landing troops at Aboukir Bay in 1801.

89 Knight, *Britain Against Napoleon*, 201 citing Henry Bunberry, *Narratives of Some Passages in the Great War with France from 1799 to 1810* (London, 1854), 94–5.

90 TNA, ADM.50.36, Admiral's Journals, Vice Admiral Lord Keith, Mediterranean, Mar. 1800 to May 1802.

91 Mackesy, *Egypt*, 234.

Planning a major military campaign today would employ a large number of multi-disciplined individuals using volumes of intelligence and the latest technology including computers, telephony, video and camera equipment, robotics, satellites and detailed meteorological forecasts. None of this was available to the small number of individuals who directed the course of the Revolutionary and Napoleonic wars. There was limited intelligence, but generally it was uncorroborated, often unreliable and usually out of date. Communications were slow. It is hard to imagine how British success was achieved. Victory was not secured by fighting alone. It was a product of Britain's island status, its naval domination of the seaways, profitable international trade and strong financial institutions which supported the national economy, and recognition that it needed the co-operation of other states and that this had to encouraged through financial support.

2

Bringing Forward Merchant Shipping for Government Service, 1793–1815

Although the duties performed by merchant shipping on behalf of the government during the wars were extremely diverse, its primary role was the delivery of the army to foreign shores and then the maintenance of its maritime supply lines. In 1814 this work accounted for 83 per cent of the total cost of the service.[1] The victualling of naval fleets, though equally important, was a much smaller part of its role. This is clearly demonstrated in chart 2.1. It shows the total number of transports engaged, together with the number allocated to each activity, at various times during the wars.

Only between 5 and 10 per cent of the transports were committed to victualling the fleet. The rest were predominately used as troop and horse ships, particularly in 1795 mainly in the West Indies and again in 1808–09, this time for the Peninsular and Walcheren campaigns. However, between 1810 and 1812 a higher proportion of transports were used as floating and mobile barracks and warehouses for munitions, supplies and provisions in support of Wellington's army in the Peninsula and the naval and military forces which blockaded the eastern coast of America throughout the war against the United States.

The Transport Board

In 1794 the newly created Transport Board inherited the responsibility for chartering and co-ordinating the activities of merchant shipping on behalf of the government. During the period 1794 to 1815 it spent over £42 million on chartering and fitting out transports for this service.[2] The Board was established as a subsidiary board of the Treasury but operated principally on the instructions of the secretary of state for war. Despite this, it was considered by many to be a branch of the navy and this perception persists today. Indeed the 18th and 31st Reports from the Select Committee on Finance in 1798 reviewed the activities of

1 HoCPP, 1813–14 (110) XI.131, 21, Estimate of the Money that will be wanted for the Several Services of the Transport Board in 1814.

2 Sutcliffe, 'Bringing Forward Shipping for Government Service', 128.

Chart 2.1. Total number of transports and the allocation between military and naval duties

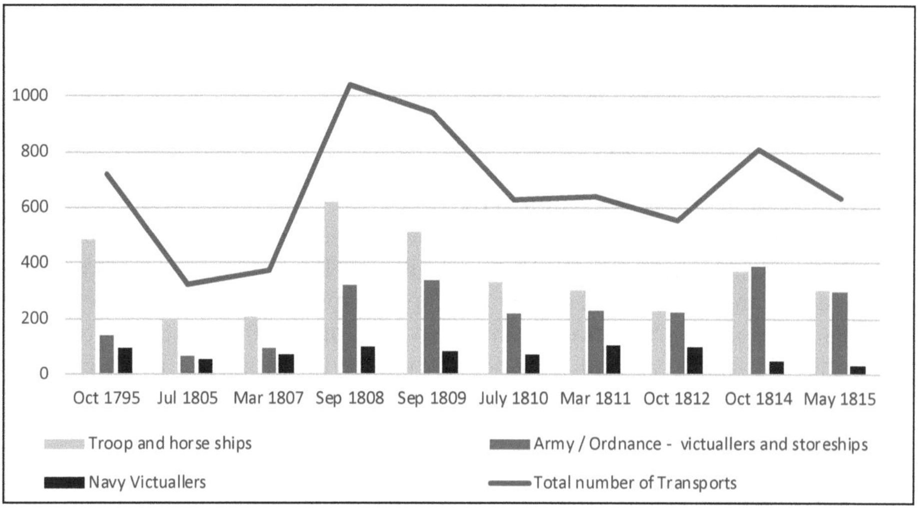

the Board under the section heading 'Admiralty, Dockyards and Transports'.[3] The Commission for Naval Revision reviewed the Transport Board's operations in its 9th Report and revisited them in its 13th Report. However, it was but one element in a complex administration system which involved the secretaries of state, the Treasury, the War Office, the Admiralty, the Navy Board, the Ordnance Board and the Victualling Board.[4]

The creation of the Transport Board inevitably ruffled feathers within the boards that were forced to concede status and responsibility, even within the Admiralty. Whereas previously the secretary of state had communicated with the Admiralty regarding Cabinet discussions concerning the need for transports, under the new regime he began to communicate directly with the Transport Office, bypassing the Admiralty, which was then advised of relevant matters by the Transport Office.[5] Admiralty officials found this irksome. Indeed the navy never stopped trying to regain control of the transport service. In 1797 Lord St Vincent, then Admiral John Jervis, Commander-in-Chief of the Mediterranean fleet, told the First Lord of the Admiralty, Lord Spencer, that he thought the Transport Board was of 'no use whatever'. Later, in 1803 when he was first Lord of the Admiralty, he boasted to Henry Addington, the Prime Minister, of 'having a very efficient Transport Board'. His view did not stop subsequent lobbying by Navy Board commissioners in attempts to win back control of the service.

[3] HoCPP, Thirty-First Report of the Select Committee on Finance on the Admiralty, Dockyards and Transports, 26 Jun. 1798, vol. 113, 3–74.

[4] Sutcliffe, 'Bringing Forward Shipping for Government Service', 111–15. The Board's relationships with the Victualling and Ordnance Boards.

[5] Condon, 'Transport Board', 69–84.

However, immediate benefits were derived from merging the responsibility for bringing forward shipping. Historic contracts were reviewed and either terminated or renegotiated to the government's advantage. Price competition was eliminated and efficiencies were gained as ships were transferred from one service to another. The involvement of commissioners who were experienced seamen in the procurement process greatly improved the quality of the ships hired. In addition, the opportunities for abuse had been much reduced by an early order that no person belonging to the Transport Board should have any vested interest in any vessel employed.[6]

The Board was not without critics in government and in the other offices of state, the army, the navy and the Ordnance Board. The criticism focused on cost-effectiveness and operational efficiency. The transport service was cost-effective but this did not stop St Vincent writing in 1802 of 'the fraud upon the public committed by the navy and transport boards, with profligate percentage on the contracts'.[7] Operationally it was frequently blamed for delays, but it was generally able to deflect responsibility to others or to circumstances beyond its control.

The position of the Transport Board in the government hierarchy created some ambiguity: it was staffed by naval officers, served the army and had been established as a subsidiary board of the Treasury. The funding of the transport service was approved annually by Parliament within the Naval Estimates, an allocation which included the pay of agents but not the cost of the Transport Office itself.

The Board was managed by commissioners. Captain Hugh Christian, the inaugural chairman, established the first office in Dorset Court, Cannon Row, London. However, Christian was promoted to Rear Admiral of the Blue in June 1795, after only a year as chairman. He was ordered to sea to command what was to become a much delayed expedition taking General Sir Ralph Abercromby's force to the West Indies.[8] His successor was the 46-year-old Captain Rupert George. Over the life of the Board there were 2 chairmen, 14 commissioners (5 sea officers and 9 civilians) and 2 secretaries.[9]

The chairman was obliged to manage a multiplicity of relationships. He was the principal channel of communication between the Board and the Treasury, the Admiralty and the secretaries of state, although most of his dealings were with the secretary of state for war. He frequently met with Pitt and Dundas to discuss the progress of the preparations for major operations and he was in regular, almost daily, contact with Castlereagh during the preparations for all the large expeditions initiated by him. In addition, he held an ambiguous responsibility to the Admiralty both as a naval officer and chairman of the Board. He was responsible for preparing annual estimates of costs, the appointment

[6] Sutcliffe, 'Bringing Forward Shipping for Government Service', 130–3.

[7] Knight, *Britain Against Napoleon*, 181.

[8] Duffy, *Soldiers, Sugar and Seapower*, 166.

[9] Sutcliffe, 'Bringing Forward Shipping for Government Service', 104–11 and 316: Commissioners of the TB.

of transport agents (who were also naval officers), providing navy victuallers, assembling convoys (for which protection was to be provided by naval vessels), the conduct of agents and transport masters, providing care for sick and hurt seamen and, finally, the Board's relations with the navy in general.

The Board also acted on instructions from the foreign secretary and the home secretary in relation to their responsibilities for prisoners of war, including the provision of cartels for prisoner of war transfers and exchanges. As the wars progressed this aspect of the Transport Board's responsibilities assumed an increasingly significant role and at times dominated its business. The Board also received instructions from the home secretary for the provision of convict transports bound for Australia. Given this multifarious range of responsibilities and relationships, it is unsurprising that from time to time there was a considerable degree of intervention in the Board's day-to-day operations, particularly from the secretaries of state for war and from the Admiralty.

The Logistical Task and the Disposition of the Transport Fleet

Managing the fleet, which expanded to over 1,000 vessels in 1808–09, balancing supply and demand and directing the fleet's disposition would be a challenge today with the assistance of computers and satellite communications. It was far more difficult when correspondence took days, weeks and sometimes months. Considerable reliance had to be placed on the resident agents and agents afloat on overseas stations. In terms of fleet size, the transport fleet was not dissimilar to the navy, although obviously the tonnage and number of seamen were much smaller.

To manage the disposition of the fleet the Admiralty maintained monthly reporting from commanders-in-chief which was consolidated into list books.[10] These list books showed the disposition of the fleet and the complement of ships by class and number of crew on each station on the first day of each month of the war. The Transport Board relied upon standard-format daily reports of ships' movements which detailed the originator of the movement order, the name and tonnage of the transport and comments on where the transports had sailed to or arrived from. This report was completed by resident agents. There were less frequent reports from floating agents. The movements were then consolidated into a register of shipping movements.[11]

By a system of regular two-way communication between the agents and the Transport Board, transports were passed from one agent to the next around the coast and then to the floating agent who escorted them to a convoy collection point and then on convoy, subsequently passing them on to other floating agents until their return to England.

Over 61,836 officers and troops were transported from Europe to the West Indies between July 1793 and June 1797, together with all the appropriate supplies

[10] TNA, ADM, 8, List books.

[11] TNA, ADM, 108/168, Register of Transports.

and provisions.[12] The West Indies, where yellow fever was prevalent, was not an attractive destination for merchant seamen. Most were very reluctant to sail once they discovered their destination and it took much persuasion and coercion by the transport agents to encourage their departure.[13] Even in the early stages of the war the challenges inherent in raising vast tonnages were exposed. In 1794 Sir Andrew Snape Hamond, Comptroller of the Navy Board, reported that trade to the Baltic and the 'speculation to the newly acquired islands in the West Indies' had employed what ships might have been available for the service.[14] However, by September 1795 there were over 300 transports involved in the West Indies operations. At the same time there was considerable transport activity in Europe: transports were evacuating troops from Holland and northern Germany; there were 174 transports landing French Royalists in the Vendée as well as victualling the supporting naval fleet that was in Quiberon Bay; and in addition there were eighty-six transports supporting naval and military activity in the Mediterranean.[15]

In July 1805 there were 322 transports of 88,835 tons in service, of which 45 ships were either in or under orders for the West Indies with a further 46 at Cork embarking troops under the command of Lieutenant-General Sir Eyre Coote that were also being sent to the West Indies. Three army victuallers were at Portsmouth ready to join this expedition. In addition there were 130 ships in or under orders for the Mediterranean. The complement was completed by 8 transports used as bomb tenders and 69 troop and cavalry ships which had not yet been allocated to specific duties; 8 others were on miscellaneous services and there were 8 defence ships.[16] Some months later there was a massive build-up of three-month ships (vessels chartered by the month for a short term) for the proposed campaign to northern Germany.

In September 1808 there were 709 transports, of which 451 were off the Peninsula having delivered Wellington and his army, whilst others were transferring 9,000 Spanish troops commanded by the Marquis de la Romana from Denmark to Santander following the Spanish uprising against the French invaders. At the same time there were 111 transports in the Mediterranean, 39 in North America and the West Indies, and 34 in the Baltic. There were also 330 transports being prepared for service, mainly to take an army under Sir David Baird to Corunna to join up with Sir John Moore's army that was pushing north from Portugal. By September 1809 the number of chartered transports had increased to 738 ships and there were another 200 'coming forward for service'. Of these, 242 transports were involved in the Walcheren campaign, 189 ships were employed on the coast of Portugal and 130 were in or on passage to the Mediterranean. Twenty-two

[12] Duffy, *Soldiers, Sugar and Seapower*, 184–92.

[13] TNA, ADM, 108/32, Transport Board Minutes, 3 Oct. 1794.

[14] TNA, HO, 28/63, A.S. Hammond, Navy Office to Evan Nepean at the Admiralty, 16 May 1794.

[15] TNA, ADM, 1/3730/311, Transport Office, Abstract of all the ships employed in the Transport Service, 3 Oct. 1795.

[16] TNA, ADM 1/3747, Transport Office, Abstract of all tonnage now in the Transport Service, 11 Jul. 1805.

ships had sailed from Portsmouth on 16 August to take Russian troops back to Russia. There were 5 ships at or on passage to the Cape of Good Hope, 3 on passage with convicts and stores to New South Wales, 12 in North America, 11 in South America and 26 in or bound for the West Indies. There were 25 in the Baltic, 26 on miscellaneous home services and 28 victualling the navy blockading squadrons off Rochefort.[17]

In July 1810 the ships involved in the Walcheren campaign had returned and many of the three-month ships had been paid off. The remaining transports had been reallocated to the Peninsula with 118 in or on the way to Spain and 242 in or on the way to Portugal to build up a reserve force should Wellington need to evacuate at short notice. There were still 105 in the Mediterranean. There were 9 in North America, 8 in the West Indies, 20 in the Baltic and 70 on miscellaneous home service.

During July 1812 there were 555 transport ships in service, including 137 transports off Portugal, 141 off Spain and 142 in the Mediterranean. In addition there were 23 in the Baltic, 50 at home preparing for service, 8 off the coast of Africa and at the Cape of Good Hope, 34 in North America and 20 in the West Indies. The number of transports increased significantly to 811 in October 1814. At that time there were 217 off the coast of North America. A significant number of transports were engaged in removing troops and supplies from France, Spain and Portugal following the collapse of French resistance earlier in the year, and there were still 154 transports in the Mediterranean. Following the Treaty of Ghent in December 1814 the focus soon switched to moving troops from Canada and the American coast and into Holland and northern Germany. These troops were initially intended to support the peace, but following Napoleon's escape from Elba in early March, troops and artillery were moved to Ostend and Antwerp in increasing numbers to join Wellington's army. It finally defeated Napoleon in the Anglo-Prussian victory at Waterloo.

The Transport Procurement Process

Procuring transports in large numbers required a systematic approach in which ships' brokers played a very important role. All the charters inherited by the Transport Board from the other boards in 1794 were hired through one broker, George Brown, who had been the principal broker in the earlier American war. The arrangement, made supposedly on the basis of security, must have been very convenient for overburdened clerks in the various boards who lacked maritime experience.[18] However, Brown's monopoly meant that he had a very significant influence on whose ships were hired and more importantly on the price. Although there is no evidence of malpractice, this monopoly cannot have been in the best interests of public service.

[17] TNA, ADM, 1/3758, Transport Office, Abstract of all tonnage now employed in the Transport Service, 2 Jul. 1810.

[18] Condon, 'Transport Service', 56.

The new Transport Board immediately set about remedying the situation by increasing the number of brokers that it traded with. Within a few years it was chartering from over 300 brokers and self-managing owners.[19] Using a range of brokers enabled the Board to assert more control over prices and made it easier to raise relatively large numbers of ships quickly. Even so the majority of the ships were still supplied by a limited number of brokers, including George Brown, Joseph Dowson & Son, James John and Thomas Dawson, James Duncan and Herring & Richardson. There is no indication of overt favouritism but there is no doubt that these brokers were the preferred suppliers, probably based on their reputations for supplying good quality ships and for trading honourably. However, those who failed to live up to the standards were dealt with harshly as Mr Herring of Herring & Richardson discovered in 1807. He was accused by the Board of price fixing. After being advised of the government's precise requirements, he attempted to persuade ship-owners to decline the Board's approaches for shipping at the going rate of 15s per ton, advising them that he would then be able to persuade the Board to pay 20s. In consequence some owners declined to offer their ships on the lower terms. The Board believed that Herring was guilty of 'retarding the service, and increasing the public expense'. He and his firm were struck off the list of approved brokers.[20]

Brokers were remunerated out of the proceeds of hire received by the ship's owner. Such commission would increase 'according to the increased price at which he contracts to let the vessels to the government'. During his appearance before the Committee on Finance in 1798, Transport Board Commissioner John Schank was asked if there would be a cost saving if the government employed its own broker, on a fixed salary, to deal directly with ship-owners. It was envisaged that he would hire transports directly, thus bypassing the traditional ship brokers. Schank believed that 'this would be attended by great advantage to the government'. He reasoned that the appointee would only be influenced by the public interest rather than by the ship-owners and would have no conflicts of interest.[21] However, when the Treasury sought the Transport Board's collective view on this matter it was advised that it was 'highly inexpedient, if not generally impracticable, to deviate from the long established mode of employing known and approved brokers, by substituting an agent of our own to perform that service'. The Board's main criterion for rejecting this proposal was that such an agent would not have sufficient contacts and knowledge of the market-place and owners, both in London and at out-ports. He would have to resort to public advertising and tender which would be exploited by some ship-owners who would be much more aware of the status of the prevailing market. Thus a government agent going out into the market to procure shipping was highly likely 'to immediately raise the price'.[22] The Board spoke from experience. In 1796 Secretary of State Henry Dundas had insisted on adopting this practice,

19 Condon, 'Transport Service', 25.
20 TNA, ADM, 108/21, TB to Castlereagh, 15 May 1807.
21 Committee on Finance (31st), 503, The further examination of Mr Schank, 7 May 1798.
22 Committee on Finance, Further Proceedings (18th +31st).

despite the Board's reluctance, arguing that it was justified by the urgency of the service. Consequently the government paid a significantly higher rate than the Board felt was appropriate. Its preferred practice was not to divulge the full amount of the tonnage required but to work with its network of brokers of proven integrity, who had comprehensive knowledge of the availability, to procure tonnage 'little by little', thus avoiding pushing up the price. Contrary to Schank's view, the Board believed that an agent of the government would be more exposed to undue influence from ship-owners than the existing brokers who were very familiar with the market.[23] The proposal was not adopted.

Transports were generally hired on the instructions of the secretary of state for war, as illustrated by this instruction from Lord Castlereagh in April 1807 when he was planning the Copenhagen expedition: 'immediately hire 16,000 tons for infantry, 10,000 tons fitted for horses and 10,000 tons as Ordnance transport and victuallers'. He described the service as 'being of a pressing nature'.[24] The proposed use for a vessel was important. It determined the size of the vessel and the fittings required: infantry transports required cabins and beds, horse ships required stalls, ordnance ships required magazine racking, and victuallers tended to be smaller vessels. Requests for such large tonnages were the exception. More frequently they were for a small number of ships for a specific purpose.

Procurement was usually by competitive tender advertised at Lloyd's Coffee House and in the waiting room of the Transport Office, where the ship-owners and brokers congregated. However, when significant tonnages were required the Board was reluctantly forced to resort to advertising in newspapers. This had obvious disadvantages because it applied pressure to the hire rate, but more importantly it gave publicity to the preparations, alerting the enemy to the government's plans. Occasionally the Board made direct contact with brokers to request submission of tenders. Tenders had to be presented in the Board's prescribed form, accompanied by the ship's customs registration document. If the tender was accepted a note was sent to the agent at Deptford instructing him to proceed with the survey. There was also a standard form for the survey which required details of the ship's registered tonnage, together with her class, height between decks and nature of sheathing (wood, copper or unsheathed) and when and where she was built. If the ship was found to be suitable for service a sea commissioner of the Board would often visit to conduct his own survey. Rupert George, the Chairman, frequently visited Deptford and other ports to examine vessels.[25] When the ship was approved the owner was required to fit her for service in line with a detailed schedule of stores and then deliver the ship to Deptford. The tender request specified the time allowed to deliver the fully provisioned ship to Deptford; failure to achieve this deadline could result in a fine of £100 or alternatively the ship could be rejected, although this

23 Committee on Finance, Further Proceedings (18th+31st).

24 TNA, WO, 6/156, Castlereagh to TB, 8 Apr. 1807.

25 TNA, ADM 108/74, TB Minute, 3 Feb 1803,'Commissioner George will inspect the *Lady Andover* of 229 tons tomorrow'.

Table 2.2. An account of the average prices of certain articles in 1795, 1800 and 1806 (Transport Office, 4 February 1807)

		1795			1800			1806		
Article	*Measure*	£	s	d	£	s	d	£	s	d
Seamen's wages	Per month	4	5	0	4	6	8	4	11	8
Cordage	Per cwt	2	8	8	3	16	0	4	0	0
Canvas no. 1, 2 and 3	Per yard		1	5		1	8½		1	9
Masts and spars	Per load	13	10	0	16	10	0	22	10	0
Bread	Per cwt	1	10	8	2	0	2	1	8	6
Beef	Per tierce	5	11	8	7	6	8	7	7	4
Pork	Per barrel	3	19	8	5	11	8	5	4	2
Insurance per annum	Per cwt	9	0	0	9	3	0	9	4	8
Value of shipping	Per ton	9	6	8	11	13	4	14	0	0

N.B. Between 1795 and 1800 there appears on aggregate to be an advance of about 20 per cent and between 1795 and 1806 an advance of about 38 per cent.

Source: Commission for Revision (9th), 370.

rarely happened. On arrival at Deptford, appropriately provisioned, the vessel was entered into pay.

Ships accepted for service were valued by the officers of the Board at Deptford where the agent, who was a naval officer, estimated the value of the rigging, cables, sails, anchors and other articles of equipment, together with the boatswain's and gunner's stores. In doing so he considered the state of wear and tear by assessing whether each item was a quarter, half or three-quarters worn. To value each article he needed to be aware of the current market prices, which changed significantly over the period. Prices increased on average by about 20 per cent between 1795 and 1800 and 38 per cent between 1795 and 1806. Naturally the cost of some items increased more than for others: the increase in the cost of masts and spares reflected timber shortages and the higher price of bread in 1800 reflected the corn shortages of the time. A range of price movements from 1795 to 1806 is clearly demonstrated in the memorandum prepared in the Transport Office in 1807, reproduced in table 2.2.[26] A shipwright valued the hull, mast, yards and pumps, and the completed valuation form was sworn on oath by the officers and sent to the Board's accountant. The owners were only advised of the valuation if the ship was subsequently lost or damaged by the enemy and the Board accepted liability. After the valuation was completed the charter party was prepared. This was the contract that set out the obligations of the parties: the ship's owners, the master and the Transport Board, on behalf of the government.

[26] Commission for Revision (9th), 370.

Transports for troops were generally fitted with cabins and beds and horse ships with stalls and beds. This work was carried out on the river Thames under the direction of the agents for transports. It was estimated that in 1812 the cost of converting a 326-ton ship for troops was £79 18s and for converting a 28-stall horse ship was £55.[27]

The ship's owner was responsible for feeding the ship's crew, but if passengers or troops were embarked the Transport Board applied to the Victualling Board for the appropriate provisions. If the transports were for provisions or ordnance stores they were temporarily transferred to the appropriate board, whose agents were responsible for the loading and stowage of the stores. When this was completed the responsibility for the ships reverted to the agent for transports who advised the Board to request that the Admiralty allocate convoy protection. When this was arranged, the agent ordered the transports to the convoy rendezvous point to await the convoy escort. Sometimes they were held there for many weeks whilst the whole convoy was assembled and made ready to sail, much to the frustration, not to mention risks to health, of the crew and troops on board. If there were horses aboard such delays caused them much distress.

Types of Hire

Transports were principally hired by one of three methods. Regular transports were hired on six-month charter parties. The owners were paid an agreed rate per ton per month, calculated on the ship's tonnage. Regular transports formed the core of the chartered transport fleet. The regular transport fleet was supplemented by ships which were also chartered by the month but usually for a specific purpose and for a shorter term, generally three months, thus they were known as 'three-month ships'. These ships were hired in large numbers when a major expedition was being planned. Although many were sent out to the West Indies between 1793 and 1795 and to the American coast in 1812 because of the shortage of regular transports, they were mainly restricted to services in European waters. Again, the owners were paid an agreed rate per ton per month, calculated on the ship's tonnage. However, if availability was tight the spot rate was frequently higher than that for regular transports. The owners of three-month ships were not protected against capture and had to rely on their own insurance cover. These ships were not required to be sheathed.[28] Finally, rather than chartering the whole ship, accommodation and storage were frequently chartered 'on freight', to convey men or supplies at a rate per ton of freight or rate per passenger on merchant ships that were taking other cargo.[29]

[27] NMM, ADM, BP/49b, Navy Office Transport Department memorandum, 26 Nov. 1829.
[28] Commission for Revision (9th), 76.
[29] Commission for Revision (9th), 76.

Regular Transports

Although wood-sheathed vessels were preferred, in the early stages of the Revolutionary war limited availability meant that unsheathed ships had to be chartered. Sheathing, and in particular copper sheathing, enhanced the ship's sailing capabilities by inhibiting hull fouling. More importantly it protected the ship against the shipworm which was particularly prevalent in the West Indies. Naval vessels had first been coppered in large numbers in 1779 but the iron bolts used to secure the copper sheets created electrolysis which caused corrosion of underwater ironwork such as the rudder irons and the heavy bolts which fastened together the keel, stem and sternpost.[30] Some years later the iron bolts were replaced by bolts made of a copper alloy and this problem was resolved.[31] Coppering had the advantages of lightness and durability but the disadvantage was cost. In 1797 the cost of coppering a 300-ton merchant ship was £629, compared with £158 for wood sheathing.[32] Coppered ships were first chartered in small numbers from 1798. By 1804 copper-sheathed vessels were preferred although wood-sheathed ones were still taken when the demand was high. From 1808 only coppered ships were chartered as regular transports and by the end of that year all the wood-sheathed regular transports in service had been coppered.[33]

The charter party for regular transports stipulated that the ship had to be armed with six guns, which, together with gunpowder, were supplied by the Ordnance Board. The ships received an allowance for additional gunpowder used for legitimate reasons. Initially ships intended to convey troops had to be not less than 5 feet 6 inches between decks. This applied even when ships were hired to carry stores, in case they might be required to carry troops at a later date. However, this stipulation had to be relaxed because of the lack of vessels meeting the requirement. From September 1794 the height of 5 feet between decks became the standard measure.[34] Two tons per man were prescribed for all troop-carrying ships bound to any port beyond Cape Finisterre and 1½ tons for very short or coasting voyages.

In 1794 the Board inherited 391 regular transports of 108,012 tons which had already been taken up by the Navy, Victualling and Ordnance Boards. The number of regular transports hired from the start of the war rose dramatically. It peaked at almost 140,000 tons in October 1795 when ships were being procured for the ill-fated Christian convoy of troops for Abercromby's campaign in the West Indies and for the Vendée adventure, which took place a few months earlier. From the peak there was a major reduction in 1796 when 304 ships totalling 74,000 tons were discharged. The simultaneous discharge of

30 Rodger, *Command of the Ocean*, 344.

31 Roger Knight, 'The Introduction of Copper Sheathing into the Royal Navy, 1779–1786', *Mariner's Mirror* 59 (1973), 299–309.

32 Condon, 'Transport Service', 73.

33 Commission for Revision (9th), 76.

34 TNA, ADM, 108/31, Transport Board Minutes, 18 Sept. 1794.

a large number of transports placed considerable strains on the administrative procedures of the Transport Board and the Victualling and Ordnance Boards, which had to receive back and account for vast quantities of unused stores and supplies very quickly to avoid demurrage charges. There was a further reduction of 30,000 tons in 1797. By December 1799 the tonnage had been reduced gradually to 30,814 despite the desperate need for transports for the Anglo-Russian den Helder expedition. That demand had been fulfilled by chartering a large number of three-month ships.[35]

The tonnage rose slightly in 1800 and 1801 in response to an instruction from Henry Dundas, Secretary of State for War, for additional transports to support Abercromby's Mediterranean campaign to Malta and then Egypt. The Peace of Amiens in 1802 eased the demand. By December that year the regular transport fleet had been reduced to thirty-five ships of 13,000 tons. This reduction was reversed in May 1803 when war was declared again. Large numbers of transports were hired in 1805, and the years that followed, to support the military operations in northern Europe, then the Mediterranean, the Peninsula, the Baltic and the North Sea. In 1805–06 transports supported the capture of the Cape of Good Hope (10 January) and regular transports were sent to South America to support the invasion of Buenos Aires. The majority of the additional vessels raised in 1808–09 for the Peninsular and Walcheren expeditions were three-month ships. However, in 1808 15,000 tons of regular transports were sent out to the West Indies to support the capture of Martinique (24 February 1809) and the surrender of Guadeloupe (6 February 1810). Tonnage increased still further in 1812–13 to support the broad range of military activities in North America, the Peninsula, the Mediterranean and northern Europe, peaking in the last quarter of 1813 at 200,000 tons. This was some 60,000 tons higher than the peak in the earlier war during the West Indies expeditions, although in that campaign there had also been considerable reliance on 'on freight' transports – merchant ships involved in the West Indies trade returning there in ballast or with unused storage capacity.

There was a steady reduction of tonnage in 1814, and then in the following year 406 ships of 127,354 tons were discharged, again placing considerable strain on the dockyards and on the administrative systems, which caused delays in unloading and paying off ships. By the time the Transport Board's responsibilities were finally handed over to the Navy Board in January 1817 the fleet had been reduced to fifty-two ships of 16,515 tons.[36]

Less than 1 per cent of regular transports were under 100 tons; 26 per cent were between 100 and 220 tons; 28 per cent were in the range 220 to 300 tons; 31 per cent were 300 to 400 tons; only 14 per cent were over 400 tons. Chartered ships over 500 tons were usually East India Company vessels. Very few of these ships were taken up as regular transports because they were expensive to hire and the company insisted on their being returned in time for the seasonal passage to

[35] Sutcliffe, 'Bringing Forward Shipping for Government Service', 29.

[36] Sutcliffe, 'Bringing Forward Shipping for Government Service', 30.

India. The Board could not always guarantee this and the penalty for failure to meet the deadline was very high.

In addition to the master, the number of men and boys employed was always defined according to tonnage at the rate of five men and one boy to each 100 tons, and for the fraction of 100 tons at the rate of one man per 20 tons.[37] This was twice the number used in coasting vessels. Owners and masters often tried to get away with using fewer men than was stipulated. However, the ships were frequently mustered by the agents for transports in an attempt to stop this practice. Sometimes transports were temporarily undermanned because of losses through death, desertion or impressments, but this met with little sympathy from the agents.[38] Ships were mulcted if they did not have a full muster.[39]

Three-month Ships and 'on Freight' Charters

Ships on three-month charters were invariably used when a large number were required at short notice for specific operations.[40] Almost 150 three-month ships were hired for the West Indies campaigns of 1793–96. In June 1800 there was a demand from Henry Dundas, Secretary of State for War, for 30,000 tons of transports which were urgently required to support proposed diversionary operations against enemy dockyards to destroy their fleets at Brest and the Spanish naval base of Ferrol. The proposed Brest attack was aborted but the majority of troop ships used in the Ferrol attack were three-month transports.[41] Again in 1805 three-month transports were chartered in large numbers to ship troops to northern Germany and out to the Mediterranean. The Transport Office abstract dated October 1806 reproduced in table 2.3 illustrates the changes in the transport fleet since February that year, when William Windham became Secretary of State for War in the 'Ministry of All the Talents'.

There had been a very significant increase in three-month charters between July 1805 and February 1806 when large numbers were brought forward to support Castlereagh's plans to send a 67,000-strong force to north Germany. At the same time he sent 15,000 tons of vessels for infantry, cavalry ships to accommodate 700 horses and riders and 3,000 tons of store ships to the Mediterranean.[42] From 506 ships of 130,655 tons on hire on 1 September 1805, the numbers increased quickly to 1,028 ships of 261,240 tons by 1 January 1806.[43] Following the decision to withdraw from northern Germany, a significant proportion of these vessels

[37] TNA, ADM, 1/3733, TB to the Admiralty, 11 Jan. 1797.

[38] Ville, *English Shipowning during the Industrial Revolution*, 96–7.

[39] Mulct – to suffer a financial penalty deducted from the charter fee.

[40] TNA, ADM, 1/3730–65, Abstracts of all ships employed in the Transport Service. From abstracts submitted on a monthly basis to the Admiralty, although few have survived, it is possible to determine the extent of three-monthly hires at certain specific times throughout the war.

[41] TNA, WO, 6/156/266, Henry Dundas, Secretary of State for War to the TB, 7 Jun. 1800.

[42] TNA, WO, 6/156/321, Castlereagh, Secretary of State for War, to the TB, 29 Aug. 1805.

[43] Castlereagh, *Correspondence* VI, 97, Memorandum of the late Equipments, Undated.

Table 2.3. An abstract showing the state of the transport service on 6 February 1806 and the amount of tonnage on monthly pay, discharged and engaged between that period and 14 October 1806

		Number	Tons
In service on 6 February 1806	Coppered	174	56,558
	Sheathed	44	12,319
	Total regular transports	218	68,877
	Three-month charters	412	88,345
	Total	630	157,222
Discharged being unfit for service	Coppered	4	1,190
	Sheathed	10	2,059
Discharged, having been engaged for the expedition to the Continent in 1805	Three-months	377	83,179
	Total	391	86,428
Taken up since 6 February 1806	Coppered	30	10,721
	Sheathed	17	4,625
	Three-month	33	7,034
	Total	80	22,380
Now in service	Coppered	200	66,089
	Sheathed	51	14,885
	Three-month	68	12,200
	Total	319	93,174

Source: TNA, ADM, 108/21/22 and TNA, WO, 1/803/75, TB to William Windham, Secretary of State for War, 14 Oct. 1806.

were discharged, but there were still 630 ships on charter on 6 February 1806. Windham was determined to reduce the cost of war, so on his insistence the discharge of another 391 transports progressed at an incredibly rapid rate. By October, even after including 80 new charters, there were only 319 transports in the service.[44] Whilst handling the discharge of so many transports during that six-month period the agents also prepared transports for 22,020 troops and provisions in 145 troop transports and 25 victuallers: 860 troops to Bermuda and the Bahamas; 1,400 to the Cape of Good Hope; 1,800 to Gibraltar; and 17,960 to the Mediterranean.[45]

When Viscount Castlereagh was reappointed Secretary of State for War he and his successors, Lords Bathurst and Liverpool, chartered large numbers of three-

44 Knight, *Britain Against Napoleon*, 232. Windham also attempted to make reductions in the cavalry, foot guards and the artillery Royal Wagon Train.

45 TNA, ADM, 108/21/22 and TNA, WO, 1/803/75, TB to William Windham, Secretary of State for War, 14 Oct. 1806.

month transports to support the Peninsular campaigns of 1808, the evacuation of Corunna in January 1809, Walcheren in 1809, the Peninsula again in 1809–14, the American war in 1812–13, the repatriation of troops, equipment and supplies from the Peninsula, France and North America in 1814–15 and sending troops to Holland in 1814–15.

The Board used vessels 'on freight' to transport troops or stores 'which can be delivered upon arrival of the ships at foreign ports', believing that it was cheaper than sending regular transports only for them to return empty. Owners of ships hired 'on freight' had to find their own return cargo. If there was a chance or expectation that the ships would be detained overseas then regular transports were used.[46] There were usually between thirty and fifty ships chartered 'on freight' at any one time.[47] However, the highest use of 'on freight' transports was for carrying troops during the West Indies campaign at the rate of £6 to £7 per man. In 1795–96 there were over 50,000 tons of ships chartered 'on freight', accounting for 24 per cent and 35 per cent respectively of the total of all tonnage chartered in those years.[48] A significant number of East India Company ships were used for this purpose, plus other ships that would probably have been registered and based in the West Indies engaged in the West Indies-to-England trade. There was an attraction in this for their owners because the vessels frequently sailed from England with only nominal cargoes to return with full loads.

The Charter Party

There were different agreement formats, one for each type of charter: regular transports, three-month ships, ships 'on freight' and transports for conveyance of convicts to New South Wales. The Commission for Revision found that since the business had been transferred to the Transport Board, the charter party agreement had been 'much improved from time to time in favour of the government by strengthening the obligations of the owner'.[49] For instance in 1799 the charter party for victuallers was revised to oblige such ships to convey a certain number of officers and troops. Up to that point ship-owners and masters had been very reluctant to do this because of the cost of providing accommodation and the inconvenience of having them on board. As a result the Transport Board had to charter additional ships at extra cost.[50]

By custom and practice the Board had always assumed the right to choose when to discharge or retain transports. When demands for transports were high the Board retained ships beyond the initial charter term; this was often against the owners' interests if there was more rewarding business to be taken up elsewhere. This practice was challenged in March 1800 by a lawyer acting on

46 Committee on Finance, Further Proceedings (18th +31st).

47 TNA, ADM, 1/3730, Abstract of all ships employed in the transport service.

48 Committee on Finance (31st), 502.

49 Commission for Revision (13th), 52.

50 Condon, 'Transport Service', 85.

behalf of a number of owners who claimed that it was not in accordance with the terms of the charter party. The Board referred the matter to the attorney and solicitor general, in whose opinion the wording in the charter party was 'doubtful'. He recommended the drafting of a revised version that clarified the right of the Board to choose when to discharge transports.[51] This was done, and in order to persuade the owners to accept the revised charter party the Board proposed to offer inducements. The pay of wood-sheathed transports would be increased to 16s per ton per month from the previous 1 January and that of copper-sheathed transports to 17s 6d if the owners entered into the new contract. In addition the freight due to 31 December 1799 would be paid up to that time subject to a reserve of six months' hire charge until their accounts could be passed and the balance paid. By this arrangement the freight was to be paid much sooner than the owners were entitled to receive it under the terms of the existing arrangements, but in future the bills would not bear interest. Up to that point interest had been paid at 3 per cent for the 90 days. The Board proposed that the owners of transports who did not consent to the new form of charter party would not receive the increased price nor benefit from the revised payment arrangements.

Invariably, some owners were not happy about this proposal and prevailed upon George Brown, who had formerly been the principal broker to the government, to call a meeting of owners of regular transports to discuss these proposals. The meeting was held at Batson's Coffee House on the 18 March 1800. After what was no doubt an explosive meeting, they resolved that whilst they were satisfied with the new proposals for payment of freight so that only six months would be in arrears at any one time, having 'maturely considered' the other contract changes they were unanimously of the opinion that the proposal could not be accepted. The owners did not think the offer of 16s by any means an adequate rate given the increase in cost of 'all kinds of stores and provisions'.[52]

When the Transport Board received this notice it wrote to George Brown to ascertain which owners had attended and the names of their transports that were in service. It was suspected that 'this document may express the opinion of a few, instead of the many who have ships in the transport service'.[53] The outcome is not clear from the correspondence but in fact all the existing charter parties were terminated on 31 December 1799 and new contracts were issued. However, the rates finally agreed were 18s for wood-sheathed and 19s 6d for copper-sheathed, and interest continued to be paid on bills. It would appear that the Board achieved the desired changes to the charter party but at a cost.

If a ship, by accident or through fault of the owner or master, became incapable of performing the service for which she was engaged, she could either be discharged immediately or, more likely due to the usual shortage of supply,

[51] TNA, WO, 1/801, TB to Henry Dundas, 25 Mar. 1800.

[52] TNA, WO, 1/800, Minutes of the meeting of regular transport owners held at Batson's Coffee House, 21 Mar. 1800.

[53] TNA, WO, 1/800, Minutes of the meeting of regular transport owners held at Batson's Coffee House, 21 Mar. 1800.

mulcted for the period that she was out of service.[54] Then the owner had to arrange for the repairs to be completed at his cost and as quickly as possible to avoid heavy deductions. All accidents and normal maritime disasters were the responsibility of the owner who either insured against the risk or carried it themselves. However, if the ship was burnt or sunk by the enemy or, in the case of regular transports only, captured by the enemy through no fault or neglect of the master, then the value was paid less a deduction for wear and tear, provided the convoy commander was prepared to certify to that effect. Ships chartered for three months were not protected against capture; the owners had to insure against this contingency.[55]

The charter party provided for advances to be made to the owners during the course of the charter, always retaining six months' pay to enable the government to indemnify itself for any claims it may have against them. The ship was finally paid off when discharged on production of: certificates from the victualling office, ordnance office or other public departments that all stores had been duly accounted for and that there were no outstanding charges; the ship's log book; and certificates evidencing that they had paid all light dues, port dues and all fees to Greenwich Hospital. Mulcts were deducted from the charter fees due and the net value was paid by thirty days' bills of exchange with 3 per cent interest per annum. Regular transport charter parties inevitably extended beyond the initial six-month period; almost 35 per cent extended to twenty-five to forty-eight months with some exceptionally extending for ten to twelve years. Many ships were chartered numerous times. The *Barrick* (300 tons, a wood-sheathed ship) was first chartered by the Board in September 1793. She was constantly in service under her master, Joseph Street, until June 1802. During that time she had been detained for a period in Brest by the French where she had been sent as a cartel to repatriate French prisoners of war. After the peace she was then re-engaged in June 1803 under Master Thomas Bailey until July 1805. She was rehired in May 1806; by this time she had been coppered. Apart from a six-month period in 1812, she was chartered up to February 1817, acquiring a new master in 1812, Thomas Dunn. The combined period of employment in the service was 249 months, which is over 20 years, and the *Barrick* must certainly be one of a very few ships that were chartered for almost the whole duration of the wars.

The potential for long charters was probably the most positive attraction to ship-owners: chartering to the government invariably meant that they were paid for periods when the ship would normally be laid up between sailing seasons. There was a strong likelihood of medium- to long-term continuous employment once the ship was in service as a regular transport. However, it did have some disadvantages. When freight rates rose, owners usually found it difficult to extract their ships from the transport service to take alternative freights at higher rates. When owners applied to have their ships discharged the request was invariably declined by the Board, as in the case of John Byron, owner of

54 Commission for Revision (9th), 77.

55 Condon, 'Transport Service', 86.

the *Elizabeth* transport of 165 tons. He requested that she be discharged from the service because he could not afford a refit due to the high price of materials. The Board rejected this request and instructed Captain Patton, the Resident Agent of Transports at Portsmouth, to have her defects made good at the owner's expense and return the vessel to service as soon as possible. The cost was subsequently offset against the hire charges.[56]

Registered Tonnage

Merchant ship owners faced a dilemma. It was in their interests to under-record the tonnage of their ships in relation to the payment of port dues, light dues and Greenwich Hospital fees, but to exaggerate it when their ships were employed by the Transport Board. Shortly after the establishment of the new Board, Lieutenant James Bowen, then the Agent for Transports at Deptford, recommended that the tonnage, as registered with the Customs Office under the 1786 Registration of Shipping Act, should be the only measure used by the Transport Board. This would replace the measured tonnage that had previously been used. He had examined the records relating to some of the ships hired in 1793 and estimated that the measured tonnages at which they were taken up were an average of 7½ per cent above the tonnages of the same ships as they were registered at the Customs Office.[57] This recommendation was accepted by the Board and from mid December 1794 all ships were taken up at the registered tonnage.[58] Bowen was destined to have a distinguished career in the transport service. In July 1803 he was appointed Commissioner of the Transport Board and in 1808–09 he was the senior transport agent at Corunna where his leadership in the evacuation of Sir John Moore's army was highly praised.

In 1807 the Commission for Revision conducted a similar exercise on a sample of seventy-two ships taken up before the establishment of the Transport Office. They found that the measured tonnage exceeded the registered tonnage by about 3½ per cent. Applying this figure to the average tonnage taken up on monthly pay from 1793 to 1806, they calculated that the total loss to the Treasury would have been £261,984 if the practice of using measured tonnage had continued throughout that period.[59] After 1794 the Board insisted upon the production of a ship's customs registration prior to her being accepted for service.

Charter Rates

Transports were hired at rates which varied in accordance with the prevailing market conditions, but increases were only implemented in agreement with the Treasury. These rates tended to reflect the availability of shipping and the urgency

56 TNA, ADM, 108/82, Mr Robert Chapman, Broker, London, to the TB, 1 Apr. 1809.
57 Commission for Revision (9th), 73.
58 Commission for Revision (9th), 76.
59 Commission for Revision (9th), 60.

of the government's requirements. Occasionally the Board had to offer longer contracts to obtain ships at the mandated rate by raising the 'certain' (guaranteed) period to twelve months rather than the usual six.[60] During the earlier American Revolutionary war (1775–83) the hire rates had increased from 10s per ton per month to 13s.[61] Freight rates had been fairly consistent in the intervening years and 13s was the going rate in 1793. The rates subsequently changed, generally upwards, although there were some short periods of rationalisation, as in 1798 and 1804 and again briefly in 1808 when freight prices reduced as a result of reductions in competing trade demand. In 1802, during the period of peace, the rate dropped to 12s and similarly the rate collapsed at the end of the war in 1815.

The Board was skilful at using the market to bring forward ships, sometimes resisting the temptation to increase the rate. Rupert George explained to the Scheldt Inquiry in 1810 that 'Most undoubtedly you may have all the ships in England if you go to the price: if the price is large enough ships may give up their engagements if that is more advantageous than the regular trade.' However, he went on to say that the higher price may well bring forward a few additional ships but the number involved would not justify the great discontent that would be generated amongst existing transport owners who were receiving a lower rate.[62] Thus raising the rate was not always beneficial.

In early 1796 the Transport Board received numerous letters from ship-owners in London, Sunderland, North Shields, South Shields, Whitby and Newcastle describing an increase in costs and requesting a corresponding rise in the rate of hire.[63] The recent corn shortages had increased the price of biscuits; other maritime costs had increased significantly, up to 25 per cent in some instances, due to shortages; and seamen's wages had increased from £2 10s per month to £4 5s. In April that year the Board was forced to concede an increase in the charter rate to 15s per ton per month. However, in the first quarter of 1798 the Board identified a reduction in demand and a weakening of freight rates, and so it reduced the rate back to 13s per ton for new hires but agreed to pay 15s if owners were prepared to copper their ships.[64] From 1798 the Board continued to offer owners of coppered ships a higher rate than for wood-sheathed vessels. This price difference was subsequently justified 'not only from the scarcity of one compared to the other but their being generally a superior class of vessels, and better adapted for particular services'.[65] Following negotiations with owners regarding the implementation of the new charter parties in 1800 the rates were increased to 18s per month for sheathed ships and 19s 6d per month for coppered

[60] TNA, ADM,108/21/22 and WO,1/803/75, TB to William Windham, Secretary of State for War, 14 Oct. 1806

[61] Syrett, *Shipping in the American War*, 251–2.

[62] Scheldt Inquiry, Sir Rupert George's evidence.

[63] Condon, 'Transport Service', 73–7.

[64] Condon, 'Transport Service', 78.

[65] TNA, ADM, 108/20/215, TB to W. Fawkner, Secretary to the Council for Trade and the Foreign Plantations, 19 Jun. 1804; TNA, ADM, 108/21/22 and WO, 1/803/75, TB to William Windham, Secretary of State for War, 14 Oct. 1806.

ships. The rate reduced to pre-war levels during the temporary Peace of Amiens but soon increased to 1800 levels after the resumption of hostilities.

In 1807 Castlereagh was again Secretary of State for War. On 8 April he demanded that the transport service immediately raise 36,000 tons of transports with the least delay: 'the board should consider itself at liberty, if tonnage cannot be raised on cheaper terms, to contract for three month ships at no more than 20/- per ton and six month ships at 25/- per ton'.[66] These transports were required for the Copenhagen expedition. The Board implemented the increase. However, on 20 April it was compelled to advise Castlereagh that despite the higher rate it had not been able to procure more than a few vessels and, furthermore, the commissioners did not believe that offering an even higher rate would be more successful because 'two Commissioners of this board have examined the river and the docks, where they saw very few ships fit for the transport service'.[67] Price was not the principal factor: there were no ships to be had because all the suitable ones were engaged in trade at that time of year. Seasonality played an important part in the availability of shipping: in January 1808, just nine months later, the Board was able to reduce the rate for new charters back to 21s and managed to hold the rate during that year despite the high demand for transports to ship troops to the Baltic and then to Portugal. The Board held the rate until January 1809 when it was urgently trying to raise shipping to evacuate troops from Corunna. It raised the rate to 25s. Then in April 1809 the Board took advantage of reducing freight rates to implement a mandatory rate reduction to 20s per ton, warning that 'any owners not accepting this rate will find that their vessels are returned to this country for discharge and are unlikely to be hired again'.[68] Ironically this was only a month before the Board was instructed to raise significant tonnage for the Walcheren campaign.

Raising shipping for the Walcheren campaign proved incredibly challenging: the rate was increased again to 25s and the Board even contemplated raising it to 30s. Following the discharge of transports returning from Walcheren the rate was reduced back to 21s. This rate held until 1813 when there was once again a very heavy demand for transports for various services, particularly to supply the army on the Peninsula. In December 1813 so extreme was the position that the Board was experiencing great difficulty raising a modest 5,000 tons for an extremely urgent conveyance of hay and biscuits to the north coast of Spain. The pressure applied by Bathurst, the Secretary of State for War, to make the shipment was so intense that the Board was persuaded to offer 30s per ton per month for all vessels fit for service that would immediately engage for two months certain. Even so, it was not able to hire sufficient ships. The urgency of this service forced the Board to hire neutral vessels 'on freight' as a last resort; it was aware that the use of foreign ships was very politically sensitive and they were used very infrequently.[69] The 25s rate continued through to January 1815

66 TNA, WO, 6/156/349, Castlereagh, Secretary of War to the TB, 8 Apr. 1807.

67 TNA, ADM, 108/21/56, TB to Lord Castlereagh, 20 Apr. 1807.

68 TNA, ADM, 108/83, TB Minutes, 20 Apr. 1809.

69 TNA, ADM, 108/24/53, TB to Earl Bathurst, Secretary of State for War, 3 Dec. 1813.

when it was reduced to 21s. As transport demand reduced, general freight rates also fell. The charter rate reduced to 19s in October 1815 and then to 15s in August 1816.

Consider the rates paid to the owners of the *Ellen* (342 tons), which was in service for over twelve years. Wood-sheathed, she was hired at the rate of 17s per ton per month on 8 March 1804. This rate was increased to 19s on 1 January 1808 and to 21s, the rate for coppered ships, on 14 August 1808 after she was coppered. However, she was mulcted £12 for 'excess time in coppering': ten days was the normal allowance for this process. During the remainder of the charter she was also mulcted £52 for excess repair time, £57 for deficiency of complement, £88 for stores unaccounted for and £1 for pilotage at Lisbon. The charter rate increased to 25s on 1 January 1809 but went back down to 21s on 1 September that year, reducing further to 19s on 1 October 1815. The owners were paid £54,116 for the whole hire period after the deductions. The changes were in line with the rate reviews with the exception of the temporary rise to 25s from April to December 1807. Neither did she benefit from the increased rate of 25s from March 1813 to January 1815.[70]

However, rates paid to ships already on hire were not always changed in line with rates for new hires. For example, in May 1804 when the rates approved by the Treasury were 19s per ton per month for coppered ships and 17s per ton per month for wood-sheathed ships, the actual range of rates being paid for coppered ships was from 16s 6d per ton to 25s per ton, with 63 per cent of the charters on the latest approved rate. Similarly the range of prices being paid for wood-sheathed ships ranged from 15s to 18s per ton. However, only 42 per cent were at the latest rate.

At times the Board preferred to hold down prices for additional ships to avoid having to pass on the benefit to existing charterers. When there was further pressure on price early in 1813 the Board advised Earl Bathurst, Secretary of State for War, that when it increased the rate in April 1812 in response to the strong competition from the heavy demand from trade, it had assured the Treasury that the increase was temporary. Thus it had subsequently reduced the hire of coppered transports from 25s to 20s per ton per month. It claimed that this reduction had already saved the government £166,615. The Board suspected that the reduction might have 'occasioned the tardiness of ship owners to offer their ships' but it was minded not to increase the rate again 'unless the exigency of the Service renders it indispensably necessary that we should hire a large quantity of additional tonnage without delay'. The Board's premise was that 'any increase, if not made general, would occasion much discontent and consequently great inconvenience, and loss to the service'.[71]

[70] TNA, ADM, 108/152/72.

[71] TNA, ADM, 108/23/97, TB to Earl Bathurst Secretary for War, 19 Feb. 1813.

'On Freight' Rates

Individuals, goods and sometimes groups of officers and troops were shipped on commercial voyages 'on freight' as a supplement to a ship's commercial cargo. On other occasions, the whole of a ship's capacity was used to transport troops at a rate per man. Rates were closely linked to the local market rates and varied depending on the destination. For example in 1797 the following rates were paid: for home services, between 15s and 20s per ton of cargo; to Lisbon and the Mediterranean, between £1 17s 6d and £3 per ton of cargo; to the West Indies, between £2 15s and £3 10s per ton of cargo and between £5 5s and £7 per man. The rates to Canada were Halifax £2 5s to £3 per ton, Quebec £3 to £3 10s per ton and Newfoundland £2 10s per ton.[72]

When the demand for transports coincided with heavy trade demand, ship-owners naturally attempted to exploit the position. For instance, in November 1798 Henry Dundas, Secretary of State for War, gave instructions to provide for the embarkation of the 9th Regiment of Foot consisting of 278 men from Yarmouth to Guernsey and of the 35th Regiment consisting of 244 men from Sunderland to Jersey. When Lieutenant Parke, Agent for Transports, visited Yarmouth to procure the necessary tonnage, he discovered a scarcity of shipping on the east coast. This had already increased the price of coals significantly in the Port of London. The owners raised the rate to £8 per man. However, the attempted opportunism backfired. Parke declined the offer as 'such unparalleled extravagance'. The Board suggested to Dundas that it might be more expedient to march these troops to Portsmouth or Southampton where there was greater certainty of procuring the necessary vessels at less exorbitant rates.[73]

Transports Lost, Damaged or Destroyed

Adverse weather conditions accounted for the loss of 223 of the 317 naval ships lost between 1803 and 1815. They were either wrecked or foundered, the majority on account of hostile natural elements.[74] Inevitably, then, given the sheer number of transport voyages, there were significant losses for the transport service of both ships and lives as a result of winter gales rather than enemy action.

There were numerous reports of havoc caused to major shipping convoys by weather conditions. Frustrating delays often occurred during the preparation stage as the convoy was being consolidated, resulting in delayed departure beyond the normal safe sailing dates, thus increasing the likelihood that the convoy would sail into adverse conditions. The best known example is the 1795 Abercromby-Christian convoy of troop and supply ships to the West Indies. It was delayed in departing until November only to be struck by gales, which drove the whole fleet back into scattered ports at the end of January with great loss of

[72] Committee on Finance (31st), 502.
[73] TNA, WO, 1/800, TB to Henry Dundas, Secretary of State for War, 7 Dec. 1798.
[74] Hall, *British Strategy*, 47.

life. It finally arrived in the West Indies at the end of the campaigning season.[75] In December 1805 eight transports carrying troops to Germany were wrecked in a storm: 664 men were drowned; the 1,552 who were not were washed up on the enemy coast and taken prisoner.[76] Wellington, then Major-General Sir Arthur Wellesley, a brigade commander, was on board the sloop *Fury* in the Downs and experienced the severe gales that had caused the dispersal of the transports. He subsequently expressed the view to Castlereagh, Secretary of State for War, that 'the season was too far advanced to send a fleet of transports to the war'.[77] Fifteen transports were lost during the Copenhagen campaign in 1807 with a loss of 402 soldiers, 60 horses and 14 seamen; 620 men and 18 horses were saved from these wrecks.[78] In March 1810 a Transport Board minute records the loss of two further transports, the *Richard* (321 tons) and the *Troy* (175 tons). The *Richard* had 130 men of the 4th Regiment on board.[79] In February 1811 the loss was reported of the *John and Jane* (241 tons), carrying part of the 2nd Battalion of the 11th Regiment to Lisbon. It was run down by HMS *Franchesse.* The *Wellington* (198 tons) was carrying ordnance stores for Malta when it was run down by the American ship *Intercourse.* Some years later the Transport Office had to advise the Admiralty of the loss of HMS *Hero* and a fleet of 120 ships, including many transports, in gales on 23 December 1811.[80] The full extent of the loss of life in sinking transports is not known. However, between 1 January 1814 and June 1816 some 1,702 souls were lost and 2,809 individuals were rescued from shipwrecked transports.[81]

Of the 284 regular transports lost, 117 (41 per cent) were considered by the Transport Board to be standard marine losses due to adverse weather or master's error. These were the owner's responsibility with no liability to the government. Only between 90 and 105 (31 to 36 per cent) of regular transport losses were attributed to capture by enemy navies and privateers. The losses were most significant in 1795–97 and again in 1813–14. Other losses included thirty-one transports abandoned and burnt during the evacuation from Holland in January 1795 and thirteen vessels which were used as fire ships in the April 1809 attack on the French at Basque (Aix) Roads.[82] However, it was not only foreign navies and privateers that masters needed to be wary of: on 1 December 1794 the *Camilla* transport (300 tons) carrying 186 French prisoners from the West Indies to Europe was seized by the prisoners. It was subsequently lost off the coast of France.[83]

75 Rodger, *Command of the Ocean*, 435.
76 Hall, *British Strategy*, 47.
77 TNA, PRO, 30/70/4/258, Sir Arthur Wellesley to Castlereagh, 16 Dec. 1805.
78 HoCPP, 1808 (10) IX.53, 2, Account of Transports employed against Copenhagen, 2 Feb. 1808.
79 TNA, ADM, 108/21/230, TB to Lord Liverpool, Secretary of State for War, 12 Mar. 1810.
80 TNA, ADM, 1/3763/84, TB to the Admiralty, 28 Jan. 1812.
81 HoCPP, 1816 (530) XIX.233, 1.
82 Sutcliffe, 'Bringing Forward Shipping for Government Service', 47.
83 TNA, ADM, 108/4b/143.

Valuation of Ships Captured or Destroyed by the Enemy

Losses due to natural maritime causes and sailing errors were not reimbursed by the government; only those due to ships being captured or destroyed by the enemy or abandoned on service were paid. This occurred in Holland in 1799. Transports could be damaged or destroyed by the navy during the course of an operation, a fate which befell merchant ships converted to fire ships used in Cochrane's attack on the French fleet at Basque Roads in 1809.

One element of the process of inspecting ships prior to engagement was the valuation by suitably qualified officers. The masts, rigging and other fittings were valued separately from the hull. It is difficult to identify the rate per ton used in the valuation, which presumably changed during the period under review as build costs increased. However, the rate per ton of two ships that were valued within four months of each other was quite different. In the case of the *Hythe* the rate was £30 whereas the *Lady Johnstone*'s valuation was £21 per ton. If a ship's loss was to be reimbursed then the original valuation was reduced by a charge for wear and tear incurred during service at a rate of 8s per ton per annum, 5s in respect of the hull, masts and spars and 3s for the rigging.[84]

The *Hythe* (148 tons, Master William Thorp) was taken into pay on 23 December 1812 and wrecked off Senegal on 23 December 1815 but was subsequently salvaged and sold. The charge for wear and tear for the twenty-six months and one day that she was on charter, at the rate of 8s per ton per annum, was £128. This, together with the proceeds of sale of £706, was deducted from the original valuation of £4,409, which gave a net value of £3,574. This sum was paid, plus interest at 3 per cent, by a thirty-day bill.[85] The *Lady Johnstone* (437 tons, Master John Richardson) was taken into pay on 14 September 1812 and captured on 24 November after only seventy-two days' service. The amount paid was equivalent to the original valuation of £9,247 less wear and tear of £35 for the seventy-two days that she was on charter, giving a net valuation of just over £9,212 to which interest of £103 was added. The full sum was again paid by a thirty-day bill.[86]

Occasionally there were disputes between a ship's owners and the Board about the valuation.[87] In one instance the Basque (Aix) Roads fire ship owners claimed that the offers made to them were totally inadequate to enable them to replace their vessels on the basis that there were 110 comparable ships for sale at that time but their values were high because the prices of new ships being built in even the cheapest shipbuilding ports of Whitby, Sunderland and Newcastle were very high.[88] They complained even though the Board had already agreed to pay an additional 25 per cent over the original valuations, the freight to the day on which they were destroyed and the wages of the masters from that day until

84 TNA, ADM, 108/34, TB Minutes, 19 Dec. 1794.
85 TNA, ADM, 108/151/184.
86 TNA, ADM, 108/151/135.
87 Rodger, *Command of the Ocean*, 555.
88 TNA, ADM, 108/174, Memorial from the owners of the late transports to the TB, 10 Jul. 1809.

the day of their arrival in London, together with the amount of their actual losses in clothes and their expenses to London. The Board considered the additional requests unreasonable and refused to pay more.[89]

Allowances and Mulcts

The Board appears to have been very fair in compensating owners for expenses incurred whilst on service. There are numerous examples of allowances granted to ship-owners by the Transport Board for: air scuttles cut in decks on troop ships to encourage the circulation of air to the sleeping decks; candles used and gunpowder expended for signalling; the loss of anchors and chain as a result of the actions of others; long boats lost in service; ballast removed; damages in action with privateers; damage sustained alongside naval vessels when transferring water and victuals. The owners of the *Admiral Gambier* were awarded £104 towards the expense of replacing three men who had quit the ship at Quebec to join a naval ship on the American lakes.[90] The owners of the *Colworth* were awarded £100 for the master's exertions in saving troops from another damaged ship off the coast of Ireland.[91]

If ships were no longer required after they had completed their mission, the owners were sometimes offered the opportunity of having them discharged immediately. They were then paid in lieu of the return voyage at the rate of four weeks' hire from the West Indies and two weeks' hire from the Mediterranean. This allowed the master the opportunity of finding another charter for the return voyage, in effect receiving double pay for that trip.

At the same time the Board was very forceful in recovering costs incurred by transports which it considered to be the responsibility of the owners. Mulcts were principally applied when a ship, by accident or through fault of the owner or master, became incapable of performing the service for which she was engaged. She would be mulcted during the time out of service and until she was reported ready for duty by an agent of the Board. This also extended to delays in sailing. Deficiencies in crew numbers were regularly mulcted as crew fell ill or deserted. Transport crews were frequently pressed into naval ships, which caused numerous delays. Another principal cause of mulcts was the master's misconduct, disobedience or negligence. There were fines ranging from £10 for a relatively minor infringement, such as failure to report arrival at a port immediately to the agent for transports, to £21 for a more serious matter such as disobeying orders. Possibly the more serious implication was that the ship's charters were more likely to be curtailed as soon as demand allowed.[92] Probably the most common cause of a mulct was the failure to account for all supplies that had been loaded on board at the start of the charter. In addition, any expenses

[89] TNA, ADM, 1/3758, unsigned note attached to the valuations list, 12 Jun. 1809. Also in TNA, ADM, 108/174.

[90] TNA, ADM, 108/152/8.

[91] TNA, ADM, 108/153/39.

[92] TNA, ADM, 108/32, Transport Board Minute, 6 Oct. 1794.

paid by agents on behalf of owners were deducted before the account was finally settled. Such expenses would include: the cost of taking in ballast, light and buoy dues, the cost of care of sick seaman at a hospital or on a hospital ship, customs duty and for damage done to navy ships at sea. Ships' log books were required to be surrendered at the end of each charter. Failure to do so resulted in a mulct of between £10 and £100, but this was frequently reimbursed at a later date when the log book was received at the Transport Office.

Speed of Preparation of Transports

The majority of demands for transports were for a specific service. Generally these ships were made available quickly and efficiently.[93] Tonnage was provided to the Victualling Board within an average of ten days from the request for victuallers for the Baltic between 1808 and 1810. Victualling ships for the Mediterranean in 1800 to 1802 had taken, on average, seven days to be made ready.[94]

The role of the Board and its agents was pivotal to the success of preparations for expeditions, but once it had procured the appropriate tonnage it had to rely on other boards to perform their own tasks efficiently. The Ordnance Board, the Victualling Board and the War Department had vital roles, and of course the weather had a very significant impact: ships took longer to prepare as the weather deteriorated.

During this process the Board was frequently called upon to provide a commentary on the state of preparation. Occasionally it prepared such reports when it was concerned about delays or when it anticipated that there might be an inquiry into delays. In these instances the motivation was to exonerate the Board from the causes of the delays. Table 2.4 is a direct copy of the report prepared by the Transport Office in August 1799, only two weeks before the commencement of troop landings, relative to the transports required to convey victualling provisions, bread and oats, horses and wagons to Holland in support of the Anglo-Russian mission known as den Helder expedition which illustrates this point. The 46,000-strong force was led initially by Lieutenant-General Abercromby and subsequently by the Duke of York. After initial successes, including the surrender of the Dutch fleet, the campaign became bogged down by enemy resistance and the force began to experience a shortage of wagons, horses and provisions due to bad weather which prevented them being landed. In October, rather than face an arduous winter in Holland, the duke negotiated an armistice to allow the evacuation of the army.[95]

The table indicates the complexity of preparing even a relatively small part of the operation. The Board received numerous instructions from Henry Dundas to prepare transports for this expedition between 20 and 29 July. The loading of the victuals was the responsibility of the Victualling Board whilst the Commissariat

93 Commission for Revision (9th), 76.

94 Davey, *Transformation of British Naval Strategy 1808–1812*, 275.

95 Rodger, *Command of the Ocean*, 463.

Table 2.4. List of Commissariat ships and victuallers destined for den Helder expedition, 14 August 1799

Ship's name	*Tons*	*Date of requisition*	*Date of appropriation*	*Time of sailing*	*Time of arrival*	*Remarks*
Commissariat for Bread and Oats						
Elizabeth	150	20 July 1799	25 July 1799	29 July 1799 from Deptford	30 July 1799 in the Downs	From the date of the application to the date of the appropriation, were only five days, and thence to the date of final preparation and sailing to their respective destinations 4 days more. We are advised that these vessels are arrived at the ports according to the dates in the preceding column.
Jenny	122	Ditto	Ditto	Ditto	Ditto	
Hope	87	Ditto	Ditto	Ditto	4 Aug 1799 in the Downs	
Ranger	75	Ditto	Ditto	Ditto	6 Aug 1799 in the Downs	
Brothers	99	Ditto	Ditto	Ditto	Ditto	
Liberty	83	Ditto	Ditto	Ditto	31 July 1799 Harwich	
Manchester	122	Ditto	Ditto	Ditto	31 July 1799 Faversham	
Commissariat for 180 Draft Horses						
Venus	239	27 July 1799	5 August 1799			Arrived at Blackwall on 5 August 1799 and from that time under the Commissariat Department.
Mercury	216	Ditto	Ditto			
John and Amy	121	Ditto	Ditto			
Penelope	312	Ditto	Ditto			
Squirrel	264	Ditto	Ditto			
Mary	244	Ditto	Ditto			

Ship's name	*Tons*	*Date of requisition*	*Date of appropriation*	*Time of sailing*	*Time of arrival*	*Remarks*
Commissariat for Wagons						
Jane	170	27 July 1799	10 August 1799	The 40 wagons were not ready till Saturday 10 August as appears from Mr Brook Watson's letter of the 8th inst. The ship is now at Northam (Devon), taking in her cargo and was expected to complete on the 12 August. By this post [14 August] it appears that this ship is on her way to the Downs with 36 of the wagons on board, not being able, on account of the unexpected accumulation of baggage, to receive the four left behind.		
Victualling Department						
Diligence	100	29 July 1799	1 August 1799	Laden and proceeding to the Downs		
Jason	79	Ditto	Ditto			
Rebecca	104	Ditto	Ditto	Loading their provisions under the orders of the Victualling Board		
John and Eleanor	80	Ditto	2 August 1799			
Newcastle Trader	122	Ditto	10 August 1799			

Source: TNA, ADM, 108/20, TB to the Secretary of State for War, 14 Aug. 1799.

was responsible for loading the other items. The bread ships were handed over for loading on 25 July and loading was completed on 29th. The whole process took nine days from the date of requisition. The horse ships were handed over to the Commissariat on 5 August. As of the 14th the Transport Board had not been advised about the rate of progress of embarkation. There had been an underestimate of the ship capacity required for the wagons and four of the forty had been left behind. The victualling ships were still being loaded. One of them, the *Newcastle Trader*, had taken longer to prepare and was not handed over to the Victualling Board until 10 August. It is clear from this table that any delays in respect of these transports were not the responsibility of the Transport Board. It had reacted quickly after receiving the requisitions to prepare the ships for handover to the other boards for loading. In addition to these vessels, the force transported to Holland consisted of 71 infantry ships of 18,576 tons with 14,861 men at 1¼ tons per man; 56 cavalry ships of 12,245 tons with 1,628 horses and men; two hospital ships; and the *Weymouth* armed transport of 1,425 tons with 1,000 men. The force totalled 33,051 tons. Another 106 ships of 22,453 tons had been prepared for the expedition but were not sent to Holland because of the decision to withdraw.[96]

When tonnage was required for major military operations, the considerable amount needed invariably meant that it took some time to assemble, prepare and load. Ships fully loaded could wait at the assembly points for weeks until the flotilla was complete and the navy took them under convoy orders.

Demurrage

If transports were delayed beyond the periods incorporated into the charter party when unloading supplies at the end of a charter, in circumstances beyond the control of the master and owner, then demurrage or further costs was paid by the government. One of the primary tasks of the transport agents was to ensure that ships designated for discharge were released as quickly as possible. This is illustrated by events reported by Captain Young, Agent for Transports at Deptford, in March 1806. He sent a list of transports that had arrived at Deptford and had returned water casks and provisions during that month. His report showed that the ships were paid for five days to report their arrival at the customs and excise offices and to unload all their stores. Due to the large number of ships being discharged at one time, delays had occurred for a number of reasons. The warehouses were full. This combined with frequent bad weather had prevented the dry stores from being offloaded. There was also a shortage of labour to unload the lighters, which in turn affected the availability of lighters. The overcrowding in the river just made things worse: ships had to moor some considerable distance from the unloading dock, increasing the time taken by

96 TNA, WO, 1/800, List of Transports appropriated to the expedition but which did not proceed under Admiral Mitchell, 14 Aug. 1799.

lighters travelling back and forth. Some lightermen had contracts to unload naval supplies, others to unload army supplies, and although the provisions were all landed at the Red House, the lightermen would only move those goods that they were contracted for, even if there was a mix of supplies on board one vessel. Additionally, lightermen refused to supply more craft until those loaded were cleared; many had not been unloaded and were lying at risk for a considerable time without any allowance for demurrage.[97] As a result none of Young's ships had been unloaded within the five days built into the charter party. Most had taken between one and three weeks longer, so Captain Young had been obliged to consent to extend the time allowed when it became clear that the delays were not caused by any neglect of the master or crew. Some of the masters were happy with the delays; they were not anxious to get their ships cleared because they had no immediate future employment for their vessels.[98]

The Board received many claims for demurrage but before accepting them it always made further investigation of the facts. The owners of the ship *Fortitude* and several other transports on passage to Portugal in December 1796 claimed demurrage during the time the vessels were detained at Falmouth, where they arrived under convoy with HMS *Seahorse*. The Board requested the Admiralty to advise if the convoy put in at Falmouth under its instructions or in consequence of adverse winds. Demurrage would be paid for the former but not the latter.[99]

Merchant ships played a paramount role in the defeat of France between 1793 and 1815. There were frequent shortages of ships, yet the Transport Board worked diligently to ensure continuity of supply. This was achieved by dealing fairly with owners, by maintaining good communications with ships' brokers and by continuing to have good market intelligence which facilitated judicious manipulation of the hire rate in line with market conditions. Despite being under considerable pressure from ministers to bring forward ships, the Board only resorted to increasing the hire rate if it was fairly sure that the decision would bring forward a significant number of additional vessels. It had discovered from experience that when availability was tight, rate increases did not yield a substantial number of additional ships; in fact rises had a detrimental effect under such circumstances. Existing owners would inevitably demand a universal price increase which would add significantly to the Board's costs if granted or damage goodwill if not.

In earlier wars the Navy Office had devised rigorous, somewhat cumbersome, procedures for bringing forward and managing shipping for government service. The Transport Board adopted many of those practices, revising them where appropriate to improve efficiency. The charter party was refined through experience to ensure that owners and masters fulfilled their unambiguous obligations and to minimise the cost and risk to the public purse.

97 TNA, ADM, 108/37, William Young, Agent to TB, 24 Mar. 1806.
98 TNA, ADM, 108/37, William Young, Agent to TB, 24 Mar. 1806.
99 TNA, ADM, 1/3734, TB to the Admiralty, 14 Sept. 1797.

Regular transports were the backbone of the transport fleet but when major military operations were planned large numbers of ships were hired, usually on three-month charters. These ships were then discharged as soon as possible. The tonnage hired peaked in 1795 during the West Indies campaigns, then in 1809 during the Walcheren expedition and again in 1813 when transports were heavily engaged in supporting the army in the Peninsula and the war against America. The Transport Board took steps to ensure that it obtained the best ships at the best value to the public service by increasing the number of brokers that it dealt with on a regular basis.

3

Competing Demands for Merchant Shipping, 1793–1815

'I have no doubt that in execution of this service, which is indispensable, you will make every exertion to prevent as much as possible any embarrassment to trade or any increase of charge to the public.'[1]

The volume of shipping required by the Transport Board changed month by month and year by year, as did the availability of the limited pool of merchant ships from which transports were procured. Whilst the Transport Board was charged with the provision of sufficient tonnage to meet government demands at the lowest cost, the constant dilemma for ship-owners was how to utilise their ships to generate the best return. Economics, not patriotism, was the principal motivation for ship-owners because military success did not earn the contractors who supplied services to the military the gratitude of the public. Victories were bought at a considerable price, and hostility emerged towards the perceived beneficiaries of the escalating military expenditure, including ship-owners.[2]

Ultimately the market determined the availability of shipping for government service, but how successful was the Transport Board in engaging with the industry to achieve its goals? When large tonnages of shipping were required for major expeditions, there were invariably insufficient ships immediately available. This chapter will identify why shortages occurred and assess the impact of government demands on the British shipping industry. For example, did the demand push up freight rates? Did it encourage the building of more ships than would be needed in the long term? Did it divert ships from trade, which might have generated additional revenue for the state?

The size of the British merchant fleet in 1793 gave the government an advantage over rival nations and it was destined to double in capacity over the next thirty years. Its main competitors were Holland and France.[3] Despite the

[1] TNA, WO, 6/156, Hawkesbury in the absence of Castlereagh, Secretary of State for War, to TB, 3 Aug. 1805, when presenting instructions to raise 18,000 tons fit for distant service, half to be got ready at Cork the other half at Portsmouth, victualled for six months, also 10,000 tons for home service to be assembled in the Downs.

[2] Gordon Bannerman, *Merchants and the Military in Eighteenth-Century Britain* (London, 2008), 140.

[3] Morriss, *British Maritime Ascendancy*, 82–3.

wars there was a rapid expansion in world trade and a corresponding increase in demand for shipping. In 1815 Britain still had the largest merchant fleet, but its closest rival, in terms of fleet size, was then the United States of America. The US had overtaken the other nations as early as 1803 when it was reported in a statement to a Committee of the House of Congress that 'the merchant tonnage of the United States was now inferior to no other country except Great Britain'.[4] This is evidenced by the changing pattern of shipping entering US ports: in 1790 only 355,000 tons of domestic shipping entered US ports from overseas; by 1801 this had more than doubled to 799,304 tons, demonstrating dramatic growth as the country took control of its own trade, supported by government protection schemes. These included the imposition of higher duties on goods entering the US if they were carried in British ships.[5] By 1812 most US trade was carried in its own shipping, even though Britain exported more to the US than it imported from there.[6]

Whilst American shipping expanded, the French merchant fleet fared badly. British naval blockades severely restricted its movements and growth. Not until 1825 did French foreign trade exceed its 1788 total, nor did France's merchant shipping tonnage regain 1788 levels until the 1850s.[7] The pre-war dominance of the British fleet was principally due to the effective adoption of protectionism, exemplified by the Navigation Acts. Protectionism restricted certain trades to British ships, imposed higher rates of duty on goods that were carried in foreign vessels and implemented higher port dues for foreign shipping.[8]

The British shipping industry had a high domestic public profile. Ship movements received daily coverage in newspapers, and throughout the wars considerable parliamentary time was devoted to trade and shipping matters. There were numerous Acts of Parliament relating to navigation including, but not limited to, the 1794 Act authorising the extension of the employment of foreigners on board British ships to up to 75 per cent of the crew.[9] The 1797

4 Robert Kilmarx (ed.), *America's Maritime Legacy: A History of the US Merchant Marine and Shipbuilding Industry since Colonial Times* (Colorado, 1979), 31.

5 Kilmarx (ed.), *America's Maritime Legacy*, 32.

6 HoCPP, 18 Feb. 1812 (63) X.1, 3, An Account of the Official values of Imports from and Exports to The United States of America from 1801 to 1810. This account showed the aggregate imports from the US (£20,105,627) and exports to the US (£65,182,261) for the ten years 1801 to 1810. This represents a ratio of 3:1 in favour of exports to the US.

7 Duffy, *Soldiers, Sugar and Seapower*, 389.

8 Palmer, *Politics, Shipping and the Repeal of the Navigation Laws*, 41–3. For other commentaries on the Navigation Acts see: Gary Walton, 'The New Economic History and the Burden of the Navigation Acts', *Economic History Review* 24.4 (1971), 21–36; HoCPP, 1847 (232) X.1, Reports of the Select Committee of the House of Commons to Inquire into the Operation and Policy of the Navigation Laws; J. Allen, *The Navigation Laws of Great Britain, Historically and Practically Considered* (London, 1848); Lawrence A. Harper, *The English Navigation Laws: A Seventeenth-Century Experiment in Social Engineering* (New York, 1939); Robert Livingstone Schuyler, *The Fall of the Old Colonial System: A Study in British Free Trade* (Oxford, 1945).

9 1794 Act for the Further Encouragement of British Mariners, 33 Geo. III, c.26.

Manifest Act mandated the compulsory requirement for a full manifest for each cargo carried on every British ship.[10] The 1798 and 1803 Convoy Acts ensured that merchant ships were legally bound to sail in convoy or forfeit insurance cover or government indemnity if captured.[11] Then an 1801 Act demanded that ensigns or colours should be borne at sea in all merchant ships.[12]

The British Shipping Register

A significant component of the Navigation Acts was the Registration of Shipping Act of 1786 which defined 'British ships'. To qualify for registration, ships had to be British owned and British built, have a deck and be of more than 15 tons.[13] The British Register of Shipping was compiled from monthly returns submitted to the registrar of shipping from customs officers at all ports in the British Empire.[14] It included ships registered not only in England, Jersey, Guernsey, the Isle of Man, Scotland and Ireland, but also in the North American and West Indian colonies.[15] Whilst the number of seamen normally engaged in sailing each vessel was recorded, it is important to observe that these figures did not reflect the number of men who were actually available at any one time.[16] Sadly the register gives no indication of whether the ships were ocean-going or coasters, nor of the trade that they were mainly involved in. So it does not readily indicate the extent of the fleet that might have met the Transport Board criteria.[17]

As shown in table 3.1, the number of ships on the register increased by 72 per cent between 1790 and 1816 and the tonnage almost doubled, suggesting

10 1797 Manifest Act, 36 Geo. III, c.40.

11 1798 Convoy Act, 38 Geo. III, c.76; 1803 Convoy Act, 43 Geo. III, c.57.

12 1801 Merchant Shipping Act, 41 Geo. III, c.19.

13 1786 Registration Act, 26 Geo. III, c.60.

14 Palmer, *Politics Shipping and the Repeal of the Navigation Acts*, 42: 'The Registry Laws formed part of the Navigation code. Where a vessel was built, the nationality of the owner, master and crew were all relevant criteria. To qualify as British a ship had to be British built and British owned. British also included colonial.' Note that in 1814 there was a major fire at Customs House in London and many of the records were destroyed.

15 The port of registration was not recorded before 1799.

16 Paul van Royan, Jaap Bruijn and Jan Lucassen (eds), *'Those Emblems of Hell': European Sailors and the Marine Labour Market 1570–1870* (St John's, Newfoundland, 1997), 17–31. The statutory Customs Register is not to be confused with the Lloyd's Register which was a voluntary register with details submitted by the ship's owners. Lloyd's Register was (and is) a classification society which rates the condition of ships to enable them to obtain insurance.

17 Several historians have attempted to assess the relative sizes of the ocean-going fleet and the coasting fleet, notably: Davis, *Rise of the English Shipping Industry*, 395–406; Ernest Fayle, 'The employment of British Shipping', in Northcote Parkinson (ed.), *Trade Winds: A Study of British Overseas Trade during the French Wars 1793–1815* (Leicester, 1948), 73; D.E. Robinson, 'The Secret of British Power in the Age of Sail: Admiralty Records of the Coasting Fleet', *American Neptune* 48 (1987), 5–21.

Table 3.1. The number of ships and their tonnage and the number of men who would normally sail in them in 1790, 1799 and 1816 (as recorded by the registrar of shipping), with a comparison of the growth of tonnage 1799–1816 and the number of enemy ships taken as prizes and registered under British ownership

	1790		*1799*			*1816*			*Growth of tonnage from 1799 to 1816*	*Prize ships on the Register in 1816*	*Prize ships tonnage*
	Ships	*Tons*	*Ships*	*Tons*	*Men*	*Ships*	*Tons*	*Men*			
England	Detail not available		11,487	1,337,181	99,309	17,442	2,152,968	134,060	61%	2,696	417,372
Jersey			61	4,611	694	77	7,992	636	73%	32	4,217
Guernsey			78	6,199	803	65	7,237	494	18%	27	4,041
Isle of Man			227	5,146	1,201	369	9,335	2,315	81%	0	
North America and West Indies			2,996	201,743	15,982	3,775	279,643	16,859	38%	942	102,231
Scotland			2,031	148,110	12,413	2,958	263,536	18,775	78%	281	31,105
Ireland			999	49,825	4,835	1,178	63,229	5,681	27%	63	6,839
Total	15,015	1,460,823	17,879	1,725,815	135,237	25,864	2,783,940	178,820	61%	4,041	565,805
Total for England and Scotland	Detail not available		13,518	1,485,291	111,722	20,400	2,416,504	152,835	62%		
Ships in England and Scotland as a % of the total			75%	86%		79%	86%				

Sources: Cobbett, Parliamentary History of England, *vols 1–24, and HoCPP, Annual Trade and Navigation Accounts. TNA, CUST, 36/5*

a modest move towards larger vessels.[18] However, John Dalley, the Assistant Registrar in the department of the Registrar General of Shipping, provided important evidence that the register needed to be treated with some caution.[19] He emphasised that the numbers did not necessarily reflect the number of ships that were operational because ships frequently remained on the register after they had been withdrawn from service. Also, inaccuracies occurred because the returns from out-ports were frequently incorrect due to the loss, capture or destruction of vessels and to changes of ownership when ships were registered at the new port without being cancelled at the original one.[20] Dalley reinforced the fact that there were insufficient seamen to man all the ships at any one time. He was convinced that the availability of seamen was the principal restriction on the availability of shipping. Nevertheless, the register does give a reasonable reflection of the size of the British fleet and the changes from year to year.

Merchant Ship Losses during the Period

It is not clear how many ships were lost during the period, nor the number removed from the register because they were no longer operational, although there have been several estimates of the losses. The secretary of Lloyd's suggested that over the whole period 1793–1815 British merchant losses were about 2 per cent a year from all causes net of recaptures, but in deep sea trades they were as much as 5 or 6 per cent, half to marine causes and half to enemy action. In the English Channel losses in 1808 were 1½ per cent.[21] If losses of English ships were at the rate of 2 per cent per annum during the period 1800–15 inclusive, this would equate to an approximate average of 290 ships per year or 4,640 over the period. This number is comparable to Winfield's assessment that the average number of British merchantmen lost between 1803 and 1814 was 440 per annum; of these 67 per cent or 294 ships were English.[22] These estimates

[18] Duffy, *Soldiers, Sugar and Seapower*, 386: 'the boom in the West Indian trade ... was one of the prime factors in the increase in merchant tonnage in the war against Revolutionary France'. See appendix 1 for the annual numbers of registered ships.

[19] Dalley Evidence.

[20] Palmer, *Politics, Shipping and the Repeal of the Navigation Laws*, 1: the register was updated in 1827 (and some 13 per cent of the ships recorded were removed). For more detailed consideration of the registration system, see the evidence of W.H. Noss, Registrar of Shipping, to the Select Committee of the House of Lords on Policy and Operation of the Navigation Laws (HoCPP, 1847–48).

[21] Rodger, *Command of the Ocean*, 559–60.

[22] R. Winfield, *British Warships in the Age of Sail, 1793–1817* (Chatham, 2005), 291 citing W.M. James, *The Naval History of Britain*, vol. 2 (reprinted London, 2001), 382 and 404 citing C.B. Norman, *Corsairs of France* (London, 1883), 453. The losses quoted are 1803: 222; 1804: 387; 1805: 507; 1806: 519; 1807: 559; 1808: 469; 1809: 571; 1810: 619; 1811: 470; 1812 not known; 1813: 371; 1814: 145. Total 4,839. Average per year for 11 years 440. Morriss, *British Maritime Ascendancy*, 86, cites P. Crowhurst, *The French War on Trade: Privateering 1793–1815* (Aldershot, 1989), 31, quoting the number of merchant ships lost to enemy privateers between 1793 and 1814 at 11,000 and an annual loss never exceeding 2.5 per

can be used to help identify the number of ships removed from the register. From table 3.1 it is clear that the net increase in the number of ships on the register of English shipping from 1799 to 1816 was 5,955.[23] To compensate for these losses, the number of ships built in English shipyards between those years was 9,200.[24] In addition, 2,696 prizes were added to the register. This would have increased the total number of registered ships by 11,896 before taking into account losses due to maritime incidents and enemy activity and to ships removed from the register. If net losses were 290 ships per year (2 per cent), in total 4,640 throughout the period, then the number of English ships taken out of service would have been 1,301, representing 0.5 per cent per annum.[25]

The recapture of ships taken by the enemy was not uncommon. It is estimated that some 10 per cent of all captures were in fact recoveries of ships that had been seized by the enemy.[26] At times the rate was much higher.[27] Recovery of captured ships was a profitable venture because the Prize Acts provided for salvage on such vessels. This was equal to one-eighth of the value of the ship, its stores and cargo when paid to naval re-captors, but one-sixth to privateer re-captors.[28]

The Tonnage of Registered Ships and of Transports

The Transport Board chartered ships of more than 100 tons but the preference was for ships in excess of 220 tons. In fact 73 per cent of all regular transports exceeded 220 tons.[29] Table 3.2 shows the average tonnage for all regular and three-month transports in 1795 and 1812. It demonstrates quite clearly that ships over 300 tons were preferred for use as troop transports. It also demonstrates that the average tonnage employed ranged from 271 tons to 296 tons, indicating that three-month ships also were generally over 220 tons.

cent of all British registered ships. Simon Ville, 'The Incidence of Loss of Merchant Vessels during the French Wars', *Mariner's Mirror* 68, 4 (1982), 450.

23 This number is determined by deducting the number of ships registered in England in 1799, shown in table 2.1 as 11,487, from the number registered in 1816 of 17,442. The difference between the two numbers is 5,955; this represents the net growth in the number of ships registered in England between 1799 and 1816.

24 TNA, CUST, 36/5.12, for 1800 to 1813 inclusive the figure was 8,225. HoCPP, Accounts Relating to Shipping, 22 Apr. 1815, shows the number of ships built in England in 1814.

25 Determined from 9,200 ships built and 2,696 prizes less losses of 4,640, giving a net figure of 7,256 compared to the recorded increase of 5,955, the difference being 1,301 attributable to decommissioned ships. N.B. Clarification of the number of ships lost and captured during the period and of ships decommissioned is still required but is beyond the scope of this book.

26 Richard Hill, *The Prizes of War* (Stroud, 1998), 200.

27 Hill, *Prizes of War*, 200, citing TNA, ADM, 1/3993 The Admiralty to the Committee at Lloyd's. Between 1 Sept. 1808 and 1 Mar. 1809 of the 72 ships that were captured by the enemy in the approaches to Britain, no fewer than 24 had been recaptured.

28 Hill, *Prizes of War*, 126 and 198.

29 Sutcliffe, 'Bringing Forward Shipping for Government Service', 31.

Table 3.2. The average tonnage of all monthly paid transports in 1795 and 1812

	3 October 1795			13 October 1812		
	No. of ships	*Tonnage*	*Average tonnage*	*No. of ships*	*Tonnage*	*Average tonnage*
Infantry ships	353	107,464	304	130	43,723	336
Cavalry ships	129	32,674	253	90	28,889	321
Navy victuallers	86	20,858	243	98	27,105	277
Army victuallers	41	8,314	203	132	29,488	223
Ordnance/baggage	100	22,292	223	61	16,701	274
Naval store ships	8	2,899	362	0	0	0
Total	717	194,501	271	511	145,906	286
Sources	*TNA, ADM, 1/3730*			*TNA, ADM, 1/3768*		

Table 3.3. An analysis of the tonnage of ships registered in England in 1790

	30 September 1790 No. of ships	*% of total*
Under 60 tons	4,347	43%
60 to 100 tons	1,659	17%
100 to 220 tons	2,556	25%
220 to 500 tons	1,360	14%
Over 500 tons	129	1%
Total	10,051	
Over 100 tons	4,045	40%
Under 220 tons	8,562	85%

Source: TNA, CUST, 36/5

The 1786 Registration of Shipping Act required all ships of more than 15 tons to be registered. Table 3.3 demonstrates that in 1790 no more than 15 per cent of the ships registered in England were over 220 tons (even in 1829 the weight profile was still very similar). The remaining 85 per cent were less than 220 tons, which was not appropriate for most of the Transport Board's requirements. This is very significant: it demonstrates, very dramatically, the challenge that the Transport Board faced. It sought to charter at least 73 per cent of its needs from a pool of 15 per cent of the ships registered, even though this relatively small proportion of the registered ships accounted for 54 per cent of the registered tonnage. However, even this does not reflect the full extent of the difficulties faced by the Board because it does not take into account ships that were without crew or had been withdrawn from service, neither does it account for the overlap of seasonal trade with major

Table 3.4. The average tonnage of ships entered and cleared at ports in Great Britain in 1792 showing origin and destination

Long-distance trades	*Average tonnage*	*Southern Europe and Mediterranean*	*Average tonnage*	*Short sea trades*	*Average tonnage*
Asia	707	Spain, Portugal, Malta, Gibraltar	126	Russia, Baltic Scandinavia	186
West Indies	233	Italy and Austria	143	Holland and Flanders	117
United States	221	Turkey, Levant, Egypt	224	France	73
British North America	147			Greenland and southern whale fisheries	270
Africa excluding Egypt	202			Ireland	75
				Isle of Man and Channel Islands	47

Source: Fayle, 'The Employment of British Shipping', 73.

expeditions. An analysis of the tonnage of merchant ships used on different trade routes in 1792 is shown in table 3.4.[30] It illustrates that traffic involved in the Asian trade used the largest ships, mainly East India Company vessels. Long-distance ships were generally over 200 tons although the average tonnage of vessels bound for Canada was 147 tons; those bound for northern and southern Europe tended to be in the 100–200 tons range. It is clear from table 3.4 that ships of less than 100 tons were generally restricted to coasting activities and short voyages in 'Home Waters' between Britain and France, Ireland and the Channel Islands.

Shipbuilding[31]

The Navigation Acts protected British shipbuilders as well as they protected ship-owners: all ships on the British register had to be British (or British colonial)

[30] Fayle, 'Employment of British Shipping', 74.

[31] For a more comprehensive review of shipbuilding in the period see Arthur Sayer, W.W. Rostow, J. Jacobson and Anna Schwartz, *The Growth and Fluctuation of the British Economy 1790–1850* (Oxford, 1953); A. Slaven, 'The Shipbuilding Industry', in Roy Church (ed.), *The Dynamics of Victorian Business: Problems and Perspectives to the 1870s* (London, 1980), 107–26.

built. This eliminated competition, not between themselves but from foreign shipbuilders. It was unlikely that British ship-owners were going to approach a foreign builder if the resulting ship could not then be registered.[32] In consequence British shipbuilding continued through the wars but at a level reduced from peacetime. The annual average number of ships built in England between 1793 and 1813 was 561 (average tonnage 72) compared to 658 (average tonnage 65) in the three years from 1788 to 1790. The rate of building had earlier peaked in 1787 when 829 ships were built. It peaked again in 1802 (721 ships) and 1803 (865 ships) after peace had been declared, the 1803 number exceeding the 1787 level. Table 3.5 demonstrates the growth in shipbuilding in 1800, 1801 and 1802 compared to the pre-war years of 1790, 1791 and 1792. It indicates that in 1800–02 there were 47 per cent more ships built with an 80 per cent growth of tonnage, suggesting that bigger ships were being built in this period.

Table 3.5. Number and tonnage of vessels built and registered in Great Britain (including Scotland) for the three years 1790–92 compared to the three years 1800–02, showing the increase in the average of the latter period over the former

Years	*Ships*	*Tonnage*	*Years*	*Ships*	*Tonnage*	*Growth ships*	*Growth tonnage*
1790	577	57,137	**1800**	845	115,349		
1791	624	58,760	**1801**	918	110,206		
1792	655	66,951	**1802**	967	104,789		
Total	1,856	182,848	Total	2,730	330,344		
Average	618	60,949	Average	910	110,114	+47%	+80%

Source: HoCPP, 1802–03 (26 (1)) VII.167, An Account of the number and tonnage of vessels built and registered in Great Britain, Assistant Register General of Shipping, 3 Mar. 1803

When war restarted the level of shipbuilding again reduced. The lowest level of building was in 1809 when only 417 ships were constructed.[33] Ville suggests that in 1776 around 40 per cent of English tonnage was launched in the north-east where low construction costs and sound design had established the area as a major centre of shipbuilding.[34] However, during the wars this dominance was defused: the shipyards of Hull produced the highest number of ships, then builders on the Thames, then Yarmouth and then the Tyne yards around Newcastle. Between 1800 and 1813 the average tonnage of each ship built in England was only 121 tons, suggesting that a large number of these were coasting vessels. Those built along the Thames were larger, averaging about 225

32 Palmer, *Politics, Shipping and the Repeal of the Navigation Laws*, 47.

33 The principal shipbuilding centres and the number of ships built are demonstrated in appendix 2 extracted from TNA, CUST, 36/5, Abstract of Ships built in several ports of England from 1786 to 5 Jan 1814. Also see Morriss, *British Maritime Ascendancy*, 85.

34 Ville, *English Ship-owning during the Industrial Revolution*, 40.

tons.[35] Demand for new ships was high. Merchant shipbuilding capacity had to be shared with the navy, but it was contained by the restricted availability of materials, principally timber, and probably more so by the availability of shipwrights, who were in great demand in the royal dockyards.[36] Indeed Henley & Sons regularly found that its favoured yards had a constantly full order book and so it had to rely generally on the second-hand market and on the ability to buy prize vessels to expand its fleet.[37]

Shipbuilders derived a considerable part of their business from the preparation and repair of transports. Messrs Dudman & Son, who had a yard on the Thames, reported that in the years between 1803 and 1812 it built 25 navy ships and 13 merchant ships, and repaired and refitted 4 navy ships, 123 transports and 155 merchant ships.[38] Another yard reported that it had survived because of the work it had done on repairing and refitting transports.[39]

The Demands of Trade

As demonstrated in chart 3.6, officially recorded imports rose from £19.6 million in 1792 to £36 million in 1815. This was an increase of 84 per cent. Imports grew modestly in the early war period and then suffered a slight reduction in 1798. However, there was significant growth between 1799 and 1802 when they peaked at almost £33 million during the Peace of Amiens.[40] Resumption of hostilities caused a reduction but imports reached £31.4 million in 1806 and peaked again

[35] Sarah Palmer, 'Shipbuilding in Southeast England 1800–1913', in Simon Ville (ed.), *Shipbuilding in the United Kingdom in the Nineteenth Century: A Regional Approach* (St John's, Newfoundland, 1993), 45–74; Morris, *British Maritime Ascendancy*, 138.

[36] Morriss, *British Maritime Ascendancy*, 165.

[37] Ville, *English Ship-owning during the Industrial Revolution*, 44.

[38] HoCPP, 1813–14 (115) VIII.1, 427, Minutes of Evidence taken before the Select Committee of the House of Commons on petitions relating to East India built shipping, Evidence of Mr Samuel Jordan.

[39] HoCPP, 1813–14 (115) VIII.1, 371, Minutes of Evidence taken before the Select Committee of the House of Commons on petitions relating to East India built shipping, Evidence of Mr Cornelius Trussit, Clerk to Thames shipbuilder, Mr Mestaer, 19 May 1814.

[40] The values quoted for imports and exports are those known as the 'official values' assigned arbitrarily to the specific articles a century before. The advantage attached to this system is that no fluctuations of price enter as a factor; the values continue to represent from year to year the proportion of trade done. 'Real values' deduced from current prices were generally much greater than the 'official values'.

Comparison of official export values with real export values						
	1805	1806	1807	1808	1809	1810
Official exports value (£000s)	34,308	36,527	34,566	34,554	50,286	45,869
Real exports value (£000s)	51,109	53,028	50,482	49,969	66,017	62,702
% increase from official exports to real exports	49%	45%	46%	45%	31%	37%

Chart 3.6. British imports and exports, 1790–1817

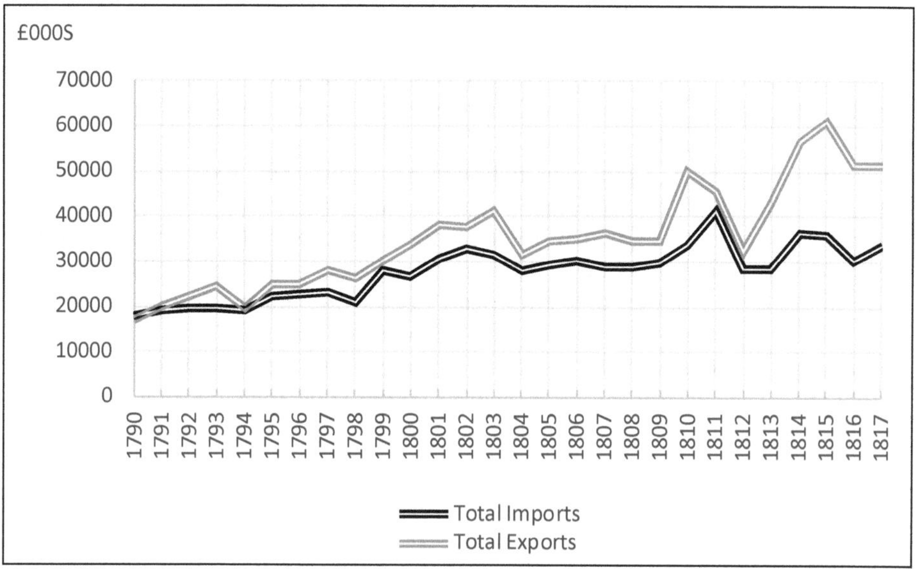

in 1811 at over £41 million.[41] The Navigation Acts and the later Orders in Council ensured that a significant proportion of imports were brought in British shipping before being re-exported to Europe. The duties raised were necessary to offset the cost of the war and the subsidies paid to support the allied armies of Russia, Prussia and Austria. The importation of corn was vital, particularly in 1796, 1801 and 1810.[42] Domestic harvests had failed so these imports saved the nation from starvation and prevented internal insurrection. Ironically the British grain shortages in 1809–12 provided Napoleon with an opportunity to generate much needed revenue by waiving the restrictions of his own Continental System to allow the export of wheat to Britain.[43]

The economic importance of the Caribbean to British trade grew substantially during the war. In direct trade the West Indies and South America increased their share from less than a fifth to nearly a quarter of British trade. When added to their contribution to the re-export of domestic manufactures of cotton cloth and refined sugar and to the African slave trade, this share had become a third of

---

HoCPP, 1812 (63) X.1, 5, An Account of the value of all imports into and all exports from Great Britain from 1805 to 1809 (18 Feb. 1812).

See also Ralph Davis, *The Industrial Revolution and British Overseas Trade* (Leicester, 1979).

41 Imports were principally corn, coffee, sugar, tea, tobacco, timber, cotton, iron, flax, hemp and linen yarn and later oils, seeds and nuts for making oil, gums and tallow.

42 *Aberdeen Journal*, 17 Mar 1800, refers to 'transports in the Mediterranean and America are to be paid a month's allowance from Government extra, discharged from Government Service and permitted to bring home immediately cargoes of wheat'.

43 Woodman, *Britannia's Realm*, 188.

British trade by 1802.[44] This re-export trade was to play a particularly important role in funding the demands of war, so much so that George III declared in 1779, 'If we lose our sugar islands, it will be impossible to raise money to continue the war.'[45] Hall notes that although Britain's colonies helped support her trade and naval power, extensive naval resources were required to preserve them from attack.[46]

Exports rose by 172 per cent between 1792 and 1815. Exports consisted of coal, iron and steel, manufactured cotton, yarn, linen and silk, together with re-exports of coffee, sugar, tea, tobacco and raw cotton. Re-exports accounted for 25 per cent of exports in 1792, increasing to 28 per cent in 1815.[47] The re-export trade in Eastern, colonial and American goods was among the most lucrative and jealously protected branches of British commerce.[48] In 1805 31 per cent of British export trade went to the United States and 38 per cent to Europe.[49] Of course the nature of trading relationships changed as allies became belligerents and then when those same belligerents became allies again, as is illustrated by the relationship between Britain and Spain. The two countries were at war between 1796 and 1802 and between 1804 and 1808. This had an adverse impact not just on trade with Spain but also on the Spanish-dominated South American trade during those periods.[50]

In 1806, Napoleon instigated the Continental System forbidding the entry of colonial produce from Britain or any of her possessions into all French and allied ports.[51] The British retaliated with a number of Orders in Council that tightened the grip on France's maritime trade which, by that time, was generally being conducted in neutral shipping. Napoleon responded by issuing his Milan Decree in December 1807. This specified that ships going to British ports or paying British dues had effectively lost their nationality and had become British property and were subject to seizure.[52] It has been suggested that the Continental System was intended to develop the economy of France at the expense not only

[44] Duffy, *Soldiers, Sugar and Seapower*, 380.

[45] Duffy, *Soldiers, Sugar and Seapower*, 385.

[46] Hall, *British Strategy*, xii.

[47] Military stores shipped out to supply the army and navy overseas were not included in the export figures.

[48] Fayle, 'Employment of British Shipping', 76.

[49] Nicholas Tracy, *Attack on Maritime Trade* (Toronto, 1991), 77.

[50] Fayle, 'Employment of British Shipping', 78–84

[51] For further study of the Continental System see: François Crouzet, 'The Continental System after Eighty Years', in Ronald Findlay (ed.), *Eli Hecksher, International Trade and Economic History* (Cambridge, MA, 2006), 323–46; Crouzet, 'Great Britain's Response to the French Revolution and to Napoleon', in Crouzet (ed.), *Britain Ascendant: Comparative Studies in Franco-British Economic History* (Cambridge, 1990), 267–85; A.H. Imlah, 'Real Values of British Foreign Trade 1798–1853', *Journal of Economic History* 8 (1948), 87–102; Imlah, *Economic Elements in Pax Britannica: Studies British Foreign Trade in the Nineteenth Century* (Cambridge, MA, 1958); Lance E. Davis, *Naval Blockades in Peace and War* (New York, 2006).

[52] Hill, *Prizes of War*. 51.

of Britain but also of French vassal states and of neutrals.[53] By 1810, however, Napoleon had effectively abrogated the Berlin and Milan Decrees by allowing the import of colonial products but at an exceedingly high import duty rate.[54] In Britain, the galvanising effect of the industrial revolution generated a rise in the consumption of raw materials, increasing dependence on overseas sources of supply and producing a corresponding growth in tonnage for imports. It also created more dependence on exports to foreign markets.[55] Indeed, despite the Continental System, exports to continental Europe doubled between 1808 and 1809 from £11.2 million to £23.7 million. In the same year exports to America, which had fallen back in 1808, recovered to earlier highs of over £10.5 million. The overall increase in the value of exports in 1809 over the previous year was a staggering 45 per cent. This surge coincided with major government demands for shipping to support the Peninsular and Walcheren expeditions, thus contributing to the difficulties experienced by the Transport Board in that year.

The Licensed Trade

The expansion of the British fleet suggests that a significant proportion of the increase in trade was carried by British ships. The statistics for ships' entrances to and clearances from British ports tell a different story. They show that a significant number of recorded entrances and clearances related to licensed foreign ships.

Licensing was an effective part of commercial control instituted by Orders in Council. There were 11,972 licences issued between 1802 and 1808. Then 15,226 were issued in 1809 alone and in 1810 another 18,356, reducing to 7,602 in 1811. The surge was to support increasing imports and exports, in particular the import of French wheat as a result of disastrous British harvests.[56] Licences were mainly issued to facilitate imports: in 1810 the issues were in the ratio five to one in favour of imports. Licensed ships were not used in the British coastal trade; this remained the sole domain of British-registered shipping.[57]

53 Tracy, *War Against Maritime Trade*, 74.

54 Nicholas Tracy, *Naval Chronicle: The Contemporary Record of the Royal Navy at War*, vol. 5 (London, 1999), 333.

55 Fayle, 'Employment of British Shipping', 25.

56 Licences were granted under various heads. The numbers included in brackets show the allocation of the 1810 licences in each category. Exports not transferable (2,085), imports not transferable (5,512), exports transferable (474), imports transferable (6,903), export and imports transferable (1,803) and exports and imports not transferable (1,574).

57 HoCPP, 1812 (323) X.69, 2, An Account of the tonnage of vessels clearing outwards and entering inwards at the Port of Liverpool during the six years ending 5 Jan. 1812, distinguishing each year, and those coastwise and those to foreign parts; and the British from Foreign 1 Jul. 1812. Liverpool Customs Office.

Table 3.7. An account of the number of vessels that arrived in Great Britain from foreign ports, 1792–1815

	1792		*1795*		*1800*	
Inwards	*British ships*	*Foreign ships*	*British ships*	*Foreign ships*	*British ships*	*Foreign ships*
Total	13,030	2,477	9,972	2,428	10,496	5,512
Proportion of the annual total in foreign ships		16%		20%		34%

	1805		*1810*		*1815*	
Inwards	*British ships*	*Foreign ships*	*British ships*	*Foreign ships*	*British ships*	*Foreign ships*
Total	11,414	4,517	13,557	6,876	16,851	4,822
Proportion of the annual total in foreign ships		28%		34%		22%

Source: Author's compilation from Cobbett, Parliamentary History, *vols 1–24, Annual Navigation and Trade Reports, TNA, CUST, 36/5.*

Table 3.7 demonstrates the number of voyages inwards, showing the proportion which was foreign shipping.[58] Of the 15,507 voyages inwards in 1792, 4,209 were from Ireland, 1,743 from France, 1,318 from Holland and 511 from the United States. After 1792 no British ships went into France and after 1794 neither did they venture into Holland: such trade was conducted in foreign bottoms. The number of inward voyages from Denmark and Norway, Sweden, Prussia and Germany continued to increase dramatically but the increases were in foreign ships and, apart from American trade, these accounted for the majority of the foreign ships licensed for inward voyages. Foreign ship-owners exploited the British owners' reluctance to commit shipping to those countries for fear of capture. The proportion of shipping inwards in foreign bottoms increased from 16 per cent in 1792 to 34 per cent in 1800–01 and again in 1809–10.

Table 3.8 might suggest that the number of foreign ships carrying British exports also increased. The voyages out from British ports in 1792 numbered 15,025 of which 7½ per cent were foreign ships; in 1795 there were 12,313 of which 18 per cent were foreign ships; in 1800 there were 16,760 of which 29 per cent were foreign; in 1805 there were 15,533 of which 25 per cent were foreign; in 1810 there were 19,338 of which 33 per cent were foreign; and in 1815 there were 22,266 of

58 These tables record the number of voyages not the number of ships, so ships on shorter routes making several voyages in the year would have each voyage counted.

Table 3.8. An account of the number of vessels that departed from Great Britain to foreign ports, 1792–1815

	1792		*1795*		*1800*	
Outwards	*British ships*	*Foreign ships*	*British ships*	*Foreign ships*	*British ships*	*Foreign ships*
Total	13,888	1,137	10,032	2,281	11,867	4,893

	1805		*1810*		*1815*	
Outwards	*British ships*	*Foreign ships*	*British ships*	*Foreign ships*	*British ships*	*Foreign ships*
Total	11,603	3,930	13,090	6,641	17,981	4,285

Source: Compilation from Cobbett, Parliamentary History, *vols 1–24, Annual Navigation and Trade Reports, TNA, CUST, 36/5.*

which 19 per cent were foreign. However, it is important to note that 50 to 60 per cent of foreign ships voyaging out from English ports during the period 1802–11 left 'in ballast', without a cargo. In 1806 and 1807, and again in 1811, almost all the foreign vessels left 'in ballast'. This demonstrates clearly that most licences were granted to facilitate imports.[59] Less than 10 per cent of British shipping departed from English ports without cargo but about 30 to 40 per cent arrived back from overseas 'in ballast'. This suggests that these vessels had made shipments to foreign ports but had been unable to find cargoes for the return voyages.

Naturally the use of neutral shipping caused resentment amongst British ship-owners. In June 1804 the government's position was attacked by William Fawkner of the Committee of the Council for Trade and Foreign Plantations. He complained to the Transport Board that as a result of the practice of permitting neutral vessels to import goods into 'this kingdom', a great number of British ships were lying unemployed.[60] Later in the war ship-owners adopted the more pragmatic view that even if the French decrees had been revoked, British ships would not venture into French-held ports for fear of capture.[61] In 1811 the merchants and ship-owners of Hull presented a petition to the Board of Trade protesting about the continued issuing of licences to foreign shipping, particularly for the trade between the United Kingdom and the Baltic States, upon which Hull had previously depended heavily. They complained that

[59] HoCPP, 1812 (321) X.63, 1, Office of the Registrar General of Shipping, 3 Jul. 1812. N.B. This refers to English ports only; the reports for Scotland do not specify with cargo or in ballast.

[60] TNA, ADM, 108/20, TB to W. Fawkner, 19 Jun. 1804.

[61] HoCPP, 1812 (210) III.1, 57, Minutes of evidence taken before Committee of the Whole House, to consider of the several petitions which have been presented to the House, in this session of Parliament, relating to Orders in Council.

foreign owners were able to demand up to three times the freight rates paid to English owners, that the trade was allowing seamen from hostile nations to obtain an 'accurate knowledge of our coasts and the principal ports', and that foreign masters were 'fraudulently surrendering their cargoes to the Danes' thus significantly increasing insurance rates. They admitted that despite this licensed trade, British shipping had been 'generally employed in other directions'. They claimed that if the Baltic trade had been available then it would have encouraged more shipbuilding. Unrealistically the petition called for an end to the granting of licences and a halt to all 'commercial intercourse' with countries where the British flag was excluded until they allowed British shipping into their ports once again. The petitioners believed that Russia, Prussia and Sweden were far more dependent upon the trade with England than England was on its trade with them. However, if continued trade was still desirable then those states should be encouraged to ship goods to intermediate depots from whence they could be transferred to British shipping.[62]

It is not surprising that the licensing rules were circumvented by foreign owners who carried false documentation or registered their vessels under a neutral flag, particularly the Prussian flag. The practice was well known: it is estimated that in 1806 there were some 3,000 sail belonging to merchants of Holland, France and Spain navigating under the Prussian flag. By these means France was able to obtain naval stores despite British blockades.[63] Even British owners took advantage of this devious practice: 50 of the 55 foreign ships licensed to import cargoes from America to Great Britain between 1 July 1810 and 8 April 1812 were British owned but were registered and sailed under foreign flags, mainly Prussian (12) and Russian (20).[64]

The Transport Board rarely used foreign ships. In 1804, in a response to the Council for Trade and Foreign Plantations, the Board confirmed that 'no neutral vessels are or have been employed in HM's Transport service'.[65] In 1809, in extreme circumstances, the Board did obtain the authority of the secretary of state to hire ships 'not having British Registers, provided they be furnished with Licences to sail under the British flag'.[66] In 1810 it reported that it had six

[62] HoCPP, 1812 (83) IX.345, 63, Papers relating to the Licence Trade, 26 Feb. 1812, quoting a letter from the Merchants and Ship-owners of Kingston upon Hill to the Board of Trade, 4 Apr. 1811.

[63] Captain A.T. Mahan, *The Influence of Seapower upon the French Revolution and the Empire* (Boston, 1892), 310.

[64] HoCPP, 1812 (83) IX.345, 63, An Account of all licenses granted for importing cargoes from any port in the continent or islands of North America, into any port of Great Britain or Ireland, in foreign ships, since the 1st Jul. 1810; specifying the description of the ships and cargoes for which such licenses were granted; the ports from which and into which such importations were allowed; and whether any, and what provisions were inserted in such licenses, as to the ships being navigated by British or foreign seamen, 9 Apr. 1812.

[65] TNA, ADM, 108/20, TB to Council for Trade and Foreign Plantations, 19 Jun. 1804.

[66] TNA, ADM, 108/21/176, TB to Sir Stephen Cotterell, 3 Jul. 1809.

foreign-owned ships in service. In December 1813 the Transport Board advised Earl Bathurst, the Secretary of State for War, that despite increasing the hire rate to 30s per ton per month, it was still unable to provide all the store ships which the Board had been requested to procure. The Transport Board understood that a great part, if not the whole, of that quantity might be speedily procured using neutral vessels that were in the river Thames, but to overcome objections from ship-owners it was proposed to employ such vessels 'on freight' only. [67]

The Effectiveness of Trade Protection

Britain, France and the United States all used trade protection laws to strengthen their own grip on the movement of goods and to contain, where possible, the market for their enemy's products. Britain moved early in the war, in June 1793, to direct naval commanders to detain all vessels loaded with flour or grain bound for French ports.[68] The right to stop and search neutral vessels for cargo belonging to the enemy or contraband goods destined for him had been recognised by numerous treaties of the seventeenth and eighteenth centuries.[69] In Britain the 1793 Act for the Encouragement of Seamen, essentially a Prize Act, established the guidelines, but it was a process that was fraught with difficulties for the boarding officer. He invariably had to deal with masters and crews who professed little knowledge of English and ships' papers and manifests, genuine or otherwise, in a foreign language.[70] In addition the guidelines changed frequently, particularly after 1806 as successive Orders in Council were issued to keep abreast with political developments. The Admiralty was concerned about the burden of decisions resting on the naval captains and issued a printed circular in February 1808 redefining guidelines.[71]

Neither Britain nor France nor America managed to achieve its objectives by the use of restrictive levers such as the French Continental System, British Orders in Council or the US Non-Importation Act, Non-Intercourse Act of 1809 and Act of 1813 outlawing the use of British passes. They were evaded by the use of neutral shipping and by belligerents sailing under neutral flags. There was widespread use of false papers because passes were relatively simple and easy to forge. There was considerable deception because corrupt officials were prepared to sign false documents regarding cargo and destinations.[72] Even Lloyd's resorted to publishing deliberately deceptive shipping lists.[73] In fact both the Continental System and the Orders in Council were crude instruments which had unintended consequences. The Russian resistance to implementing the Continental System was one contributory factor to the French attempted

[67] TNA, ADM, 108/24/53, TB to Earl Bathurst, 3 Dec. 1813.
[68] Mahan, *Influence*, 233.
[69] Hill, *Prizes of War*, 10.
[70] Hill, *Prizes of War*, 21.
[71] Hill, *Prizes of War*, 49.
[72] Crowhurst, *War on Trade*, 35.
[73] Tracy, *Attack on Maritime Trade*, 73.

invasion of Russia, which ultimately failed. The impact of Britain's Orders in Council brought the country into contention with the US.[74]

The success of the Orders in Council was ultimately dependent upon the determination of naval captains to stop and detain merchant ships which appeared to be acting contrary to the rules of the day; but while these officers enthusiastically apprehended neutrals for prize money, they were also acutely aware of the consequences of detaining innocent vessels. In such instances they became liable for costs incurred in the legal process and in retribution. This led Earl St Vincent to write, 'Where one captain makes a fortune by the capture of neutrals, ten are ruined; [no one else] bearing any part of the onus.' However, it does appear that an informal mechanism was developed to underwrite such costs when the captain had detained a vessel, in good faith, on suspicion that it was not what it seemed.[75]

Blockade was also used to enforce Orders in Council. When a blockade was properly in force no vessel of whatever flag, whatever the cargo could lawfully enter or leave the blockaded port. However, the legality in terms of determining prize law depended upon three criteria. Had the blockade been notified by proper authority? Was it being effectively enforced? And had an attempt to break it actually been made?

The Availability of Shipping for Government Service

Given the number of ships on the British register it would appear that there ought to have been sufficient to satisfy the demand from both trade and the government. For instance, at the peak of demand in September 1808 when the government hired 1,012 ships of 250,917 tons (average 247 tons) there were 22,646 ships of 2,324,829 tons (average 102 tons) on the British register. The demand for transports represented only 4 per cent and 10 per cent of the ships and tonnage respectively. However, this was not the reality because 14 per cent of the registered ships were registered in North America and the West Indies and were unlikely to be available to the Transport Board except as 'on freight' ships. They were used in large numbers as such in 1795. Irish-registered ships, which accounted for another 6 per cent, were not used in great numbers by the Transport Board except for troop movements between England and Ireland. The remaining 80 per cent of registered ships were in England, Scotland and the Channel Islands. According to John Dalley, Assistant Registrar of Shipping, approximately 29 per cent of these were either no longer seaworthy or could not be manned due to the shortage of seamen.[76] After deducting those ships, only 57 per cent of vessels on the register were effectively available for coastal

[74] Tracy, *Attack on Maritime Trade*, 81. Although Tracy also suggests that the US had an ulterior motive in the desire to achieve the annexation of Canada.

[75] Hill, *Prizes of War*, 21 and 92. Hill suggests that St Vincent in his letter to Grenville in 1806 was exaggerating but nevertheless making a valid point. He suggests that income from the Droits fund was used by the Admiralty to underwrite these costs.

[76] Dalley Evidence.

and international trade and to satisfy the government's demand. However, given that 73 per cent of transports hired were over 220 tons (see page 76), those ships on the register under 220 tons should be removed from the available pool, thus reducing the 57 per cent mentioned above to 9 per cent (10 per cent if the unseaworthy element was reduced to 15 per cent) of British-registered shipping over 220 tons and operational but not necessarily available. This restricted pool had to serve international trade as well as government transport requirements.

Chart 3.9 demonstrates the proportion of the pool of operational ships over 220 tons that was taken up for transport service during the course of the wars. As already indicated, two scenarios are demonstrated: the higher proportion is based on John Dalley's assumption that 29 per cent of the fleet was non-operational; the lower proportion assumes that the non-operational ships represented only 15 per cent of ships over 220 tons, thus more ships were operational and the pool larger. However, varying the proportion of operational ships does not materially impact the outcome. In summary, chart 3.9 demonstrates that at times of peak demand the government was using well over 30 per cent of the fleet of ships over 220 tons, possibly peaking at 39 per cent in November 1808 (at 29 per cent non-operational). This is considerably higher than has previously been recognised.[77]

Chart 3.9. The percentage of the transports on hire compared with the number of available merchant ships over 220 tons at 15 per cent and 29 per cent non-operational levels

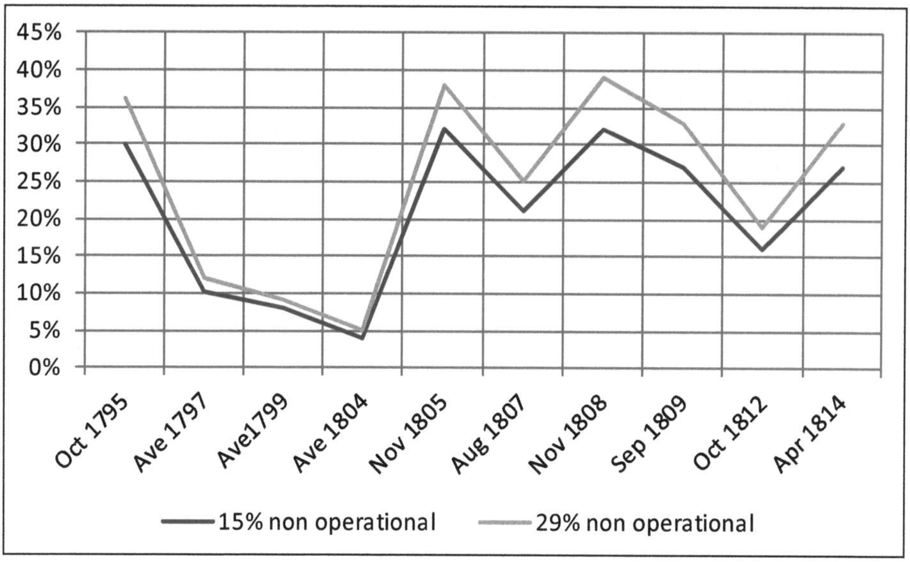

[77] This proportion of available appropriate shipping taken up by the Transport Board is somewhat higher than previously estimated, see Morriss, *British Maritime Ascendancy*, 349; Ville, *English Shipowning during the Industrial Revolution*, 153; Rodger, *Command of the Ocean*, 434; Duffy, *Soldiers, Sugar and Seapower*, 184.

This must have had a material short-term impact upon the availability of shipping for trade. In particular at those times when the government procured considerable tonnages of colliers mainly for use as horse ships and in 1799 and 1809 and again in 1813, there is evidence that 'the enormous quantity of tonnage already in service is said to render shipping difficult to be obtained for the purpose of trade, and to have already materially affected the price of coals'. Freight rates for other commodities including sugar rose significantly when the government were preparing major campaigns.[78]

However, it is significant to observe that during the Revolutionary war this position was not quite as extreme as in the later period. Even in 1795 when large numbers of ships were needed for the West Indies expeditions, the government chartered East India Company ships, as well as others that were involved in West Indies trade, 'on freight'. Many of these ships would have been registered in the West Indies. Trade would not have been disrupted because most of the ships would probably have made those passages 'in ballast', without cargoes. At the same time, owners of ships not normally involved in that trade had some difficulty persuading their masters and crews to go to the West Indies for fear of contracting the deadly yellow fever. The government also resorted to using ships of less than 220 tons. After 1795 the number of transports taken up was much lower than it was to be post-1805. In addition, owners of ships that before the war had been involved in trade with France and other European ports, and even the US, found that trade had reduced significantly. For them the government's requirement for transports was a very welcome alternative use for their ships.

According to John Dalley, vessels employed in the foreign trade averaged two-and-a-third voyages per year and those employed in the coastal trade about five voyages per year.[79] Ships employed in the West Indian trade generally made one outward voyage per year, leaving Britain in the autumn to pick up the sugar harvest in May or June of the next year, arriving home in August or September in time for some of the cargoes to be transhipped to the Baltic before the ice formed in the Baltic ports. Those on the North American run made two voyages.[80] In European trades several voyages inwards and outwards each year were more common. Even so there were delays in waiting for convoys, in the resulting slow passages and in extended turnaround times in ports caused by a large number of ships arriving at one time. These delays suggest that about one-and-a-half round voyages a year in the southern European trades and two in the Baltic and Scandinavian trades were achievable.[81]

[78] Hall, *British Strategy*, 43 and Castlereagh, *Correspondence* VI, 256, Memorandum respecting the Expense of Transports, 1 Apr. 1809.

[79] Dalley Evidence. However, Dalley did not clarify whether he excluded ships on voyages to and from Ireland in his assumption. These ships must have made many trips to and from Ireland during the course of a year and would have heavily biased Dalley's estimates.

[80] Hope, *A New History of British Shipping*, 239.

[81] Fayle, 'Employment of British Shipping', 74.

Table 3.10. A comparison of the number of ships whose trade was displaced by the impact of war with the demands of the Transport Board

	1792	*1795*	*1800*	*1805*	*1810*	*1815*
Voyages outwards of British ships	13,888	10,032	11,867	11,603	13,090	17,981
Exclude to Ireland, Channel Islands and Isle of Man	6,965	7,031	8,081	7,270	8,000 (est.)	10,752
Exclude to France (ships tended to be less than 100 tons)	1,317	7	5	0	0	737
Voyages outwards excluding to Ireland or France	5,606	2,994	3,781	4,333	5,090	6,492
Number of ships using Dalley's 2.33 voyages per annum	2,406	1,285	1,623	1,860	2,185	2,786
Ships displaced by war 1792 (actual for the year)		1,121	783	546	221	-380
Number of ships taken by Transport Board		717		322	627	
Number of displaced ships still unemployed after Transport Board demand was taken up		404		224	short-age	

Source: Author's compilation from Cobbett, Parliamentary History, *vols 1–24, Annual Trade and Navigation Reports, Summary of Transports employed from chapter 2, Commissioners of Naval Enquiry, Fifth Report The Sixpenny Office and by deduction.*

Using Dalley's judgement it is possible to assess the impact of the wars on the English and Scottish shipping fleets. This is demonstrated in table 3.10. The starting point is the number of voyages outwards of British ships to foreign ports in 1792. There were 13,888. The 6,965 voyages to Ireland and the Channel Islands have been removed from this total. The 1,317 voyages to France, which ceased at the onset of war, have also been deducted because the ships on this run were generally under 100 tons and therefore not particularly suited to Transport Board requirements. If Dalley's estimate of 2.33 voyages per year for each ship is assumed then the remaining number of voyages in 1792 represents 2,406 ships. In 1795 the equivalent number of ships was 1,285, suggesting that 1,121 ships had been displaced by the impact of war. In October that year the Transport Board was hiring 717 ships. The inference is that the Transport Board hires were effectively using some of the ships that otherwise would not have been employed. The same is true in 1805 when the number of ships displaced by war was 546 and the number of ships hired by the Board in July that year was 322. It was not until the end of 1805, when ships were being procured for the northern Germany expedition, that the situation changed and the Transport Board hired more ships than the total whose trade had been displaced by war.[82] Although the Transport Board did experience some shortages before 1805, these are more likely to be attributable to seasonality, the supply of seamen or the ships not being in the right place than to the actual availability of active shipping.

Ship-owners facing a similar situation in earlier wars resorted to applying for letters of marque to enable them to engage their ships in privateering. Starkey, in his excellent book *Pirates and Privateers,* demonstrates that 'during the 1790s and 1800s relatively few private men of war were fitted out'. He illustrates that in the Revolutionary war British privateering was on a much smaller scale than might be imagined. He estimates that during the peak monthly activity rates there were on average 27 deep-water privateers operating during the period 1793 to 1815 (plus 42 privateers, smaller vessels who operated mainly in the English Channel) and during the later Napoleonic war there were on average 10 deep-water privateers and 91 'Channel' privateers.[83] The number of ships involved in privateering, particularly deep-water vessels, does not have any material impact upon this assessment.

However, the above assessment does seem to indicate that, up to 1805–06, the British shipping industry benefited from the demands of the Transport Board which replaced the loss of trade due to the closing down of access to enemy-controlled ports. If the Transport Board had not hired those ships, they would almost certainly have been unemployed, despite the fact that as some ports closed to British merchant ships, others in Northern Europe opened up. Table 3.10, based upon the recorded inward and outward voyages from British ports, includes sailings to newly opened destinations.

Given such apparent spare capacity it might be imagined that charter prices would have reduced. Unfortunately it is very likely that manpower availability was being absorbed elsewhere, which would have meant that some ships did not have crews. On balance, charter rates generally held steady during the period 1793–1800 despite a significant increase in costs, notably in seamen's wages which rose considerably. Overall cost increases are estimated to be 20 per cent between 1795 and 1800.[84] The Transport Board did recognise the pressure that ship-owners were under in 1796 and agreed to increase the rate of 13s per month per ton (which had existed since before 1793) to 15s per month per ton. The Board further increased the hire rate in 1798 to compensate those owners who coppered their ships. The rates did rise generally in 1800 because of the increased transport demand combined with growing trade and additional pressures created by the

[82] Sutcliffe, 'Bringing Forward Shipping for Government Service', 87–8.

[83] David Starkey (ed.), *Pirates and Privateers: New Perspectives on the War on Trade in the Eighteenth and Nineteenth centuries* (Exeter, 1997), 132.

[84] See page 48.

corn shortages. Rates fell back at peace but increased to the 1800 level when war recommenced in 1803.

With excess capacity it would be expected that hire rates would decrease, but there were extenuating circumstances which prevented this during the period 1793–1805. Therefore there is strong evidence to suggest that by its demand for shipping for military purposes the government probably staved off a damaging downturn in the shipping industry during the Revolutionary war, thus deferring it until the end of the Napoleonic war. However, as alluded to above, throughout both wars there were two further important issues which impacted short-term availability of shipping: the supply of seamen and the seasonality of certain trades.

The Availability of Seamen

Seamen were in short supply. There was strong competition from the navy, particularly in 1801 when the number of its seamen rose above 130,000 and again from 1807 onwards when the total rose above 130,000 to peak at 147,000 in 1813.[85] To counter this, an Act of 1794 allowed up to three-quarters of the crew of a British ship to be foreigners; the level had previously been capped at a quarter.[86] Not all ship-owners took advantage of this. In Henley's ships, foreigners never constituted more than about half of the crew; maybe they were considered less suitable, but Henley's judgement was that it was unsafe to proceed with an almost entirely foreign crew.[87] There were never enough seamen to crew all the available merchant ships despite the numerous reasons why they preferred the merchant service, such as level of pay and regularity of payment, discipline, safety and length of service, which were all more favourable than in the navy.

Ville has been able to extract considerable statistical data on wages of seamen during the wars from his study of the Henley papers. After a period of relative stability in the late 1780s, wages rose steeply following the outbreak of the war in 1793 from their pre-war levels of £2 10s per year. They continued to rise in the next few years, especially during the labour shortage of February and March 1795. Lower wages in 1796 and 1797 reflected the return of many vessels from the campaigns in the West Indies. Campaigns in Malta and Egypt at the end of the 1790s pushed up wages once more, only to be followed by a reversal during the temporary peace. They rose again as the navy recruited to rebuild the fleet after the Peace of Amiens in 1803, whilst another upturn in 1809–10 coincided with high demand for seamen from trade, government and the navy. With the return of peace seamen's wages collapsed in response to the reduced demand for shipping.[88]

Whilst pay varied from trade to trade and port to port and season to season, it is quite clear from table 3.11 that though the continuity of employment was likely

[85] Rodger, *Command of the Ocean*, 639.

[86] Hope, *A New History of British Shipping*, 256.

[87] Ville, *English Shipowning during the Industrial Revolution*, 96.

[88] Ville, *English Shipowning during the Industrial Revolution*, 101.

Table 3.11. A comparison of seaman's pay in the navy and in the merchant service

		Navy (1)			Merchant in Mediterranean trade (2)			Merchant in Baltic trade (2)			Merchant in Canadian timber trade (2)			Merchant in West Indies trade (2)		
		£	s	d	£	s	d	£	s	d	£	s	d	£	s	d
1797 rate	Able seaman	1	9	6	4	0	0	4	15	0				4	5	0
1807 rate	Able seaman	1	13	6	4	10	0	4	17	6	5	0	0	4	10	0

Sources: Author's compilation from Rodger, Command of the Ocean, *626, and Ville,* English Shipowning during the Industrial Revolution, *165–7.*

to be less because there were periods of unemployment during winter months, pay in the merchant service was considerably more than in the navy.[89] The pay for five months' employment in the merchant service was equivalent to a year's pay in naval service. Thus it becomes clear that a personal protection from impressment, although not always effective, was 'the most precious document any man living in a coastal area could possess'.[90] To persuade seamen to join the navy the Admiralty paid bounties. It was not unusual for seamen to run from merchant ships and transports to join a naval ship and take advantage of the bounty. This prompted occasional protests from the Board to the Admiralty, as in 1803 after five seamen deserted to join the navy.[91]

Dalley pointed out during the course of his evidence to the Commission of Naval Enquiry that it was impossible for all the registered tonnage to be ready for use at any one time because of the restricted availability of seamen. When asked to estimate how many seamen were actually employed in the merchant service, he explained that the registrar did not record that information so he estimated the number based on the register of ships in September 1801. At that time the number of ships registered in England and Scotland was 15,046; these required 124,478 men to crew them if they were all afloat at the same time. He estimated that there were 37,783 men in the foreign trade, 35,970 in the coasting trade and 14,628 in the fishing industry. This amounted to a total of 88,381 men.[92] This represented 71 per cent of the number required to man all the ships, indicating that in 1801 there we never more than 71 per cent of the fleet at sea at any one time.[93]

[89] Morriss, *British Maritime Ascendancy*, 231–44, a very detailed review of 'Recruitment into the naval service' and 'The impressment of seamen'.
[90] Ville, *English Shipowning during the Industrial Revolution*, 99.
[91] TNA, ADM, 1/3744, TB to the Admiralty, 20 Oct. 1803.
[92] Commissioners of Naval Enquiry, Fifth Report into the Sixpenny Office, 214.
[93] This fact has led to frequent misinterpretation of the statistics, for example in Woodman, *Britannia's Realm*, 142: 'by the Peace of Amiens [the British merchant fleet] was manned

The Seasonality of Trade and of Military Expeditions

Weather patterns dictated that shipping was a seasonal trade. Invariably the times of greatest demand for trade shipping coincided with the most favourable times for military expeditions and the consequent demand for transports. Long-distance trade voyages were planned to avoid the most serious adverse weather conditions, such as winter storms in European waters, hurricanes in the Atlantic and typhoons in the Indian Ocean. Convoys to the West Indies aimed to leave before the end of October and generally set off back to England in April to avoid the summer hurricanes. In European waters April to November were the most favourable months, the season possibly ending a little earlier for Baltic passages. Clearly, bringing forward and preparing a large number of ships presented a challenge at any time in the year, but the seasonal demands of trade complicated the situation. Seasonality thus affected the preparation of expeditions. Unlike the government, ship-owners planned months ahead: they had to negotiate for cargoes and plan voyages well in advance of the sailing seasons so that no time was lost to generate returns. In contrast the government's demands were not planned around the anticipated availability nor did they take into consideration trade seasonality. In May 1794, prior to the establishment of the Transport Office, the Navy Board experienced great difficulty in procuring transports in the River to convey troops to Europe and the West Indies. The Baltic convoy had sailed a month earlier and few ships were expected from the 'East Country' for more than five or six weeks. Trade with the recently acquired islands in the West Indies had also employed a considerable number of ships.[94] Although the 1795 Abercromby and Christian West Indies expedition was ordered in May, partially due to delays in raising shipping it did not sail until 16 November and suffered the ignominy of sailing straight into very strong westerly gales which scattered the convoy and drove them back into port. In 1799, 20,000 tons of transports for den Helder expedition were ordered on 28 June; the landing was in August. In June 1800 Dundas instigated the preparation of 30,000 tons of transports for the 1801 Aboukir Bay expedition. In 1805 the major troop convoys for the Mediterranean sailed on 19 April. In April 1807 Castlereagh ordered the Transport Board to assemble 36,000 tons of transports to support the Copenhagen expedition. A few months later, in June 1807, the Board had to advise that 'the difficulty in procuring transports on the present occasion has been owing to the unusual scarcity of unemployed shipping at this season of the year'.[95]

In earlier years the shortages had tended to be experienced at the beginning of the season but from 1807 shortages occurred throughout the season. In April 1808 Castlereagh demanded 34,807 tons for Sir John Moore's expedition to the

by 144,558 seamen'; and in Morriss, *British Maritime Ascendancy*, 227 where the numbers in the table column headed 'paid 6d duty in the merchant service' are actually from 1799, the numbers of men on the British Shipping Register that would be needed if all the ships were in commission at the same time.

[94] TNA, HO, 28/63, Navy Office to the Admiralty, 16 May 1794.

[95] TNA, ADM, 108/21/57, TB to W. Fawkner, 8 Jun. 1807.

Baltic. In May 1808 a significant number of transports were ordered to ship troops for Wellington's campaign in Portugal. In May 1809 Castlereagh instructed the Transport Board to raise shipping for the Walcheren expedition. All these demands coincided with the peak seasonal demands of trade. All were generally required at short notice.

Discharging Chartered Shipping

Several demands for large volumes of shipping followed shortly after instructions, from either the Treasury or the secretary of state for war, to demobilise a significant portion of the transport fleet to save money. In 1796, following delays in preparing transports bound for St Domingo, Henry Dundas was forced to concede that 'The wish expressed by Mr Pitt and myself for the reduction of transports at home to six thousand tons appeared to have led to a mistake.'[96] Again in January 1807, just prior to the instruction to raise shipping for the Copenhagen campaign, the Treasury demanded the discharge of transports. Then in April 1809 the Board received instructions directing that 8,000 tons of three-month ships be paid off; this was just one month before the instruction to bring forward ships for the Walcheren expedition.[97] If those ships had been retained, it might have speeded up the preparations for the campaign but the demands for short-term cost saving were paramount.

Freight Rates

Adjusting the freight rate was the principal mechanism available to the Transport Board to counteract the competition from trade in times of shortage. Occasionally it resorted to extending the committed term of the charter instead of moving the rate. However, just as Knight and Wilcox suggest that Victualling Board members were cautious commissioners, so the same might be said of the Transport Board.[98]

The Board was both cautious and prudent, almost always obtaining a number of tenders for each service and increasing freight rates only under pressure. However, it seized on opportunities to reduce them with alacrity. To increase competition the Board quickly moved from using only George Brown, the sole broker that it inherited in 1794 from the Navy Board, to using a large number of brokers. Chart 3.12 compares the movement of the Transport Board hire rates for regular transports with the freight rates in other shipping trades. It demonstrates that the Transport Board followed the market rather than establishing the going rate until 1806 –07 when its high demands forced its rate up against most trades except the Honduran mahogany market. At that time the Board's rates were

[96] TNA, WO, 6/156/183, Henry Dundas, Secretary of State for War, to the TB, 2 Jan. 1797.

[97] TNA, WO, 6/156/477, Castlereagh to TB, 5 Apr. 1809.

[98] Roger Knight and Martin Wilcox, *Sustaining the Fleet 1793–1815: War, the British Navy and the Contractor State* (Woodbridge, 2010), 28.

Chart 3.12. Comparison of the freight rate movements of regular transports with those of various trades: 1793=100

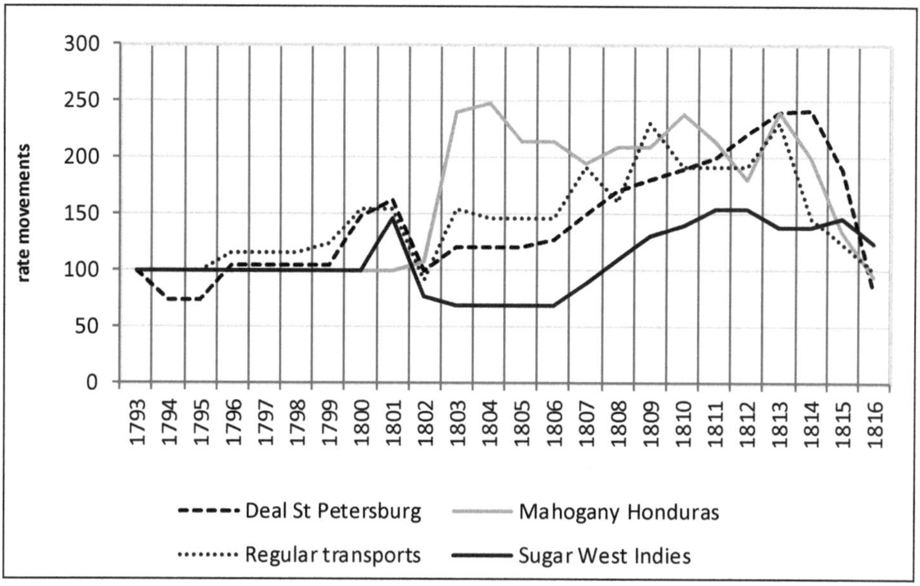

rising faster than in some trades. This probably indicates why the Board was able to charter enormous tonnages of three-month transports during this period.

One of the more significant hire rate increases occurred in 1800 when, due to the difficulty of raising ships, Henry Dundas, Secretary of State for War, authorised an increase in the rate for coppered ships from 16s per ton per month to 19s 6d. Later, in April 1807, it was increased from 19s to 25s. In June the Transport Board reported that:

> the Commissioners have found much difficulty in procuring a sufficient quantity of shipping for the service for which they have recently been directed to provide; ... they have no doubt, that so great a proportion of the trading tonnage of the country being taken out of its proper course, must have very considerably raised the rates of merchants' freight; but to what extent this has actually taken place, this Board has no means of ascertaining.[99]

After the Walcheren expedition in 1809 the regular transport rate was reduced to 21s. The three-month rate continued at 25s and even rose to 30s in 1813–14, by which time the regular rate had moved back to 25s per ton per month. This explains why the *Alice*, a ship owned by Michael Henley & Sons, was paid 30s from February to August 1814 whilst at the same time another of the company's ships, the *Polly*, was paid at the rate of 25s throughout 1808–15 despite the various

[99] TNA, ADM, 108/21/57, TB to William Fawkner, 8 Jun. 1807.

rate movements highlighted above. The *Polly* was a regular transport whilst the *Alice* was a three-month transport.[100]

During the period 1807–13 the Transport Board was paying almost at the top of the freight scale to obtain shipping, which reflects the shortage of shipping for both trade and the higher level demands of the Transport Board for the numerous expeditions of the period – Copenhagen, the Baltic, the Mediterranean, the Peninsula, Walcheren and the Peninsula again. However, increasing the hire rate did not always bring forward additional shipping. In 1809 Sir Rupert George told the Scheldt Inquiry Committee that even significantly raising the price would not have made any difference to the ability to raise ships for the Walcheren expedition. He claimed that the increase would only have brought forward a limited number of ships due to seasonable availability, and it would have initiated demands for the higher rate from existing transport owners. The ultimate cost would have been greater than the benefits derived.[101]

Ville's study of Henley & Sons led him to conclude that the Board's increasing demands for shipping 'pulled up freight rates'. There is evidence that brokers and ship-owners exploited their position to push up rates by withholding shipping from the Board in anticipation of obtaining higher rates to fulfil demand.[102] There is no doubt that the government demand for shipping did increase freight rates, making British shipping less competitive than its international rivals. There was already some concern about British competitiveness. In 1796 the minutes of a Special Committee of the Court of Directors of the East India Company recorded that:

> It is a fact that becomes each day more evident, that unless English ships are obtained for the Indian trade, on nearly as moderate terms as foreign flags can furnish freight then the greater part of that trade will be transferred to the ports of other nations.[103]

The company proceeded to negotiate price reductions with the owners, making a significant saving in return for greater security in the tenure of the charter parties.

There is further evidence of this lack of competitiveness. According to evidence presented to Parliament, British shipping had experienced a few good years between 1807 and the later part of 1810. Due to that unprecedented demand, freight rates had risen significantly and become overpriced; neutral vessels were cheaper and had taken some trade away from British ships. But by 1812 the position had changed: the shipping market was depressed and ship-

[100] Ville, *English Shipowning in the Industrial Revolution*, 128.

[101] HoCPP, 1810 (12), VIII.I, Papers relating to the Scheldt Expedition, Evidence of Sir Rupert George.

[102] Ville, *English Shipowning in the Industrial Revolution*, 128, TB impact on freight rates.

[103] HoCPP, 1715–1800 (106), IX.223, Resolutions and Proceedings of the Court of Directors of the East India Company from 10 Mar. 1796, respecting taking up ships for the Company's Service by fair and open competition.

owners attributed this partly to the reduction in the Board's freight rate in 1809.[104] This ignores the fact that the Board could only have reduced the rate because rates in the market were falling. However, it does indicate the reliance of some ship-owners on government demand. Unsurprisingly, general freight rates fell back even more dramatically when the Board's demands, in concert with the demand for war goods to supply the military and naval forces, reduced in 1802 and again after 1812.

The volume of shipping available to the Transport Board changed constantly. Not all of the vessels on the British Shipping Register were readily available to the Board. The ships registered in North America and the West Indies and Ireland were beyond its normal reach. Dalley estimated that almost a third of the ships registered in England, Scotland and the Channel Islands were not operational, being no longer seaworthy or unmanned due to the shortage of seamen.[105] More importantly, only 15 per cent were over 220 tons, which was the size favoured by the Transport Board for troop shipping but which was also in great demand for international trade. This reduced the pool of available ships dramatically and explains why the Transport Board experienced great difficulty in procuring transports for major expeditions, particularly during the Napoleonic war. At peak times demand equated to 30–39 per cent of the available suitable ships.

With the possible exception of 1795, in the years prior to 1805 there ought to have been enough ships to supply all the Board's requirements.[106] Indeed there are indications that up to 1805–06 the Board's activities sustained the national fleet, pumping millions of pounds into the industry and the wider economy at a time when the industry was excluded from business in so many of its traditional markets. Besides sustaining the fleet, the work of repairing and refitting transport service vessels sustained many shipbuilders during difficult trade periods when there were few commissions for new ships. However, there is also strong evidence to indicate that by taking ships from trade it had a material impact on freight rates which led to increased costs and shortages of commodities, particularly coal.

The Transport Board competed strongly for sufficient capacity but the level of demand changed frequently, resulting in a number of government instructions to reduce the fleet to cut costs during the period between 1807 and 1809.[107] Such was the pressure to reduce costs, a lack of forethought and medium-term planning meant that occasionally instructions to discharge transports were followed relatively quickly by the receipt of conflicting demands for significant additional

[104] HoCPP, 1812 (210) III.1, 577, Minutes of evidence, taken before the Committee of the Whole House, to whom it was referred, to consider the several petitions which have been presented to the House, in this session of Parliament, relating to the Orders-in-Council, 1812, Evidence of Mr Buckle, 577.

[105] Dalley Evidence.

[106] See pages 91–2.

[107] HoCPP, 1807 (115) IV.105, 1, 10 Aug. 1807, Return of the amount in tonnage of the Transports in His Majesty's Service.

volumes. This caused frustration and difficult challenges for the Board but there is no evidence that it materially delayed any expedition.

The restricted availability of seamen and the seasonality of trade conspired to make the Transport Board's task more demanding when record numbers of transports were required to support military operations.[108] The Transport Board had to pay high prices to obtain ships. In 1809 even that measure had its limitations since there were no more ships available except foreign ones and the Board was very reluctant to use this resource. Yet despite these considerable difficulties, over the wars as a whole there is no doubt that the strength and depth of the British merchant fleet gave the country a significant military advantage in the conflicts against Revolutionary and Napoleonic France.

[108] The degree of ship-owners' willingness to release shipping also had an impact. This will be discussed in a later chapter.

4

Economy versus Efficiency

'The transport service is liable to a great variety of unforeseen or unavoidable incidents and delays which, though independent of the Board itself, might blemish the impact of honest and unwearied energies.'[1]

The Inevitability of Delays

Each government administration was anxious to reduce the costs of war. Occasionally, attempts to minimise expenditure resulted in decisions that lessened the efficiency of the transport service – in particular, instructions to discharge transports only to find that they were required again within a few months. The ensuing difficulty of re-procuring and preparing the transports for service caused frustration, extra effort and sometimes additional cost. Retention of transports might well have speeded up the preparation of some expeditions. These situations exposed the lack of forward planning and exacerbated the combined effects of ministerial interference, misunderstandings, miscommunication, departmental obfuscation and confusion that were invariably prevalent during the build-up to most military expeditions. At the port of embarkation the agents for transports were responsible for bringing order to an often frenetic situation. Delays were almost unavoidable given the number of departments involved, each with its own priorities and all operating under Treasury scrutiny for economy, combined with the inevitable reliance upon good fortune and favourable weather conditions. Such delays were hardly a surprise, but rarely were they anticipated or were mitigations planned. When the preparations for expeditions did not run as smoothly as was expected there was invariably a degree of manoeuvring to place the responsibility for failure elsewhere.

The Transport Board was in an invidious position given its central role in this logistics process; other departments were not slow to criticise it. That criticism, which was usually directed through the secretary of state for war, invariably led to demands to conduct an inquiry into the circumstances relating to the complaint. Study of the directions issued to the Board by its political masters and of the outcome of some of these enquiries reveals much about the complexity of the process of preparing for an expedition: it highlights the huge endeavours

[1] TNA, ADM, 108/19/1, TB to Henry Dundas, Secretary of State for War, 29 Dec. 1796.

required and demonstrates the quality of the achievement. This chapter will consider the impact of political direction and the outcome from three enquiries, which will illustrate how delays occurred and the effects thereof.

West Indies and Portuguese Expeditions, 1796

First, an inquiry initiated by Henry Dundas, Secretary of State for War 1794–1801. He had been extremely irritated by the delays to the West Indies convoy led by Rear Admiral Christian in 1795.[2] So when there were delays to much smaller expeditions to the West Indies and to Portugal in the following year, Dundas's frustration boiled over and resulted in him taking the unusual step of bypassing the Admiralty to direct the departure of the convoy. This was, in his words, 'an extraordinary deviation from the established regulations of the service'. It was an unprecedented occurrence which he described as 'a measure so extraordinary'. He was determined to find out what had caused the delays by questioning all the departments involved in the preparations 'in order that blame may not be imputed indiscriminately ... and to prevent the reoccurrence of similar evils and complaints'. He demanded that the Board report on its role in this matter, trusting 'that both the Board and its subordinate agents stand exculpated, and deserving of that confidence which a liberal government will never withdraw, but upon strong and palpable grounds of misconduct or neglect'. At the same time he requested consideration of a complaint that numerous officers, some men and baggage had been left behind when the convoy finally sailed, due to there being insufficient accommodation on the transports.[3]

Needless to say, in true civil service style, the Transport Board totally refuted any allegation that responsibility for the delay in the embarkation of troops for the West Indies and for Portugal was attributable to it or its agents: 'Not a day nor an hour have been lost; that no possible exertion, within the reach of this Board and its agents has been omitted; to carry into the fullest effect the orders communicated to us.'[4] Instead the Board tactfully blamed the delays upon ministerial interference. When ordering the shipment of 3,000 troops to the West Indies on 7 October 1796, Dundas himself had specified which ships were to perform the service. They were teak-built East India Company ships, but unknown to Dundas they were 'in a very deficient state'; three of them were actually in dock in the Thames undergoing repairs. This delayed their sailing from Deptford until 28 and 31 October and 11 November, but then they experienced further delays due to contrary winds. When they finally arrived at Portsmouth it was realised that they could only receive 2,014 of the 3,000 men, so the armed transports *Coromandel* and *Weymouth* had to be prepared to take

2 For a narrative of the Abercromby-Christian expedition see Duffy, *Soldiers, Sugar and Seapower*, 159–252 and Roger Morriss, 'High Exertions and Difficult Cases: The Work of the Transport Agents, 1795–1797', in Helen Doe and Richard Harding (eds), *Naval Leadership and Management, 1650–1950* (Woodbridge, 2012), 95–107.

3 TNA, WO, 6/156/180, Henry Dundas to the TB, 24 Dec. 1796.

4 TNA, ADM, 108/19, TB to Henry Dundas, 29 Dec. 1796.

the remainder. Preparations and embarkations were completed and they were ready to sail on 12 December, nine weeks after the initial instruction; of this period, three to four weeks were attributable to the essential repairs.

Dundas was also unhappy about delays in the preparation of troop ships for St Domingo that he had ordered on 26 November 1796. The Board was obliged to remind him that on 5 November, when he and Mr Pitt had met Captain George, the chairman had been instructed to 'discharge all vacant tonnage at Portsmouth beyond 2,000 tons' to reduce the 'expense' of transport and to retain only 4,000 tons in the Thames. George had taken action immediately and consequently on the 26th (three weeks later) very few ships were ready to be allocated to the St Domingo service. The 4,000 tons of transports in the River had already been taken up by the Victualling Board which had, in fact, required 7,700 tons for provisions for the fleets on the Leeward Islands and Mediterranean stations. In response, Dundas accepted that 'the wish expressed by Mr Pitt and myself for the reduction of transports at home to 6,000 tons, appeared to have led to a mistake'. He explained they had only considered the forthcoming requirements for troop transports; they had not made allowance for the demands of the other departments including the Victualling Office, 'of which we could form no accurate opinion'. He did suggest that the Board, in concert with the other departments, ought to have made 'similar estimates upon the best material, which can be formed'.[5] This clearly demonstrates several deficiencies on the minister's part: a lack of planning even over a relatively short three-week period and a failure to communicate precisely the parameters of the instruction to discharge transports. It exposed a weakness both in the communications between the Board and the Victualling Board regarding shipping intentions and in the Board's preparedness to question an instruction from the minister.

Nevertheless, to ensure that the St Domingo service was prepared as quickly as possible, George had instructed Captain Patton, the Agent at Portsmouth, to give preference to that service above all others. Patton advised that he had appropriated four ships for the reception of the troops. Unfortunately, because the crew of those transports had been impressed by the navy, the refitting and provisioning of the ships had been delayed. When these ships eventually left Portsmouth for Cowes to embark troops, the commander of the forces under embarkation thought that the ships were overcrowded, even though at 1,235 tons for 636 men they were almost at the usual proportion of 2 tons per man. This meant that an additional ship had to be quickly brought into service to placate him. It was not unknown for army officers to interfere with the embarkation process. A difficult officer could create many additional problems for the transport agents. The transports were completed in nineteen days and were ready to sail on 15 December.

Dundas agreed to write to the Admiralty about pressing men from transports, requesting that their lordships prevent any further interruption to the service.

5 TNA, WO, 6 / 156/ 180, Henry Dundas to TB, 24 Dec. 1796, TNA, ADM, 108 /19, TB to Henry Dundas, 29 Dec. 1796 and TNA, WO, 6 / 156 / 183, Henry Dundas to the TB, 2 Jan 1797.

The Admiralty confirmed that to avoid future delay and inconvenience, the officers of the impress service had been ordered not 'upon any account whatever' to impress crews of any transports upon production of a protection from the Admiralty that 'stated the number and description of the men employed to navigate such a vessel'. The Board explained that examination of the ship's charter party and the ship's register would enable the impress officer to determine how many men should be on board.[6] The Board argued that it was impossible to deliver a description of the men because seamen engaged while the ship was in port changed continually. It argued that impress officers should be ordered not to take any men where there was no apparent excess of complement. Neither should men be taken from the transport's boats or from those working ashore where each individual seaman could produce his ticket of leave for the day, signed by one of the agents at the port.[7] Although the Admiralty agreed on numerous occasions to clarify with its officers the position on impressments, in reality it knew that it could not prevent officers from pressing seamen from transports. Thus it tended to turn a blind eye to such problems.

Dundas was equally unhappy about the delays in preparing the cavalry transports for Portugal which were ordered on 1 November. The instruction required the Board to bring some foreign regiments which were in Guernsey and Jersey to Portsmouth for onward passage to Portugal. Six transports were selected to collect the troops and bring them to Portsmouth. Three transports were sent to Guernsey, arriving back at Spithead on 22 November as directed. However, the three that had been sent to Jersey received orders from the naval commander there to proceed to Falmouth to await the appearance of the fleet, rather than to Spithead. He had not considered that they were not provisioned to sail on to Portugal, nor was one of them, the *Flora*, fitted for the conveyance of troops on a foreign voyage. She was only intended to be used for the short run to Portsmouth. When the Board learned this it immediately directed Lieutenant Motley, the Agent Afloat with these ships, to ensure that the *Flora* was suitably equipped and all the ships were provisioned for the voyage to Portugal. At the time the Transport Board was writing its response, the Lisbon fleet was at Falmouth awaiting sailing orders from the Admiralty. The Board pointed out that the original instruction had said that there were 1,051 men to be shipped out but in fact there were actually 1,289, an excess of 238 men, increasing the demand for tonnage by almost 500 tons. This situation was not unusual: the transport agents regularly discovered that when the troops arrived for embarkation there were more than had been expected, necessitating the urgent preparation of additional tonnage.

To add to the Board's woes, one of the transports selected for the Portugal service, the *Dover*, suffered a series of misadventures which illustrate the sheer unpredictability of the service. She had recently arrived back from the West

6 The charter party stated the number of men or alternatively the ship's register would show the registered tonnage and the number of men was based on the ratio of five men and a boy per 100 tons.

7 TNA, ADM, 1/3733, TB to the Admiralty, 11 Jan. 1797.

Indies and consequently required refitting in sails and rigging. Not only that, her master had to be replaced as a result of bad conduct and a new crew had to be found because the original one had absconded. Consequently she had been detained at Gravesend for refitting and to await a full complement of men. Whilst there she received damage from two ships both 'running foul of her', necessitating repairs and causing further delays. Then, on her passage round to Portsmouth in fog and bad weather she 'struck upon the Owers and beat off her rudder' and sustained other damage, all of which was repaired within a few days. The final problem occurred on 15 December when in response to Dundas's order that the convoy sail immediately it ran into foul winds which drove the larger part of the fleet into Portsmouth, Falmouth and Torbay. Excluding the problems with the *Dover*, the preparations had taken six weeks to organise.

The preparation of the East India ships took longer than might have been expected due to the necessity for repairs, but the Portugal preparations had taken about the length of time that might have been predicted. However, the work to prepare the St Domingo service took nineteen days and this was particularly speedy. The larger East India ships sailed again on 21 December but the Admiralty had delayed the sailing of the other transports because of the weather conditions and they were still at Falmouth on 29th, the date of the Board's report. These circumstances do not seem sufficient to warrant Dundas's 'extra-ordinary actions'. His actions must have been motivated by other, unidentified events.

In response to the complaints from the army officers who had wanted passages, Captain Charles Patton, the Agent at Portsmouth, was able to demonstrate that many applications from officers had been made through the proper channels and these had been dealt with quickly and effectively. Some applications, however, had bypassed the established procedures. They had been made directly to him without first obtaining approval from the secretary of state and the Board itself. Some of these officers, as well as others who had applied through the correct channels but too late, had not been embarked. There had been more applications from officers, who required cabins, than had been anticipated. The agent had not been able to provide enough cabins at very short notice. The Board was confident Dundas would agree that however responsive the Board might be, it was essential 'that proper and timely communications should be made to us of the extent of their numbers and demands; or it will be difficult, if not impossible, to prevent such disappointing delays, as must inevitably arise from sudden and irregular applications'.[8]

The agent had discovered that despite his best endeavours to accommodate these late applications by making 'liberal and general offers' to the victuallers and ships 'on freight' to take in as many of these officers as could be conveniently accommodated, the masters of those ships had been extremely reluctant to accept any officers upon any terms. The agent had thus to find additional transports. Fortunately, the Board had sent round four troop ships. They were not particularly intended for that service because it was believed that sufficient

[8] TNA, ADM, 108/19, TB to Henry Dundas, 29 Dec. 1796.

transport had been allocated, but in fact, without them the remaining officers and stores could not have been accommodated.

Dundas supported the Board's position regarding the complaints from the officers for whom no passage to the West Indies had been reserved: 'the blame of disappointment lies with the officers themselves, who appear in most instances not to have made their application in time and in many not till several days after the time fixed for the departing of the convoy'. He planned to discuss the issue with the Duke of York to devise some measures to be taken to prevent reoccurrence.[9] Subsequently, in 1799, the Board changed the terms of the charter party to oblige victuallers to take a reasonable number of passengers as well as provisions, thus providing additional accommodation in each convoy.

The Board told Dundas that the tonnage demanded for ordnance stores and provisions for these expeditions had exceeded all precedents and had not been anticipated by either the other boards or the Transport Board. The volume of shipping required had only been realised at the last moment. Consequently, the 'demands were sudden and unexpected but happily, they have all been put on board'. The Board requested Dundas to ensure that for all future services, the departments needing tonnage should anticipate their requirements in advance and it would be the Transport Board's responsibility to 'make the necessary provision'.[10]

Dundas's response to the Board's review of events was prompt: 'His Majesty has been pleased to receive your report in the most gracious manner and to signify his entire satisfaction at the distinct justification therein contained of your conduct in the service to which it related.'[11] Thus he exonerated the Transport Board. Nevertheless, these events had exposed many embarrassing deficiencies in the government's administrative system, including a lack of planning in ordering the discharge of ships that were needed only a few weeks later; ill-informed ministerial interference in selecting ships which needed repairs and refits and not being clear about their troop-carrying capacity; lack of communication between departments about anticipated demands in respect of the numbers of troops to be shipped; the extent of officer accommodation required and the volume of supplies to be shipped; uninformed interference, possibly the infringement of procedures, by army officers and naval commanders; and the self-interest of the navy as evidenced by the activity of impressment gangs.

Major-General Spencer's Expedition to Sicily, 1807

In November 1807 Castlereagh requested the Board to prepare transports for 7,000 troops intended to sail under Major-General Spencer to Sicily by way of Lisbon. Preparations took six weeks; the fleet sailed on 20 December. Unsurprisingly, when the convoy finally sailed it ran into gales shortly after departure. Spencer,

9 TNA, WO, 6/156/183, Henry Dundas to the TB, 2 Jan. 1797.

10 TNA, ADM, 108 /19, TB to Henry Dundas, 29 Dec. 1796.

11 Source as note 9 above.

together with half the fleet, returned to Falmouth; the remainder pressed on for Gibraltar. He did not sail again until 21 February.[12] Lord Hawkesbury wrote on behalf of Castlereagh on 19 December, demanding an explanation for the delays from the Transport Board. His letter seems to demonstrate a complete lack of understanding about the complexity of preparing a fleet of the seventy transports that this expedition would have required and the time needed to do so. Again the secretary of state must bear some responsibility for the delays because he had specifically designated which ships were to be used for the service. Those particular vessels had only recently arrived from South America, the Mediterranean and Copenhagen; in consequence of long voyages, accidents and 'uncommonly tempestuous weather', they had required significant repairs to their hulls, masts and sails before they were ready to be replenished with provisions and the troops embarked.[13] Clearly Castlereagh had not taken this into consideration.

Captain Patton, the Resident Agent at Portsmouth, was first advised of the details of the ships that had been allocated to the service on 10 November 1807. At the same time the Board advised him that the quantity of beds recently ordered had been increased from 2,000 to 10,000 to support this expedition, illustrating the extraordinary demands put upon contractors at relatively short notice.

Some of the ships needed their copper sheathing replaced, some needed repairs, others needed their ballast changing and all wanted caulking in the upper decks and bulwarks as well as significant refitting in their rigging and sails. But due to the lack of availability of dry docks and the shortage of skilled labour, it was not possible to refit so many ships at the same time at Portsmouth. The only dry docks at Portsmouth were in the dockyard, and since the navy refused to allow transports to be repaired in the dockyards, ships had to be beached to facilitate repairs. The queue of ships waiting to be beached was increasing, so some were ordered around to the Thames for repair.[14] To speed up the process at Portsmouth, Patton had engaged shipwrights from Bursledon and Fareham and had even tried to procure some from other ports including Cowes and Cork, but without success.

The Portsmouth victualling agent was also experiencing considerable difficulties. His local suppliers were fully extended and they were encountering significant problems in keeping up with his demands. In addition, he had not enough men or vessels to supply both the naval ships and the transports. Instead he had been obliged to employ all the vessels at his disposal in supplying the navy's ships. He had not been able to comply with Patton's earlier requests for provisions and water for some transports which had arrived from the Baltic with the King's German Legion on board. Nor had he been able to supply water for the transports for Ireland even though some of them had less than three days' provisions on board.[15]

[12] Mackesy, *War in the Mediterranean*, 263–4.

[13] TNA, ADM, 108/21/75, Hawksbury to TB, TB to Castlereagh, 19 Dec. 1807.

[14] TNA, ADM, 108/21/75, Captain Patton to TB, 11 Nov. 1807.

[15] TNA, ADM, 108/21/75, Captain Patton to TB, 20 Nov. 1807.

Meanwhile, pressure was mounting on Patton. He had been advised that another 3,246 men were expected to arrive in Portsmouth shortly, and because all the barracks on shore were full, transports had to be ready to receive them.[16] Inevitably, when there were to be major troop shipments the port towns became inundated with soldiers. The local barrack facilities could not cope with such large numbers. Army commanders preferred to embark the troops directly onto transports which were then moved out to Spithead, reducing the opportunity for absconding. Castlereagh demanded a status report from the Board and it further added to the demands on Patton by insisting that he send, by return, a complete list specifying the state of forwardness of each ship and advising when it would be ready to receive troops. Captain Patton was struggling to attend to the competing demands on his time: 'It is impossible without neglecting the business of these transports to complete the weekly return in the usual form.'[17] More agents were appointed to assist him, including Captain John Halstead. Patton made him responsible for the transports at Spithead and put the transports in the harbour under Lieutenants Richard Chessman and James Morgan and the horse ships under Lieutenant Edward Tregurtha. By 27 November many of the transports were ready to receive provisions and water. A few were still in the carpenters' hands and were expected to be completed by the time the others were provisioned. The onus was then on the Victualling Board to complete the supply of provisions and water. On the same day the Transport Board wrote to the Victualling Board to remind it yet again of the urgent need to provide provisions and water.[18]

Meanwhile, Patton had realised, perhaps a little late, that there was not nearly enough bedding in store to supply the 7,000 troops that were to be embarked if they were all to have new rather than washed bedding. The Board advised him that the contractor had been instructed to increase the supply of new bedding urgently. It was intended that the troops were to be supplied with new bedding as far as possible, with the deficiency made up of washed bedding.[19] The transports with the German troops on board which had arrived from the Baltic were required for Spencer's troops but would need to be fumigated and fitted with new bedding prior to the arrival of the troops who would embark on them. The general in charge of troops at Portsmouth had not received any orders to disembark the German troops; on the contrary he had been directed to keep them on board because 'they cannot be quartered on shore for want of room'.[20] Therefore Patton had to make arrangements to fumigate these ships whilst the troops were still on board, making the task extremely difficult and very unpleasant for the troops themselves.

16 TNA, ADM, 108/21/75, Captain Patton to TB, 20 Nov. 1807.
17 TNA, ADM, 108/21/75, TB to Captain Patton, 25 Nov. 1807.
18 TNA, ADM, 108/21/75, TB to Captain Patton, 27 Nov. 1807.
19 TNA, ADM, 108/21/75, TB to Captain Patton, 27 Nov. 1807.
20 TNA, ADM, 108/21/75, Captain Patton to the TB, 29 Nov. 1807.

More troops were preparing to march to Portsmouth.[21] The Transport Board advised Patton to remove the troops under orders for Ireland who were already embarked on six-month ships (which were now required for this service) and put them on three-month ships.[22] This was more economical; the three-month ships could be discharged sooner if they were no longer required.

The agent victualler had received directions from the Victualling Board, prompted by the Transport Board's letter. He had been ordered to complete the transports with water and provisions 'with all possible despatch'. Unfortunately, by the same post he had received an Admiralty order to replenish the stores of seven sail of the line with water and provisions 'with all possible speed'. The naval ships took precedence but the agent victualler did agree to employ some vessels to supply the transports with troops on board at Spithead that were out of water and provisions and were surviving on a daily supply collected in their own boats.[23]

Finally, on 2 December the general in command of the troops at Portsmouth received directions from the Department of War to land the troops that were under orders for Ireland as well as the German divisions. But he had requested that, in order to make room for them, 450 men who were joining the expedition might be embarked on board the three coppered transports – *William* (259 tons), *Ralph Nicholson* (357 tons) and *Jupiter* (341 tons) – which were the only transports provided with water and provisions. Despite this, on 6 December the German Legion was still on board. Needless to say, the men were complaining bitterly, having been at Spithead for over four weeks.

In the meantime there were other services to be dealt with. For instance, Patton was ordered to allot transports for a company of German Light Artillery with their guns, wagons, ammunition and stores. One of the ships allocated for the mission was the *Richard*. On her previous voyage those on board had suffered an outbreak of opthalmia. Since then she had been well cleaned, fumigated and examined by a surgeon.[24] He had confirmed that she could be used, but by that time the crew had deserted so replacements were required.

Despite the arrival of additional vessels there was still a shortage of regular transports and the Board was forced to consider sending three-month ships with the fleet. By 2 December at Portsmouth, 'ready and making ready' were coppered transports for 5,133 men and sheathed transports for 815 men, not including the *Ann* (365 tons) or the *Christopher* (297 tons). The former was in dock and the latter still on Gosport Reach for repairs to damage received when she struck rocks at St Aubin's Point, Jersey.[25]

Major-General Spencer arrived at Portsmouth on 11 December bearing the message that the secretary of state 'was particularly desirous that there was no further delay'. Unfortunately, Thomas Reeks, the Agent Victualler, chose

[21] TNA, ADM, 108/21/75, Captain Patton to the TB, 29 Nov. 1807.
[22] TNA, ADM, 108/21/75, TB to Captain Patton, 30 Nov. 1807.
[23] TNA, ADM, 108/21/75, Captain Patton to TB, 30 Nov. 1807.
[24] Opthalmia: severe inflammation of the eye, highly contagious.
[25] TNA, ADM, 108/21/75, Captain Patton to TB, 2 Dec. 1807.

that day to advise Patton that he had neither pease nor butter to supply the transports, and that he was short of several other articles.[26] The Transport Board immediately ordered the substitution of rice and sugar for pease.

To add to Patton's woes, Portsmouth was experiencing 'an uncommonly low high tide' which was preventing the *Ann* from floating out of the dry dock and the *Christopher* from being floated off Gosport Beach.[27] He was planning to substitute the *Aid* (330 tons) for the *Christopher* if she could not get off in time and send the *Richard* in place of the *Ann*. The *Enterprize* (328 tons), returning after repair in the River, had been delayed by the wind but she was expected to arrive shortly. To complete the embarkation Patton recommended that several smaller ships should be used, although these would not normally be used as troop ships. As one of them, the *Eliza*, was not ready, extra hands had been put to work on her.[28] In the meantime the Board had arranged for five transports to be loaded with entrenching tools, engineers' stores, brass ordnance, ammunition and stores at Woolwich. They were to join the fleet at Portsmouth.[29]

At Portsmouth the victualling agent was experiencing further problems: besides the shortage of pease and other supplies, he was now faced with the dilemma that the loaded victualling craft had become grounded in the extremely low water at the victualling wharf. This was not the first time delays had occurred for that reason.[30] Patton's view was that the agent victualler 'has done everything in his power to forward the service'.[31]

On 16 December the agent victualler was expecting a vessel with pease to arrive from the Downs on the next day. For this reason, and because he did not know in what proportion to substitute the rice and sugar, he had put off supplying the transports with rice and sugar.[32] However, Patton was determined that if the ship did not arrive with the pease as expected then the rice substitute would be supplied.

Prior to receiving the letter from Castlereagh's office on 19 December, the Transport Board had sent Commissioner Captain George Henry Towry to review and finalise the preparations. He confirmed, probably not unexpectedly, that even with all the 'untoward circumstances', all the troops were embarked on board before the supply of the provisions and water could be completed to allow them to proceed to sea. He judged that because of the time taken to provision the ships, bringing forward more three-month ships or crowding the troops at 1½ tons per man would not have expedited their sailing. He also confirmed that Major-General Spencer 'expresses himself highly pleased with Captain Patton's exertions', and Towry agreed that 'he has in my opinion used all the means in

26 TNA, ADM, 108/21/75, Captain Patton to TB, 11 Dec. 1807.
27 It is not clear which dry dock. If it was in the dockyard then it is not clear how Patton had been able to persuade the dockyard officers to allow this ship to be repaired there.
28 TNA, ADM, 108/21/75, Captain Patton to TB, 16 Dec. 1807.
29 TNA, ADM, 108/21/75, TB to Captain Patton, 9 Dec. 1807.
30 TNA, ADM, 108/21/75, Captain Patton to TB, 9 Dec. 1807.
31 TNA, ADM, 108/21/75, Captain Patton to Admiral Montague, 9 Dec. 1807.
32 TNA, ADM, 108/21/75, Captain Patton to TB, 16 Dec. 1807.

his power to forward the arrangements'.[33] He reinforced the point that there were insufficient facilities to repair a large number of transports at the same time at Portsmouth. Adverse weather conditions had compounded the problems and had made it particularly difficult to repair ships that were moored at Spithead, owing to the distance of Spithead from the harbour. Towry confirmed that the transports would have been able to sail at the end of the previous week had the Victualling Board been able to complete the loading of water and provisions. Again the Board asserted that 'no delay has occurred within the power of this Board; whatever delays have occurred have been partly due to the weather and to causes not related to us' but to 'the department to which this part of the service belongs in water and provisions'.[34]

Troops Bound for Portugal, 1813

In February 1813 Lord Bathurst, Secretary of State for War, demanded an explanation for the delay in sending cavalry transports from Plymouth to Cork from where they were to convey 410 artillery horses to Lisbon. This order had been issued on 5 January. It was Bathurst's understanding that the ships were still at Plymouth despite the horses being needed urgently in Portugal for the Duke of Wellington's army. This was just one of a number of instructions that Bathurst had issued at the time that required horse ships. To supply these he had ordered horse ships back from Portugal, as well as authorising the Board to procure additional transports for 1,200 horses. Bathurst claimed that he had been advised on 26 January that the horse ships from the Tagus were beginning to arrive and that although some had been allocated to other services, there were transports ready at Plymouth for 528 horses. These were sufficient for the conveyance of the artillery horses from Cork. However, he had been informed by the Admiralty that no application for a convoy to proceed from Plymouth to Cork had been made until 1 February, even though the wind at that time had been perfectly fair for the voyage. On the 1st the wind had changed and the vessels were still detained at Plymouth, incurring 'a very injurious delay in sending the horses to Portugal'. Bathurst accused the Board of being dilatory between 26 January and 1 February.[35]

The Board denied responsibility for any delay. It also denied providing any report on the 26th that the horse ships were complete. 'Your lordship will observe that we could not possibly have reported on 26 January that ships for 520 horses were ready at Plymouth; nor can we find that any report to that effect either written or verbal was made from this office.'[36] It suggested that, on the contrary, because of the daily reporting to the minister the actual state of the service must have been very apparent to him. Bathurst was not happy with this response. He reminded the Board that its 'Statement of particular services under order' dated 26

[33] TNA, ADM, 108/21/75, Commissioner Towry to TB, Portsmouth, 18 Dec. 1807.
[34] TNA, ADM, 108/21/75, TB to Lord Hawkesbury, 31 Dec. 1807.
[35] TNA, WO, 6 / 159, War Department to TB, 11 Feb 1813.
[36] TNA, ADM, 108/23/97, TB to Earl Bathurst, 19 Feb 1813.

January reported that there were 'at Plymouth ordered to Portsmouth 520 stalls'. On receipt of that statement Colonel Bunbury, the Military Undersecretary, had sent a messenger to the Transport Office to suggest that instead of sending all the ships to Portsmouth thus taking extra time, sufficient number for the conveyance of 410 horses should go direct from Plymouth to Cork. The messenger had been told that indeed this was already in hand and that the ships in question were already ordered to proceed direct to Cork.[37]

There was no delay in applying for convoy, the Board argued; it had applied to the Admiralty for a convoy from Plymouth to Cork on 23 January. The agent at Plymouth reported on 27 January that HMS *Leonidas* was to lead the convoy. The Board had assumed that the horse ships would proceed with the convoy. Orders to that effect had been given to the agent. However, by a letter from him dated 30th, and received by the Board on 1 February, it appeared that although the transports were expected to be complete by 31 January, the *Leonidas* would not be ready to sail until 3 February. Therefore the Board judged it advisable to apply immediately for a special convoy for these horse ships to send them sooner, hence the application for convoy protection dated 1 February. This incident illustrates the breakdown in communication which occasionally occurred within and between government departments. The Board appears to have been taking criticism that might have been more properly directed at the Admiralty.

However, this was not Bathurst's only complaint. He had 'noticed the loss of time in providing transports for certain detachments of infantry ordered to be embarked for the Mediterranean'. He recollected that the tonnage for those troops had been ordered on 7 January but he had been embarrassed by receiving advice from His Royal Highness the Duke of York, the Commander-in-Chief, that 'no transports are yet forthcoming'.[38] The Board was anxious to explain that the difficulty it had experienced in procuring additional ships had not been caused by the reduction in the hire rate from 25s to 20s per ton, which it had implemented in the previous April. This reduction had been in line with the assurance that the Board had given to the Treasury when it authorised the last increase that the rate would be reduced as soon as the state of trade permitted. Although this may have 'occasioned the tardiness of ship-owners to offer their ships when latterly wanted', the reduction had already saved the government £166,615. The Board hoped that Bathurst would approve of it not offering an increased rate of hire at that time 'as any increase, if not made general, would occasion much discontent and consequently great inconvenience, and loss to the service'.[39] The Board claimed that other services had been prioritised on Bathurst's instructions, thereby delaying the execution of that particular instruction. It was expected that sufficient tonnage would be ready to embark those detachments whenever the wind allowed the ships to go round from the Thames to Portsmouth. Bathurst did not agree with the Board's approach to

37 TNA, WO, 6/159, War Department to TB, 11 Feb. 1813.

38 TNA, WO, 6/159, War Department to TB, 11 Feb. 1813.

39 TNA, ADM, 108/23/97, TB to Earl Bathurst, 19 Feb. 1813.

holding the charter rate down: 'However desirable an adherence to the rules you have adopted to the hiring of transport ships may be in ordinary cases I cannot consider it good economy to occasion any in the sending out of troops and stores to these destinations under the present circumstances.'[40] Bathurst's will prevailed; the charter rate was increased back to 25s a few weeks later.

During the preparation of the horse ships the Board had experienced a recurrent problem. There were no stores of forage at Portsmouth or any of the other main ports. When a large supply of forage was required it was usually ordered round from the River. On this occasion it was sent to Portsmouth but the ships from the Tagus that were to receive the forage were forced to put into Plymouth due to contrary winds, and there were no forage supplies at Plymouth. Bathurst had been surprised to learn that there were not adequate supplies of forage as well as 'every necessary means of re-equipping transports' at each of the principal ports. He instructed the Board to immediately establish magazines of forage at four of the principal ports: at Cork for 500 horses for three months, at Plymouth for 300, at Portsmouth for 1,200 and at Yarmouth for 500. The Board was to ensure that these stores were replenished according to consumption.[41]

Economy versus Efficiency

The need to minimise the cost of the wars was clear to every administration. Cost was the predominant feature of most of the parliamentary scrutiny regarding the progress of the conflicts. In certain cases it inhibited efficiency. One example of this was the reaction to Castlereagh's proposal, early on in his tenure as secretary of state for war, to maintain a 'discretionary' military force, together with the necessary transports, prepared and available for immediate action. The subsequent Whig administration dismissed this approach on the basis of cost.

In August 1807 the size and cost of the transport fleet featured in a series of clashes in the House of Commons between the recently reappointed Castlereagh (supported by George Canning, the new Foreign Secretary) and the ex-minister William Windham and his colleagues. The debate revolved around the level of preparedness that the new Portland administration had inherited from the 'Talents' administration. The previous ministers were forced to defend their actions in reducing the size of the transport fleet, which rendered it incapable of providing conveyance for the cavalry that might have been offered to support the Swedes or undertaking the speedy preparation of sufficient horse transports to ship Sir David Baird's force to the Cape.

Castlereagh accused Windham of eliminating the disposable fleet of transports, which he had built up and which might have been used for those purposes, on economic grounds in an attempt to save £40,000 per month. He claimed that 'this demonstrates that the policy between the two governments is so different'. Windham considered the accusation that the fleet reduction

[40] TNA, WO, 6/159, War Department to TB, 22 Feb. 1813.
[41] TNA, WO, 6/159, War Department to TB, 22 Feb. 1813.

had been made from financial motives to be 'fallacious'; rather it had been in recognition of the state of Europe at the time. In Castlereagh's view the 'difficulty of providing cavalry transports, besides the time and trouble of fitting up', should have been recognised and the transport fleet should not have been reduced so significantly.[42]

Despite the level of interest in this expensive transport service, the information presented to Parliament detailing the cost was very restricted. It was included under the Heads of Naval Estimates. For the year 1814 the total estimated cost of the transport service was £2,980,623 which was allocated under two headings. To support the army, £2,488,458 was allocated for 'the hire of Transports for the Conveyance of Soldiers and Horses, and of Stores belonging to the Ordnance, Commissary General, Storekeeper General and Barracks Departments; For Provisions occasionally supplied to Soldiers on board Transports; Building Cabins and Horse stalls, Forage for Horses and Pay of agents afloat'. Then a sum of £492,165 was allocated to supporting the navy, 'for the hire of Transports employed as Navy Victuallers and Navy Store Ships, and Pay to Agents Afloat'.[43] This allocation of costs quite clearly re-emphasises the point that the service operated principally to support the army rather than the navy. Other than that it did not give MPs any real indication of how the funds were spent. Within Parliament there was a tendency to focus on the headcount, pay and pension commitments of the various departments, which were the minutiae, rather than how the significantly larger estimates were spent. Morriss suggests that Parliament and the Treasury only became more focused on these costs from about 1811 when they became more easily accessible because they were published in Hansard for the first time.[44]

It is thus hardly surprising that, with the exception of the secretary of state for war, few parliamentarians had much appreciation of the costs and complexity of the transport service. This is illustrated by an exchange of letters (in 1806) between William Windham, the Secretary of State for War, and the Prime Minister, Lord Grenville, about the possibility of sending troops to support Prussia. Windham, who was against the idea, disingenuously suggested that the cost of hiring transports for 10,000 cavalry for three months would be £2.4 million, knowing, however, that the cost of shipping troops was considerably less than that of shipping the same number of cavalrymen. The latter incurred four or five times the cost of the former. Grenville might have been expected to be aware of that. Instead he replied that the cost was 'so very beyond any idea I could have formed'. This caused Grenville to reconsider the proposal

[42] Hansard, House of Commons sittings on 31 Jul., 7 Aug. and 13 Aug. 1807. Ironically the same charge might have been laid against Castlereagh himself in 1809. This relates to his instruction to discharge all horse ships returning from Walcheren and Lisbon shortly before his resignation as secretary of state.

[43] HoCPP, 1813–14 (110) XI.131, 21.

[44] Morriss, *Culture*, 70 and 203 and P.K. Crimmin, 'Admiralty Relations with the Treasury 1783–1806: The Preparation of Naval Estimates and the Beginnings of Treasury Control', *Mariner's Mirror* 53 (1967), 63–72.

to send troops to Prussia.[45] The point is that Grenville, despite his extensive government experience (he had been foreign secretary between 1794 and 1801), did not appear to appreciate the difference in cost between shipping troops and shipping cavalry.

Nevertheless the transport services and the Transport Board, along with all other government services, did receive considerable scrutiny by parliamentary commissions. Although the Board was not in existence when the Commission on Fees was in progress, comment on transport activities featured prominently. The chairman, together with Commissioner James Bowen, was interviewed by the Commission for Naval Revision during the preparation of the 9th Report. This commission did not find fault with the Board although it did suggest a revision to the management structure to accommodate the effective administration of its broader accountabilities following the integration of the Sick and Hurt Board's responsibilities. Commissioner John Schank and William Harding, the accountant at the Transport Office, were interviewed by the Committee for Finance in the preparation of the 18th Report. In addition the chairman was called before the Scheldt Inquiry to give evidence.

The Absence of Strategic Planning

Although the Board played a vital role at the principal crucial points of the war, it was far removed from any war planning process, which was determined at Cabinet level. It was continually forced to seek directions from the secretary of state regarding the further use of transports. Should they be retained or discharged? The Board's knowledge of future projects was limited to that which the secretary of state was prepared to share. It is clear that Castlereagh rarely if ever consulted the Transport Board before a major Cabinet decision was taken, following the practice of all previous secretaries of state for war, including Henry Dundas. Whereas the Admiralty and senior military officers were canvassed for ideas, views and opinions on proposals, the Transport Board was probably considered to be too lowly for the same treatment: instead it received instructions. However, once preparations were in progress, as Sir Rupert George told the Scheldt Inquiry, he met with the secretary of state for war almost every day to discuss progress.[46]

Castlereagh clearly understood the difficulties inherent in the transport procurement process. On one occasion he told the House of Commons that 'the interval between the wish to obtain transports and the power of obtaining them was much greater than was generally to be imagined'.[47] However, the impact of the weather and the inability of all the departments involved to perform in

[45] Hall, *British Strategy*, 136. In fact the three-month cost at 19s per ton per month of transports for 10,000 troops at 2 tons per man would have been £57,000. For 10,000 cavalry at 8 tons per horse the cost of transports would have been almost £240,000 not £2.4 million.

[46] HoCPP, 1810 (12) VIII.1, 39, Scheldt Inquiry, Sir Rupert George's Evidence, 20 May 1810.

[47] HoCPP, Hansard, 7 Aug. 1807.

harmony during the preparation phase were often underestimated. The Walcheren expedition in 1809 is a classic example of the how the various British administrations invariably reacted to external political and military events rather than working to a grand plan. Although proposals to capture Flushing and Antwerp had been considered from time to time by the Cabinet, most recently in December 1808 and early January 1809, it was not until 22 April when the ministers learned of Austria's declaration of war with France (which had occurred a few weeks earlier) that the plan was implemented.[48] On 20 May the Transport Board was instructed to hire shipping. This was about the same time that the Earl of Chatham was being sounded out to lead the campaign and that advice on the proposed campaign was requested from the Admiralty and senior military officers.[49] Sir Rupert George disclosed in his evidence to the Scheldt Commission that he had not been advised of the plans until 20 May. Given that this was the peak of the maritime trade season and there were already pressures on the industry which meant licensed foreign ships were being used in large numbers, and that this was a time when the Board had already engaged a record tonnage of shipping, one might wonder why he had not been involved in the process sooner.[50]

To emphasise the point that the government was reactive to events rather than proactive, even on 5 April 1809, only six weeks before Chatham was appointed, Castlereagh had not been contemplating an early adventure and had given the instruction to discharge 8,000 tons of three-month ships, which were sufficient to carry 4,000 troops, thereby reducing costs.[51] Given that the cost of 8,000 tons would have been at least £8,000 per month, from an economic perspective saving £24,000 for three months was an appropriate decision. Yet if these vessels had been retained, fewer additional ships would have been required and the procurement process would have been conducted much more quickly.

This was not the only instance when interventions from ministers advising the Transport Board to cut the transport fleet had repercussions because a significant tonnage was required shortly afterwards. This was certainly the case when Dundas and Pitt instructed Rupert George to reduce the fleet dramatically, as described above. Then in January 1807 the Treasury issued directions that 'all vessels not required for current service should be discharged'.[52] This was followed, in April 1807, by Castlereagh's order to the Transport Board to assemble 36,000 tons of transports to support the Copenhagen expedition. This caused some difficulty for the Board. It had to report that there was an unusual scarcity of unemployed shipping at that time.[53] In December 1807 Castlereagh issued the

48 Hall, *British Strategy*, 176.

49 Castlereagh, *Correspondence* VI, 250–90.

50 Curiously there is no evidence of any written instructions until 1 Jul. 1809 when the Transport Board was advised of the urgent need to bring forward a significant amount of shipping.

51 TNA, ADM, 6/156.

52 HoCPP, 1807 (117) IV.109, 1, Return of Amount in Tonnage of Transports in His Majesty's Service on 24 March 1807 including letter from the Treasury to TB, 22 Jan. 1807.

53 TNA, ADM,108/21/57, TB to W. Fawkner, 8 Jun. 1807

instruction to 'proceed to discharge all horse ships that are employed over and above the tonnage that may be required for 1,000 horses'.[54] This was only four months before a request for 10,793 tons of horse ships.

Just as Castlereagh had reversed Windham's instructions to reduce the transport tonnage immediately after he took office in 1807, so his instruction to discharge all unappropriated cavalry horse ships, given at the end of September 1809, damaged the Board's ability to raise transports for 1,817 horses to Lisbon ordered by his successor Lord Liverpool on 27 December, just three months later.[55] Then in February 1812 Liverpool, having reviewed the latest return of all the ships in the transport service, decided that a proportion could be immediately discharged 'without prejudice to the public service'. The Board was instructed to discharge all unappropriated infantry and horse transports, retaining only 7,000 tons of infantry shipping and cavalry transports for 300 horses.[56] Six weeks' later there was an urgent request for transport to Lisbon for 200 horses. At the same time Liverpool instructed the Board to dispatch an order to return to Plymouth all the horse ships in the Tagus.[57] None were returned. Liverpool ought to have been aware that, to be effective, such instructions needed the authority of his office and should have been issued simultaneously by him to Wellington and by the Admiralty to Admiral Berkeley. A week later further requests for horse ships were received. The Board was obliged to seek authority to convert some of the troop ships at Portsmouth to horse ships to meet this demand.[58] By 15 April 1812, no horse transports had returned from Lisbon so the Board had to advise Liverpool that 'Cavalry tonnage now coming forward is still insufficient for all the horses preparing for Portugal.'[59]

The lack of information on future requirements ensured that the Board was generally reluctant to make decisions relating to the discharge of ships without first checking with the secretary.[60] In addition the Board was expected to report in considerable detail on transport movements, dispositions and availability to the secretary of state of the day.[61] This applied particularly when major expeditions

54 TNA, WO, 6/156, Castlereagh to TB, 25 Dec. 1807 then TNA, ADM, 6/156, 10,793 tons of horse-ships ordered 21 Apr. 1808.

55 TNA, WO, 6/157, Castlereagh to TB, 26 Sept. 1809 and Liverpool to TB, 27 Dec. 1809.

56 TNA,WO, 6/159, War Department to TB, 4 Feb. 1812.

57 TNA, WO, 6/159, War Department to TB, 19 Mar. 1812.

58 TNA, WO, 6/159, War Department to TB, 21 Mar. 1812.

59 TNA, WO, 6/159, War Department to TB, 21 Mar. 1812.

60 TNA, ADM, 108/22/4, TB to Liverpool, 14 Aug. 1810, seeking direction on the appropriation of transports. TNA, ADM, 108/22/16, TB to Liverpool, 25 Sept. 1811, considerable increase in tonnage unallocated following recent arrivals from Tagus; can any be discharged?

61 TNA, WO, 6/156, Castlereagh to TB, 13 May 1807 and TNA, ADM, 6/156, Castlereagh to TB: 'you will make a return of all transport tonnage in the form herewith enclosed'. TNA, ADM, 6/156, 'Daily return of three months ships to be transmitted', 16 Feb. 1808. TNA, ADM, 6/157, Lord Liverpool requests a daily report on the state of transports, 7 Nov. 1809. TNA, ADM, 1/3751, Dundas to TB, 5 and 10 May: 'Daily return of troop embarkations for foreign service required'.

were in preparation; weekly, sometimes daily, reports of the state of preparations and the disposition of the fleet were demanded. Sir Rupert George advised the Scheldt Inquiry that following the initial instruction to raise shipping he had 'daily communication with the Secretary of State, indeed several times in the day I reported on the progress'.[62]

Ministers occasionally interfered to the extent of selecting the vessels to be designated for certain tasks, sometimes without full awareness of the implications of those choices. This is clear in the case studies above: first Dundas appointed four East India ships for an operation without being aware of their condition and how long it would take to repair and then prepare them; then Castlereagh fell into a similar trap when he too nominated ships that needed repairs and refits before they were again ready for service. Dundas and Castlereagh tended to get very involved in the detail of operations.[63] Dundas appears to have been disdainful of the Transport Board. He was overbearing and assertive, and one might wonder about his grasp of the complexities of the transport service, despite his extensive experience on the Board of Control of the East India Company and as Treasurer of the Navy. Castlereagh was a younger, ambitious minister who in particular seems to have ensured that he received wide-ranging advice from senior army officers during the planning stages of campaigns.[64] Yet he does not appear to have sought the advice of Sir Rupert George during those deliberations. He must have been aware of the potential impact of bad weather on expeditions. Nevertheless, he was clearly prepared to bear the consequences when he allowed the 1805 northern Germany expedition and Sir David Baird's 1808 Corunna expedition to progress so late into the year when there was a strong chance that the fleet would meet gales and the army would have to endure winter conditions. He also permitted the 1809 Walcheren campaign to sail when there was a strong likelihood that the climate on Walcheren Island would be very unsuitable and an outbreak of fever might be expected. In each case the timing was beyond his control and political expediencies dominated his decision-making.

Further Causes of Delay

When vast numbers of ships were brought forward very quickly the Thames became exceedingly crowded. The lack of dock space at Deptford, the principal victualling yard, and at Woolwich, where no more than two ships could be alongside at any time, meant that provisions, supplies and ordnance materials had to be transhipped using lighters and small vessels to transports that were often forced to moor or anchor some considerable distance, sometimes several miles, from the warehouses. Indeed this was one reason why the Board was keen to pursue the opportunity to purchase Dudman's dock in 1798. This double-

[62] HoCPP, 1810 (12) VIII.1, 39, Scheldt Inquiry, Sir Rupert George's Evidence, 20 May 1810.

[63] TNA, ADM, 1/3743/390, TB to Admiralty, 16 Feb. 1803. TNA, WO, 6/156, Hawkesbury in the absence of Castlereagh to TB, 3 Aug. 1805. TNA, WO, 6/156, Castlereagh to TB, 20 Jul. 1807.

[64] Castlereagh, *Correspondence* I, 257–73.

handling was inefficient and considerably extended the time taken to prepare ships for service in the River. The situation at Portsmouth was little better. Ships were generally moored at Spithead which was some distance from the harbour, and in inclement weather the small boats experienced great difficulty reaching, mooring alongside and then performing the arduous task of transferring provisions and troops in the swell of the Solent. Similar comments could be made in respect of the Nore and the Downs where ships were anchored in large numbers awaiting provisions, troops and convoy escorts. At Portsmouth the victualling wharf suffered from lack of water during low spring tides and it could only manage a limited number of ships alongside. More facilities for mooring alongside would have considerably speeded up the preparation process.

Similarly, repairing transports was made more difficult by the Navy Board's refusal to allow merchant ships to be repaired in the royal dockyards, and at Portsmouth the only dry docks were in the navy yards. Transports at Portsmouth had to be sent around to the merchant yards on the Thames for extensive repairs. This incurred extra time, and poor weather or adverse winds frequently delayed their return to Portsmouth. In inclement weather it was difficult to repair ships at anchor at Spithead due to the distance from Portsmouth harbour.

Despite frequent requests for the vessels' return, commanders of both army and navy stations overseas did everything within their power to retain transports on their station to be used as store ships and to be held in case an emergency evacuation was required. Transports that had been overseas for a long time or had experienced particularly bad weather often sustained major damage. They usually needed significant repairs or a refit before they were again ready for service.

Speedy preparation demanded that the Ordnance and Victualling Boards also performed effectively. This was not always the case. There were numerous breakdowns in communication and working relationships between the Transport Board and the Ordnance Board that led to delays. From time to time the Victualling Board did not have the capacity to meet the extraordinary demands placed upon it to provision both the naval fleet and troop transports and store ships that were to convey troops to overseas theatres of war.

The second case study illustrates that delays were caused by the Victualling Board's agent running out of pease and other essential supplies because of the huge demands made and the contractors' inability to supply more when they were already working to capacity. Neither did he have any spare vessels to deliver supplies and water to the transports because he had given priority to the naval fleet. Often the number of troops that finally turned up at the embarkation point was greater than the number expected, which forced transport agents to scramble to find additional ships.

The Admiralty occasionally found the demands of providing convoy protection overwhelming and in 1806 it instructed the Board to advise it at the beginning of each month how many convoys were expected during the coming month. Enemy naval activity disrupted plans from time to time, less so in the Napoleonic war, when British naval blockades were particularly effective, but the point was illustrated in August 1805: vessels carrying troops under Sir David Baird to the Cape of Good Hope were detained for over two weeks after embarking at Cork when Castlereagh learned that a squadron of three French sail of the line had escaped from Rochefort and its location was unknown.[65]

Finally, the weather often conspired to undo the best laid plans. Contrary winds frequently prevented ships from sailing on time or gales at sea forced them back into port, often having sustained significant damage.

The Impact of Delays on Expedition Preparations

Syrett identified major shortfalls in the planning of campaigns during the American war. He found that the Cabinet in general and the secretary of state for war in particular tended not to give due consideration to the timescales necessary for raising and preparing sufficient transports to support the operation.[66] So did this situation improve in the wars of 1793 to 1815?

The most notorious delay affected the Abercromby-Christian convoy in 1795. A large number of transports were ordered on 23 July. The plan – to send 30,000 troops plus supporting officers, artillery and ordnance – was finally approved by the Cabinet between 14 and 16 August. The volume of shipping for the ordnance supplies was not determined until 7 September. In the event 18,742 officers and men with their supplies and provisions sailed on 16 November in 137 transports, hired East Indiamen and West Indies-owned merchant ships and naval ships. Without this support the force would probably have required more than 200 transports of over 55,000 tons.

The number of men in the force that sailed was less than originally envisaged. Not only were transports taking some time to complete, so too were the recruitment and preparation of the military force. The bulk of the transports had arrived at Portsmouth ready for the embarkation of the troops on 20 October, twelve and a half weeks after the instruction to bring forward large numbers of transports at the height of the trade season. The fleet sailed some three and a half weeks later. Less than thirty-six hours after departing, the convoy sailed into the eye of a dramatic storm and the gale-force winds drove it back to port. A second attempt to sail, on 9 December, was again foiled by gales and storms and was forced to return home. The convoy did not sail again until March 1796.

It is difficult to clarify the impact of these delays. Certainly the original campaign plans had to be revised. Duffy says that despite the setbacks and the immense loss of life, this expedition 'saved the war'.[67] The delays resulted from the combination of a multitude of circumstances, not only the impossibly high demands placed on the Transport Board to raise a vast tonnage of shipping in less than two months in the middle of summer, but also the difficulty in raising the troops. Neither of these objectives was achieved by the original planned deadline of 15 September.

65 Castlereagh, *Correspondence* I, 141.
66 Syrett, *Shipping and the American War*, 243–8.
67 Duffy, *Soldiers, Sugar and Seapower*, 371.

Looking at the more significant expeditions, a study of the elapsed time between the Transport Board taking receipt of the first instruction to consolidate a fleet of transports and the point when the transports were ready for troop embarkation has yielded the following results: den Helder (1799), the first landings were 8 weeks after first instruction; Northern Germany (1805), 257 transports were ready for embarkation within between 10 and 13 weeks; South America (1806), the fleet of 58 transports sailed within 15–18 weeks; Sicily (1806), Major-General Spencer's army sailed within 7 weeks; Copenhagen (1807), 377 transports sailed after 16 weeks following some delays awaiting the political decision to proceed; the Baltic (1808), 34 transports were embarked within 3 weeks; Corunna (1808), 66,000 tons of transports sailed within 6 weeks; and Walcheren (1809), 40,000 troops sailed in 10 weeks.[68]

Following the failure of the Walcheren campaign, the subsequent inquiry was anxious to determine if delays in preparing transports had contributed to the calamitous outcome but assessed that this was not a major factor in the disaster. The failure had primarily resulted from weak planning, particularly in anticipating the impact of the expedition on the health of the troops and the breakdown in communications between the senior commanders in the army and navy. Ironically the preparation time for Walcheren was ten weeks which was the shortest period of all the major expeditions considered above.

These events indicate that ministerial failings in the American war, as identified by Syrett, were again prevalent during these wars. As can be observed from the commentary above, in the expeditions considered, the time between the Transport Board receiving the first instruction to procure shipping and the date of sailing was fairly consistent: between 10 and 16 weeks.

Extended preparation of major expeditions tended to attract speculation in the newspapers. In many cases press coverage alerted the enemy to the preparations at a level of detail that the government may have been anxious to avoid. Wellington referred to this in 1809 when planning his return to Portugal, urging that cavalry and artillery complete, with the full complement of horses, should be sent ahead of the general and staff officers. In his words, 'as soon as the newspapers shall have announced the departure of officers for Portugal the French armies in Spain will receive orders to make their movements towards Portugal'.[69]

The determination of each government administration to minimise the cost of war had an adverse effect on the efficiency of the Transport Board when it came to preparing for major expeditions. Castlereagh's plan to have a disposable army on standby, together with the appropriate number of transports to convey them overseas, was eminently sensible. This would have eliminated or reduced, to some degree, the constant hiring and discharging of transports, therefore speeding up the preparation for expeditions. However, the additional cost

[68] Sutcliffe, 'Bringing Forward Shipping for Government Service', 171.

[69] Gurwood, *Dispatches* III, 183: Memorandum on the Defence of Portugal, 7 Mar. 1809.

was not insignificant and thus Windham disbanded this fleet, even the horse transports. Thereafter the constant high demand for transports meant that there were not enough vessels to have a fleet in waiting. The army was probably under similar pressures and it is unlikely that a standby force was available.

Government strategy tended to be reactive to external political and military events, so the planning process was not as refined as would be expected now. Whilst more senior boards and individual military officers were consulted for advice and opinions, the Transport Board was merely instructed, even though its role was an essential element in the preparations of military campaigns. However, once preparations commenced the chairman was involved in frequent, sometimes daily, conversations with the secretary of state for war. Even so the Board's day-to-day awareness of future demand for transports was very restricted. It was not in a position to question ministerial instructions to reduce the size of the transport fleet, some of which were in retrospect ill-timed or premature. The Board was subject to considerable ministerial interference in the detail of transport operations.[70] These interventions were frequently uninformed.

The Transport Board's pivotal role in the preparation of military expeditions exposed it to regular criticism. The Board was frequently called upon to defend its performance by the secretary of state for war, who had to answer criticism for delays in Parliament. Inevitably, preparing a large number of transports for an expedition was not a speedy process. Transports were often at sea for extensive periods and on return home generally needed significant repairs, particularly to copper sheathing and rigging. Once the ships were in a good state for service, the further preparations required the co-ordinated co-operation of the Victualling Board, the Ordnance Board, the War Office, the Commissariat and the Admiralty, in addition to the many private contractors who had to increase output of provisions and supplies at short notice. The shortage of dockside wharfing meant that a considerable part of the preparations had to be done remotely from the dockyards, requiring double-handling and making the process subject to the vagaries of the weather. Given the complexity of the process, the enormity of the task and the numerous potential causes of delay, one might wonder how any major expedition was successfully prepared. However, an analysis of the time taken to prepare various expeditions shows a pattern: it demonstrates that between ten and sixteen weeks were required from the first instruction to the Transport Board to the date of sailing. It is difficult to understand why this was not taken into account as a material aspect of the campaign plan, alongside a consideration of the seasonal impact on the availability of merchant shipping.

70 There was less interference from Hobart, who was the Secretary of State from March 1801 to April 1804, spanning the Peace of Amiens. There were no major campaigns during this period. His contacts with the Transport Board related principally to convict transports. Likewise, Camden, who was in office for less than a year between May 1804 and July 1805, although that was a critical time for the build up of transports bound for the Mediterranean, refrained from becoming too involved in the details of the service.

5

Conjunct Operations and Amphibious Assaults

Naval and Military Co-operation

> 'If anyone wishes to know the history of this war, I will tell them that it is our maritime superiority that gives me the power to maintain my army while the enemy is unable to do so.'[1]
>
> *Duke of Wellington, 21 September 1813*

Although victory at sea alone could not win the war, the British navy played a very significant role in the ultimate defeat of Napoleon, principally by achieving and retaining hegemony at sea. This allowed the relatively unfettered movement of troops and military supplies by sea, much faster than the French were able to achieve by land.[2] It also ensured the security of the vitally important British maritime trade whilst restricting the trading capacity of Britain's enemies.[3]

During the Revolutionary war and in the early years of the Napoleonic war the naval strength of Britain and its enemies was finely balanced. The British navy concentrated on destroying or neutralising enemy naval forces through successful fleet actions and relentless concentration on blockading their principal naval bases.[4] Naval domination was almost complete after the decisive Battle of Trafalgar on 21 October 1805. In comparison, the weaknesses in the army had

1 Hall, *Wellington's Navy*, 111, and Duffy, 'Festering the Spanish Ulcer', 28, both citing Hamilton (ed.), *Letters and Papers of Admiral of the Fleet Sir T. Byam Martin*, 409, Duke of Wellington to Rear Admiral Thomas Byam Martin, 21 Sept. 1813.

2 Knight, *Britain Against Napoleon*, 241 and 457: 'for a dozen years every one of Napoleon's moves in the Mediterranean was countered by Britain's ability to move troops and supplies faster by ships than he could move them by land'.

3 Blockades were not totally effective. Enemy fleets did escape, particularly when the blockading ships were driven off station by severe weather conditions. It should also be noted that whilst naval fleets could be blockaded, privateers were more difficult to contain. Generally operating independently, they began to pose a more serious threat to small naval ships and merchant vessels.

4 These included: the Glorious 1st of June 1794, commander Admiral Richard Howe; Cape St Vincent (against the Spanish fleet), 1797, commander Admiral Sir John Jervis (later Earl St Vincent); Camperdown (against the Dutch fleet), commander Admiral Adam Duncan, 1797; Aboukir Bay (also known as the Battle of the Nile), 1798, led by Rear

been regularly exposed during the period. Though there were successes in the West Indies they came at a high cost of life.[5] All the European expeditions ended in failure. Indeed, until Lieutenant-General Sir Ralph Abercromby's campaign in Egypt in 1801 it was naval successes that sustained the morale of the British public and fuelled the political resolve to continue the war during this period.

Nelson's victory at the Nile did not remove the French from Egypt; neither did Copenhagen nor Trafalgar affect the French strategic position in Europe. Complete victory over Napoleon's forces would only be achieved on land and then only with the support of allies.[6] If it was to be achieved, the army and the navy needed to engage in conjunct operations. These were most successful when inter-service rivalries were suppressed. However, there are numerous well-documented illustrations of tensions in army–navy relations. During the Revolutionary war, inter-service relations were particularly strained. In 1795 an infantry officer serving aboard a warship as a marine was dismissed from his ship by a naval court martial for insubordination to the captain and the army backed the officer. Controversy ensued and the case finally reached the king before being dismissed. This incident led to 'the Admirals' mutiny' when the Duke of York, Commander-in-Chief of the army, issued regulations for soldiers serving afloat that appeared to remove them completely from naval discipline. Spencer managed to defuse this anger by persuading the duke to withdraw the new regulations.[7] Another serious dispute arose in the following year surrounding the appointment of the naval commander-in-chief to support Sir Ralph Abercromby on his first major independent command to the West Indies.[8] The impressionable Earl Spencer (First Lord of the Admiralty, December 1794–February 1801) created a dilemma by appointing Rear Admiral Hugh Christian in place of the more senior Admiral Sir John Laforey, who was less well regarded by the army, being an old man 'perhaps a little less active than one could wish'.[9] Faced with opposition from other members of the Admiralty Board and senior admirals, he vacillated. Laforey would remain and Christian would be under his command, but Spencer failed to advise Dundas and Abercromby of the change of plan. When they learned of it they tore Spencer's new plan apart, and the general declared that he might not accept the command if Laforey was involved. Pitt was very concerned, and despite Dundas confirming that Spencer

Admiral Horatio Nelson; Battle of the Straights, 1801, a night action commanded by Rear Admiral James Saumarez; Copenhagen, 1801, led by Admiral Sir Hyde Parker.

5 Knight, *Britain Against Napoleon*, 75: between 1793 and 1801, 89,000 rank-and-file soldiers served in the West Indies; 43,747 (almost half) died. If discharges and desertions are included then the losses were 62,250 (70 per cent).

6 Charles John Fedorak, 'The Royal Navy and British Amphibious Operations during the Revolutionary and Napoleonic Wars', *Military Affairs* 52 (Jul. 1988), 141–6.

7 Rodger, *Command of the Ocean*, 435.

8 Duffy, *Soldiers, Sugar and Seapower*, 163–70, for a comprehensive assessment of the proceedings. N.B. Abercromby was related to Henry Dundas through his mother who was part of the Dundas clan.

9 Duffy, *Soldiers, Sugar and Seapower*, 167.

had complete freedom of choice, Spencer stood Laforey down. Senior naval officers were incensed; they resented this interference and were alarmed that it might create a precedent by suggesting that the army possessed a veto over the appointment of admirals.[10]

During the course of these exchanges, Abercromby wrote, 'Soldiers and sailors are not generous, disinterested people. They would willingly cut each other's throats for fame or profit. They should not be trusted to co-operate amiably.'[11] That tensions existed is to be expected. They mainly occurred under stressful conditions as a result of the personal characteristics of the individuals involved. Co-operation between the lower ranks to achieve a common purpose was equally important and, bereft of political implications, was regular practice.

Political tensions within each administration impacted expeditionary planning, effectiveness and co-operation. For example, the fractured relationship between the inexperienced Earl Spencer at the Admiralty, who lacked the confidence to challenge his cautious Board, and the powerful Secretary of State for War, Henry Dundas, invariably led to expeditions being planned as military operations by the War Office, usually without Admiralty involvement.[12] They may have been labelled conjunct operations but there was no expectation that the navy would participate once the army had landed, apart from securing supply lines. However, though not planned in London, there were numerous occasions before Trafalgar in 1805 when naval officers seized the moment and took marines and seamen ashore to support the army in attack or the local population in defence. Nelson, Sir Home Popham, Sir Sidney Smith and Samuel, Lord Hood all fought ashore.

In August 1793 Hood was the naval commander of the fleet blockading Toulon, the major naval base on the southern French coast, where approximately half of the French naval fleet was moored. French Royalists trapped in the town appealed to him for British support against the advancing French government troops. Hood attempted to organise the defence of the town, but with insufficient troops he was unable to do so effectively. By December it had become clear that the defenders would not be able to halt the advancing government force. The British withdrew, taking with them several thousand refugees and three French ships of the line, but in the haste to escape only nine other ships of the line were destroyed; thirteen more and five frigates were undamaged. However, the destruction of the timber stocks that had been stored in the dockyard dealt a crippling blow to the French navy.[13]

Nelson led several forays ashore. In 1794, as a captain attached to the Mediterranean fleet under Admiral Lord Hood, he blockaded the ports of Corsica. During the siege of Bastia he led a party of 1,100 troops and 250 seamen when they manhandled fifteen heavy guns, including two 13-inch and two

10 Rodger, *Command of the Ocean*, 435.

11 Duffy, *Soldiers, Sugar and Seapower*, 167.

12 Knight, *Britain Against Napoleon*, 83; Rodger, *Command of the Ocean*, 432.

13 Knight, Roger, *The Pursuit of Victory: The Life and Achievements of Horatio Nelson* (London, 2005), 164–6; Rodger, *Command of the Ocean*, 427; Knight, *Britain Against Napoleon*, 69.

10-inch mortars, over rough terrain to a position overlooking the town.[14] Their artillery support strengthened the siege, forcing the town to surrender. An attack on Calvi on the other side of the island was similarly successful. In 1796 at Elba Commodore Nelson was involved in a combined operation during which his squadron was to support a detachment of troops under Major John Duncan. The objective was to occupy Elba and avoid it falling into French hands to be used as a base to launch an attack on Corsica. The troops were landed and Nelson's ships stood by to destroy the town's fortifications, although fortunately the governor surrendered the island without bloodshed. In 1797, Nelson led a rather foolhardy expedition to capture a Spanish treasure ship anchored off the town of Santa Cruz on Tenerife. Though he lacked reliable intelligence, the plan had been expanded to include a landing and an assault on the town. The first attempted landing failed miserably but Nelson was stubbornly determined to try again. On the second occasion the landings were strongly contested. Nelson was shot in the right arm: he and his force withdrew immediately. A number of troops did manage to gain entry to the town but were forced to capitulate after a brief skirmish, though they were allowed to return to their ships on the understanding that the British would leave Tenerife and not attack the Canary Islands again. Losses and casualties were heavy: 153 men were killed and drowned and 110 were wounded.[15]

Following the French invasion of Egypt in early July 1798 and the subsequent decimation of the French naval fleet in an attack led by Nelson in Aboukir Bay on 1 and 2 August, Napoleon and his army of 35,000 men were trapped and his plan to progress to India was stalled. He decided to return to France via Damascus, Constantinople and Vienna. In January 1799 he marched east with half his force; his artillery was to follow by sea. He succeeded in taking Gaza and Jaffa. The next target was Acre, where Captain Sir Sidney Smith had recently arrived in command of two ships of the line.[16] Smith had been instructed to observe the progress of the French army as it advanced north up the Syrian coast. Not content with merely observing, Smith captured the vessels that were carrying the heavy artillery Napoleon intended to use in the attack on Acre. Smith brought this artillery ashore at Acre where it was used, with some irony, to defend the town against Bonaparte's troops. Smith, like Nelson, thrived on direct confrontation with the enemy. He led a body of 800 marines and seamen ashore to support the Turks who were defending the old walls of the town, resisting numerous onslaughts from the French. Then Bonaparte learned that Acre was about to be reinforced by 20,000 troops from Damascus. He sent out a force to intercept these reinforcements, who were successfully beaten back, and then launched two more attacks on Acre. The final attack was partially successful: the French

[14] Knight, *Pursuit of Victory*, 169. There is confusion about who was in command ashore. The army commander appears to have tried to stamp his authority. There was rivalry at least for control over the seamen in the party.

[15] Knight, *Pursuit of Victory*, 242–54.

[16] Acre was controlled by the Ottomans. It is now known as Akko, in northern Israel, and is at the northern extremity of Haifa Bay.

penetrated the walls of the town but were then driven back. French losses were very significant, and Napoleon finally recognised that he could not take Acre because of the resistance of the British navy. On 20 May 1799 he withdrew his troops and retreated to Cairo.[17]

Captain Sir Home Riggs Popham had a chequered naval career and is primarily remembered for the invention of a system of signals. Through his role as the Duke of York's naval liaison officer in Flanders in 1794 he gained early promotion and the ear of the Secretary of State for War, Henry Dundas. In 1798 he proposed and led a raid to destroy the locks at Ostend at the seaward end of the ship-canal to Bruges, which was an important inland port and an assembly point for invasion forces. The raid was successful in that the lock gates were destroyed. However, most of the troops taking part could not re-embark because a gale sprang up and marooned them ashore where they were captured. Subsequently he led the naval element of the successful expedition to the Cape of Good Hope in 1805. He then made a serious misjudgement, believing that he would be able to stimulate unrest and insurrection in the Spanish colonies in South America if he could take Buenos Aires in Argentina. Acting without authority from the Admiralty, he persuaded the military commander at the Cape, Major-General Sir David Baird, to provide 1,400 troops to attempt a landing and transported them across the South Atlantic. He had initial success but after a few months ashore he was forced to withdraw. Popham was called home to face a court martial for leaving his station at the Cape of Good Hope without authority. He escaped with a censure.[18]

Trafalgar proved to be a watershed in large fleet naval battles, though there were two big ship actions after 1805.[19] The era of great sea battles may have been over but there were still several fleets to be neutralised.[20] These included the Spanish fleet at Ferrol, Cartagena and Cadiz, the Portuguese and Russian fleets at Lisbon and in the Baltic, and the surviving French navy at Brest and Toulon. The Spanish ships were removed from French control as a result of the 1808 Spanish revolt and the squadron of seven Russian ships of the line which was in the Tagus at that time was captured. The Portuguese fleet was despatched across the Atlantic to Brazil when Napoleon invaded in November 1807.[21]

[17] Tom Pocock, 'Smith and the Siege of Acre 1799: Sea Power Decisive on Land', in Hore (ed.), *Seapower Ashore*, 26–38.

[18] He claimed that he had discussed this expedition with Pitt in London before he had sailed. Pitt had since died. Dundas confirmed that there had been some discussion on the subject but that no specific orders had been issued.

[19] In February 1806 Vice Admiral Sir John Duckworth was successful off San Domingo. In 1809 Captain Lord Cochrane, under the command of Admiral Gambier, led a fearless fire ship attack on the French fleet in the Basque Roads.

[20] Richard Harding, 'Expeditionary Armies and Naval Power: The North German Campaign of 1805–6', in Helen Doe and Richard Harding (eds), *Naval Leadership and Management, 1650 to 1950* (Woodbridge, 2012), 64.

[21] Duffy, 'Festering the Spanish ulcer', 15.

The British approach to the war changed following the appointment of Viscount Castlereagh as Secretary of State for War in 1807. In stark contrast to the lack of direction and action of the Ministry of All the Talents, during his thirty months in office he despatched six large expeditionary forces into Europe to support allied opposition to Napoleon.[22] The navy's primary role was to provide security at sea for military and merchant shipping through intensive blockades of enemy bases and effective convoy management. The impact of an effective blockade, both economically and militarily, was very significant. In Europe effective naval blockades on Brest, Toulon and the Scheldt severely restricted the movement of the French and their allied navies. Brest was virtually abandoned as a shipbuilding centre by Napoleon in 1810 because of the blockades.[23] Between 1812 and 1814 the British navy devastated the effectiveness of the US navy and destroyed the nation's economy.[24] However, maintaining a blockade was expensive. In 1826 it was estimated that the annual cost of sustaining a blockade involving only two ships of the line with combined crews of 2,784 men was £160,740. Of course during the war blockades involved considerably more ships, but it was naval capacity rather than cost that determined the extent of blockades.[25]

All expeditions, of any duration, depended upon regular replenishment of recruits, horses, munitions and provisions which were delivered by transports under naval escort. Good convoy organisation was essential; this was an established naval skill dating back to the end of the seventeenth century.[26] Several Acts of Parliament made sailing in convoy mandatory for merchant ships.[27] Nevertheless the temptation to leave the convoy, particularly as it was approaching its destination, in order to be the first to land goods or on the home voyage to avoid the crew being pressed into naval service, was enormous. Prosecutions for leaving convoys did occur: William Evans, Master of the *Douglas* out of Liverpool, was imprisoned for leaving a convoy and the ship's owners were also prosecuted; and the master of the *Coquette* who had left a convoy bound for St Thomas was prosecuted, convicted and spent two months in Marshalsea prison.

During the Revolutionary war, when resources were scarce, convoy protection was an operational distraction for the Admiralty. Convoy duty was generally

22 Copenhagen, Sept. 1807; Baltic, May 1808; Portugal, Aug. 1808; Corunna, Nov. 1808; Portugal, Apr. 1809; Walcheren, Jul. 1809.

23 Rodger, *Command of the Ocean*, 562. French ship building was transferred to Antwerp.

24 Arthur, *How Britain Won the War of 1812*, chapters 4–7.

25 NMM, ADM, BP/46B, an account of the whole annual expense of the cost of a blockade, Navy Office, 28 Apr. 1826. The ships were HM ships *Ramillies* (crew, including 97 officers, 1,468) and *Hyperion* (crew, including 91 officers, 2,784 men). The cost comprised pay £75,558, victualling £65,974, materials £9,642, allowances and travel £7,719 and clothing for seamen £3,846.

26 Rodger, *Command of the Ocean*, 91.

27 Although the convoy system was in place in 1793, the Acts for the Better Protection of Shipping of 1798, 1803 and 1815, commonly known as the Convoy Acts, provided for financial penalties for owners who did not observe the laws.

despised by naval officers because it deprived them of the opportunity to take prizes and because it could 'try the patience of a saint'.[28] Nelson perhaps best summed up the attitude of naval officers to convoys when he wrote in 1781, 'They [the merchant ships] behaved as all convoys that I ever saw, did shamefully ill, parting company every day.'[29]

Transport agents sailed with every sizeable convoy to assist the convoy commander in managing the transports. On distant services there was generally one agent to ten transports and on short or Mediterranean services one agent for every fifteen or twenty transports. On services in home waters there was no set number.[30] Sailing with convoys was restrictive for ship-owners and frustrating for ships' masters. Nevertheless, the benefits of the convoy system were obvious. In a sample of 132 convoys between 1793 and 1797, four were attacked. Out of 5,827 ships escorted, 398 straggled from their convoys and 35 were lost – less than 1 per cent.[31] From 1803 to 1815 it has been estimated that only 0.6 per cent of all ships that sailed in convoy were lost compared with 6.8 per cent of stragglers.[32] Between June and December 1809 some 2,210 merchant ships were escorted through the Great Belt to the Baltic, mainly in large convoys, without loss from the large number of privateers who lurked in the area. In 1814 the chairman of Lloyd's said, 'Effectual protection can only be given to British commerce by a rigid adherence to the convoy system.'[33]

The demands upon the Admiralty for convoy protection grew as the expansion of international trade was matched by increasing military demands for transports. Whilst transports often sailed with the scheduled convoys, it was not unusual for troops or supplies to be required urgently, making additional specific convoys necessary. The demand for transport convoys became so acute that the Admiralty had to insist that the Transport Office supply a monthly list of all transports due to sail during the coming month, signifying their cargo and destination. The large number of convoy requests meant that the Admiralty was compelled to synchronise warship movements with convoy protection, so that naval vessels sailing to join new stations or returning home for refits or on reassignment to other duties were generally obliged to shepherd convoys heading in the same direction.[34] From the summer of 1808 until the spring of 1814 one Admiralty List shows that 404 convoys sailed from Britain to various Iberian destinations to support the Peninsular war; they ranged from small movements of one or two ships to large flotillas of ninety or 100 vessels – in total some 13,427 voyages, not including return voyages or those to and from other destinations overseas.[35]

[28] Hall, *Wellington's Navy*, 119.
[29] Knight, *Pursuit of Victory*, 67.
[30] Commission for Revision (9th), 204, Interview with Captain Bowen.
[31] Rodger, *Command of the Ocean*, 559.
[32] Hall, *Wellington's Navy*, 121.
[33] Rodger, *Command of the Ocean*, 559.
[34] Hall, *Wellington's Navy*, 112.
[35] Hall, *Wellington's Navy*, 112.

Chart 5.1 shows the number of convoys and ships convoyed between 1809 and 1814, excluding the convoy protection of military expeditions; it is clear that convoy management was a very significant element of the navy's role in the later years of the wars, peaking in 1809 when there were 298 overseas convoys (inwards and outwards) involving 10,217 ships and again in 1813 when there were 366 convoys with 16,288 merchant ships.[36] It is little wonder that the Admiralty resisted all attempts to restrict the size of its fleet when so many ships were required for this service.[37] This partially explains the significant increase during the latter stages of the war in the number of smaller naval ships which were ideal for convoy duty as well as for service in the rivers of Europe.[38] Despite the navy's best efforts, Wellington found many reasons to complain about the administration of the convoy system but such criticism only served to demonstrate his army's absolute dependence on these services and reflected the significance of their role in all expeditionary warfare.

Chart 5.1. Number of convoys and ships convoyed 1809–14 (excluding military convoys)

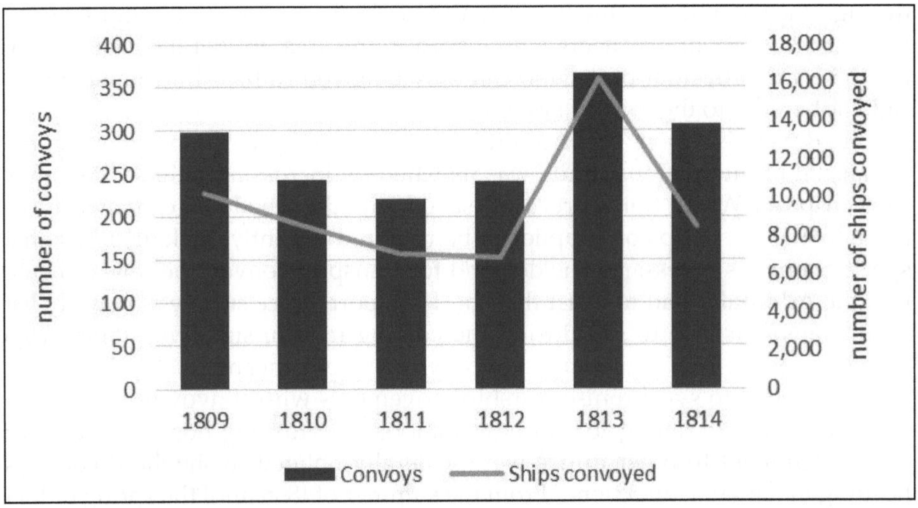

Thus the navy developed a dual role, firstly suppressing enemy trade and defending British trade (mainly against French, Danish and later American privateers), and secondly protecting and supporting military expeditions. The latter involved escorting the armada of transports conveying troops, horses, ordnance and supplies; supporting landing operations; securing and protecting

[36] TNA, ADM, 7/64, Admiralty notifications of convoys to Lloyd's. This list does not include purely military expeditions where the insurance of merchant ships was covered by government indemnity.

[37] See chapter 6.

[38] Knight, *Britain Against Napoleon*, 240. By the end of the war successive administrations had ordered the building of 174 brig sloops and 87 smaller gun brigs, while 50 captured enemy brigs joined the British fleet.

communication and supply lines; artillery bombardment of coastal defences to support a landing or an attack on a strategically important military target; disrupting enemy supply lines and communications; supplying intelligence; and occasionally initiating and participating in guerrilla attacks ashore. Though the navy's standing had never been higher than after Trafalgar, opportunities for glory and wealth, particularly for the younger officers, seemed to be drying up. Adventures ashore must have provided an outlet for the more audacious young naval officers and seamen who were deprived of participation in sea battles.

Not until the long Peninsular war did the army – and in particular Wellington, despite his frequent criticisms – appear to appreciate the real value of naval support. Not only did the navy protect supply lines and communications but it provided enormous additional advantages. The navy pursued limited, but very effective, guerrilla interventions. This is illustrated by the actions of Captain Thomas, Lord Cochrane who in 1808 created havoc on the Mediterranean French and Spanish coasts.[39] He landed marines and seamen at various locations from his ship *Imperieuse*, a 38-gun frigate attached to Collingwood's Mediterranean fleet, destroying numerous forts and gun batteries. He also disabled many semaphore stations which were very important to the coasting trade of France, causing it to be temporarily suspended. The landing parties captured French code books but left burnt documents that persuaded the French that they had been destroyed, so the codes remained unchanged. Cochrane was thus able to read French messages and pass on intelligence. For some time he made the road that ran along the Catalonian coast unpassable for French troops, cavalry and supply trains. In the process he demonstrated that such actions could tie up enemy forces that were urgently required elsewhere by forcing the diversion of 2,000 troops who were marching towards Spain to quell the Spanish uprising.[40] He made contact with local guerrilla leaders and encouraged resistance against the French occupying forces; under his leadership they captured sufficient military stores to arm an active local guerrilla unit.[41]

In November 1808 a French column of 6,000 men marched into Catalonia to put down the rebellion. Their first target was Rosas. Cochrane recognised that such a large force could not be defeated. Nevertheless, he landed seamen and marines to support the local defence of the town. He established his force in a hill fortress, Fort Trinidad.[42] Through meticulous planning against the forthcoming attacks, Cochrane and his small party played a significant role in helping Spanish troops delay the inevitable conquest of the town for nine days, killing hundreds of French and Italian troops in the process. When the town fell, the Spanish surrendered. Cochrane withdrew his party from the fort; it was situated on the cliff-top so they were able to escape to the beach below using rope ladders. Cochrane and his ship's gunner were the last to leave the fort, having lit the fuses of previously laid

39 Knight, *Britain Against Napoleon*, 301.

40 Lavery, *Nelson's Navy*, 312.

41 Richard Woodman, *The Sea Warriors* (London, 2001), 239; Donald Thomas, *Cochrane, Britannia's Sea Wolf* (London, 1978), 132–3.

42 Also recorded as Fort Trinity in some accounts.

charges. As the *Imperieuse* and its accompanying ships sailed out of Rosas Bay, the main tower of Fort Trinidad collapsed after the charges exploded.[43]

The threat of amphibious assaults launched from naval ships forced the French to detach thousands of badly needed troops to patrol beaches, garrison ports and man coastal batteries. In 1810 two allied squadrons based at Ferrol and Corunna tied down some 20,000 French troops along the Biscay coast; a further 20,000 were needed to invest the naval base at Cadiz.[44]

The Admiralty began to emphasise the importance of proactive naval support for the Peninsular army. Naval officers were instructed to find ways of 'carrying on a desultory and distracting kind of warfare against the French on the coast'.[45] Commodore Sir George Collier was ordered 'to protect allied communications and intercept those of the enemy, to co-operate actively with the guerrillas and to act in every way with Wellington to facilitate his operations'.[46] Captain Sir Home Riggs Popham was appointed specifically to join Admiral Keith's Channel fleet to render all possible assistance to the northern Spanish resistance in order to commit French troops there, thus weakening their forces elsewhere. His instructions from the Admiralty dated 13 July 1812 were: 'You are above all to endeavour to contribute in every way to Lord Wellington's objects; and in all your operations ... you are never to forget ... that a furtherance of his Lordship's views is a duty upon all parties.'[47]

It has been suggested that the navy became the 'bedrock' of Wellington's defence of Lisbon and his preparations for an emergency evacuation in 1810–11.[48] During this period Wellington and Vice Admiral George Berkeley co-operated very closely. Both men surveyed potential embarkation sites. Setubal, 19 miles south of Lisbon, was initially considered as an evacuation point because it was the best harbour in Portugal for small vessels, but it would have required almost 10,000 men to defend the embarkations.[49] After considerable investigation Peniche, a harbour a few miles north of Lisbon, was selected, once Berkeley was satisfied that the anchorage and loading points were adequate. Wellington planned to build barracks for 5,000 men, including 2,000 artillery, and magazines so that the embarkation point could be garrisoned and defended if evacuation was required. However, in a letter to Liverpool in April 1810 Wellington indicated that he actually preferred St

43 Thomas, *Cochrane*, 134–42; Hall, *Wellington's Navy*, 48-50.

44 David Gates, *The Spanish Ulcer: A History of the Peninsular War* (Cambridge, MA, 1986), 29.

45 Duffy, 'Festering the Spanish Ulcer', 23. Citing 1810 Admiralty instructions to Rear Admiral Popham in Hugh Popham, *A Damned Cunning Fellow: The Eventful Life of Rear Admiral Sir Home Popham* (Tywardreath, 1991), 194.

46 Duffy, 'Festering the Spanish Ulcer', 25.

47 Hall, *Wellington's Navy*, 200, quoting C. Lloyd (ed.), *The Keith Papers*, 265–8.

48 Duffy, 'Festering the Spanish Ulcer', 22, citing David D. Horwood, 'British Seapower and its Influence on the Peninsular War', *Naval College Review* 31 (1978), 59–67.

49 Gurwood, *Dispatches* IV, 58, Wellington to Vice Admiral George Berkeley, 8 May 1810. Gurwood, *Dispatches* IV, 54, Wellington to Charles Stewart, 6 May 1810.

Map 4. Lisbon region showing the river Tagus, Torres Vedras lines, Setubal and Peniche

Julian as the point of embarkation because of its closer proximity to Lisbon.[50] In May 1810, just as Berkeley was able to report to Wellington that 'everything is prepared for us either to go or stay. I expect that the embarkation will be made with utmost order', the Admiralty invited him to take the Plymouth command. In view of the timing in relation to the possible evacuation, but mainly because of his good working relationship with Wellington, Berkeley declined the offer.[51]

In June 1810 Wellington advised Berkeley that construction work on the fortifications of the Torres Vedras lines was almost complete. These comprised 53 miles of man-made defences in two lines, positioned to take advantage of the rugged terrain north of Lisbon. They stretched from the Tagus to the Atlantic coast. Wellington needed some essential support from the navy and requested Berkeley to supply some copies of Popham's telegraph vocabularies, after having removed the naval codes and ciphers to avoid them falling into enemy hands. He also asked for a number of officers of marines or 'steady midshipmen' or even non-commissioned seamen and marines to operate the telegraph stations on these defence lines because he did not believe that he had time to train the army to be effective in their use and was fearful that 'mistakes and blunders' would occur. He estimated that he needed one signalman and a seaman at each station and naval support in the form of gunships at each end of the lines – at Alhandra on the Tagus to protect the right flank and off the coast between Mafra and Maceira on the left to deter the enemy from pushing a column along the beach and thus sneaking around the defences.[52] With this vital navy support, the army was eventually able to pass a message from one end of the line to the other, a distance of 29 miles, within 7 minutes.[53] The gunboats on the Tagus were put under the command of Lieutenant Berkeley, the admiral's nephew. They supported the right of the army near Alhandra and engaged with enemy reconnoitring parties on numerous occasions to the extent that Wellington later praised their efforts: 'They have been of great service to us.'[54]

Berkeley wrote to the Admiralty to underline the level of support that the navy was providing for the army, enclosing a map of the Tagus (map 5) which demonstrated the naval dispositions in early November 1810. Berkeley's naval detachment consisted of nearly 400 seamen and 200 marines. General Fane and a regiment of chasseurs and cavalry had crossed to Mugen on the south bank of the Tagus where they were supported by a number of flatboats with howitzers and carronades. Two 12-pounders with 200 seamen under the command of Captain Houston are shown at positions C and D on the plan. The remainder of the gunboats under Lieutenant Berkeley with some heavy guns remained to protect the right of the British lines at Alhandra along with the *Growler* gun-brig, shown as positions A and B on the plan. Seamen were also manning batteries in the village of Alhandra, which was in front of the Vedras line. The plan also shows

50 Gurwood, *Dispatches* III, Wellington to Liverpool, 2 Apr. 1810.

51 Gurwood, *Dispatches* IV, 58, Wellington to Berkeley, 8 May 1810.

52 Gurwood, *Dispatches* IV, 87, Wellington to Berkeley, 15 Jun. 1810.

53 Hall, *Wellington's Navy*, 94–5,100.

54 Gurwood *Dispatches* VI, 526, Wellington to Liverpool, 20 Oct. 1810.

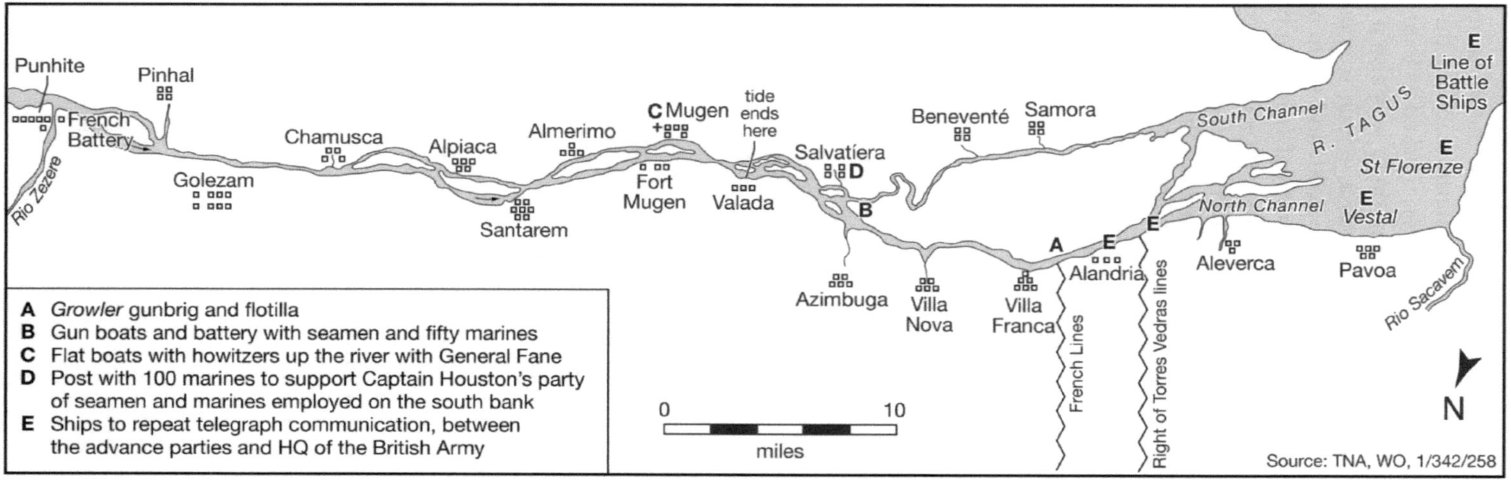

Map 5. Plan demonstrating naval support in the Tagus in November 1810. Source ADM, TNA, 1/342

the locations of the ships positioned to repeat the telegraph communications from the advance parties to army headquarters and to the flagship anchored with the line-of-battle ships in deeper water, marked as E on the plan. Although it is not specifically identified on the plan, there was also a battery and post of seamen and marines on the island of Lizeria, which was used for grazing the stock of cattle and as a bridge between the north and south banks of the river.[55]

On 14 November the French began to retreat from the banks of the Tagus. Berkeley was advised by telegraph from the ships up the river in the evening. Wellington immediately ordered General Hill, who commanded the right, to advance and requested Berkeley to support his flank up the river. Lieutenant Berkeley proceeded as requested with a flotilla of gunboats. Berkeley also sent Rear Admiral Sir Thomas Williams with boats and pontoons to organise any river crossing that might be required by building flying bridges.[56] By the 22nd the enemy had dug into a strong post at Santarem. The British and combined forces were assembling at Valada where some divisions crossed to the south side of the Tagus supported by the flotilla under Sir Thomas Williams.[57] The French retreat continued. By mid-December 1810 they had retreated north towards the Mondego. Berkeley still had seamen and boats at Mugen where there were four flatboats. The bridge boats and vessels for transporting the army were at Salvaterra, but because bad weather had set in and sickness was beginning to show itself among the seamen, he recalled the greatest part of the brigade of seamen and all the marines at Valada and Salvaterra, on the basis that they could be returned on one tide and it would provide the opportunity to clean and repair both the heavy gunboats and the flatboats to make them ready for further action. In the meantime the navy assisted in the defence on the heights on the south side of the Tagus, opposite Lisbon, with a 1,000-strong battalion of marines from the squadron under Major Westrope. They were deployed on the understanding that they would be not be located in positions that would prevent them from being ready to embark in their respective ships if that was required urgently.[58]

Wellington had earlier requested support from Berkeley when he was having difficulty providing troops to defend Peniche, the potential evacuation site, whilst also wanting to create a mobile, disposable force capable of moving to any point to act offensively against the enemy if the opportunity arose. He requested the use of a brigade of seamen and a battalion of marines to man forts ashore, thus relieving sufficient troops for these alternative services.[59] Berkeley acquiesced. He was only too happy to second his seamen and marines, but when he requested an additional shore allowance for his seamen, Wellington declined; Berkeley withdrew his men. Wellington referred the matter to the secretary of state who finally agreed that lieutenants should receive 10s per diem, midshipmen 6s, petty officers 1s 6d and

55 TNA, ADM, 1/342, Berkeley to Admiralty, 10 Nov. 1810.
56 TNA, ADM, 1/342, Berkeley to Admiralty, 20 Nov. 1810.
57 TNA, ADM, 1/342, Berkeley to Admiralty, 22 Nov. 1810
58 TNA, ADM, 1/342, Berkeley to Admiralty, 14 Dec. 1810.
59 TNA, ADM, 1/342, Wellington to Berkeley, 10 Nov. 1810.

seamen 1s.[60] However, this came too late, because when Berkeley advised the Admiralty of Wellington's request, it refused to allow seamen to be taken ashore for this purpose, believing that the ships should be fully manned and ready to sail immediately if they came under threat. It was happy for the marines to be employed ashore so long as they could be re-embarked quickly.[61] Berkeley was obliged to withdraw the seamen and to ensure that he only committed marines, as long as he could be reasonably sure of their speedy re-embarkation.

Given that about 4,000 British merchant seamen (from the transports moored in the Tagus in anticipation of the evacuation) were under-employed, discipline issues were inevitable. In August 1810 Berkeley was complaining of the 'bad conduct' of transport crews and the Admiralty recommended that such men should be pressed into the navy as punishment. However, the pressing of men from transports was a direct violation of the charter party and thus a breach of contract by the government. The impact of ships losing crews might have been significant and the government could have been liable. Berkeley wanted to find a way to bring them under military discipline where they would be subject to court martial. His concern was that 'Crimes which are committed afloat under the British flag are ignored under the laws of this country [Portugal] as they consider them vessels of war under naval power.'[62] The Admiralty submitted the matter to its lawyers for advice. The solicitors' opinion was that unless these men were taken under naval or military discipline there was no remedial action that could be taken.

In late November 1810 Berkeley's immediate problem was that some transport crew members had been accused of theft from regimental baggage, which was stored on various transports with only one soldier to guard it.[63] The Portuguese authorities considered transports to be part of the naval establishment and thus did not judge the restrictions which applied to merchant ships to be applicable: they refused to interfere in this matter. If the crime had been committed in England then the accused might have been tried by the Court of the Admiralty. Berkeley was frustrated, believing that if the men were sent back to England to be tried it would be no deterrent, whereas the sentence of a court martial could

[60] Gurwood, *Dispatches* IV, 270 and 455, Wellington to Berkeley, 9 Sept. 1810 and 9 Dec. 1810.

[61] Gurwood, *Dispatches* IV, 406, Wellington to Berkeley, 10 Nov. 1810.

[62] TNA, ADM, 1/342/190, Berkeley to Admiralty, 17 Aug. 1810.

[63] TNA, ADM, 1/432/283, Richard Poulden, Resident Agent in the Tagus, to Berkeley, 26 Nov. 1810.

He also describes another similar incident: 'Another daring robbery has taken place on board the *Defence* transport No. 6, Thomas Suckfield, Master. Between 11/12 o'clock on 20 November by two of the said transport's crew. Robert Thin and John Lewis removed property under the charge of Sergeant Meacham who was guarding the baggage of 40th Regiment. Seamen who were supposed to be on watch took the sergeant's chest said to contain £120 in valuables from under his hammock, made off and deserted in the jolly boat. Robert Thin had since been caught with part of the property on his person and confessed and is now in custody on *Barfleur*.'

be instantly inflicted and its effect could be seen and felt by the surrounding fleet. He confessed that he was 'at a loss to know how to proceed and how these men are to be dealt with'.[64] Though transport crews no doubt had their share of villains who took advantage of such situations, there is no evidence of systemic misbehaviour.

Wellington himself agreed that he was always happy to acknowledge the cordial assistance that he had received from Berkeley and the naval force under his command, and whilst Berkeley was on station Wellington's relations with the navy were workmanlike.[65] Then in 1812, in accordance with his instructions from the Admiralty to support Wellington, Sir Home Popham took a squadron of ships with two battalions of royal marines to the north coast of Spain to organise diversionary operations in concert with guerrillas and other Spanish troops by attacking Castro, Bilbao and other places on the coast. This activity tied up a considerable body of French troops who might otherwise have been deployed to reinforce the main army. Popham's marines captured 454 pieces of ordnance including 100 36-pounders and 291 24-pounders together with hundreds of muskets, shot and ammunitions, 8 ammunition wagons and 3 forage carts.[66] These actions received Wellington's praise but he was later disconcerted when Popham encouraged publication of an account of his success in diverting the attention of the enemy. He wrote, 'I mention this because I know that the French act a good deal upon any information from our papers.'[67] Yet when Popham advised Wellington that he was planning to return to England, the latter complained to Popham that without his diversionary activities 'I will have on my hands more of the enemy than I can manage.'[68]

In June 1813 Wellington felt obliged to complain to Earl Bathurst, Secretary of State for War, after he had been advised by Sir Charles Stuart that due to a lack of naval protection a number of transports had been captured off the coast of Portugal. In addition, a convoy of ordnance transports had been delayed at Lisbon for want of convoy protection to Santander. He bemoaned the fact that for the want of convoys his army had been obliged to use French ammunition which was of a smaller calibre than the standard British issue. Wellington made it clear that his army would not be able to remain in northern Spain without well-stocked magazines and stores. For this he needed supplies and specie to purchase additional goods locally. Bathurst had explained that the Admiralty was under great pressure; it had a significant part of the fleet and men tied up off the American coast following the outbreak of war a year earlier. This did not stop Wellington from declaring that 'For the first time I believe that it has happened to any British Army, that its communication by sea is insecure; and certainly we have not money to purchase in the country all that we need.'[69]

64 TNA, ADM, 1/342/283, Berkeley to Admiralty, 27 Nov. 1810.

65 Gurwood, *Dispatches* IV, 414, Wellington to Berkeley, 12 Nov. 1810.

66 Gurwood, *Dispatches* VI, 9, Wellington to Commodore Sir Home Popham, 4 Aug. 1812.

67 Gurwood, *Dispatches* VI, 71, Wellington to Earl Bathurst, 12 Sept. 1812.

68 Gurwood, *Dispatches* VI, 85, Wellington to Commodore Sir Home Popham, 8 Nov. 1812.

69 Gurwood, *Dispatches* VI, 548, Wellington to Earl Bathurst, 24 Jun. 1813.

There were more complaints. A few months later the French were being allowed to have constant communication between their bases at St Jean de Luz and Bayonne and their forces at San Sebastian because the blockade of San Sebastian was not being constantly maintained. Wellington's view was that restricting French coastal trade between the country's Mediterranean and Atlantic coasts would significantly slow down their ability to build up magazines to supply their army because it would take much longer if it had to be done overland.[70] This developed into another inter-service feud. Admiral Lord Keith, stung by the criticism, blamed a lack of co-ordination in the army's demands for convoys. Wellington retorted that his army 'requires to have constant communication by sea' but, unlike normal convoy arrangements, these services did not always require a great convoy: 'The maritime communications of an army cannot be carried on in the same mode as the commercial communication.'[71] He believed that the navy had failed to recognise this distinction, and to illustrate this, he pointed out in November that a single ship loaded with great coats for the army had been waiting at Oporto for a convoy since August. Those coats were then required urgently as the temperature had dropped significantly. Not only that, the specie that he needed badly was delayed in the Tagus. To get the money by an overland route would take two months.[72]

The army also had urgent need of tents to replace those destroyed in recent storms and other general supplies. Such items were often required at short notice and could be conveyed in single ships, whereas the navy's policy was to wait until there were sufficient ships to form a sizeable convoy to justify the commitment of a naval escort. He did not wish to call forward more supplies than were necessary because this caused other problems; the harbours were generally small and unloaded transports clogged them up, making it difficult for other ships to manoeuvre around them. Yet there were limited storage facilities ashore to facilitate these transports being unloaded.

Unfortunately, to Wellington's dismay, naval presence in the waters surrounding the Peninsula was the responsibility of various commands. The northern Spanish station was under the remote command of the Channel fleet.[73] Then there was a Portuguese station based at Lisbon. It had been established in 1807 under Admiral Charles Cotton who remained in charge until December 1808 when Vice Admiral Berkeley took over; he was succeeded in June 1812 by Vice Admiral Sir Thomas Byam Martin. Meanwhile the Mediterranean fleet, under a separate command, was responsible for the south and eastern coastlines of the Peninsula.[74] It is understandable that this division of responsibility caused

70 Gurwood, *Dispatches* VI, 691, Wellington to Earl Bathurst, 19 Aug. 1813.

71 Gurwood, *Dispatches* VII, 239, Wellington to Earl Bathurst, 1 Nov. 1813.

72 Gurwood, *Dispatches* VI, 575, Wellington to Earl Bathurst, 2 Jul. 1813.

73 Channel fleet commanders-in-chief: Vice Admiral Lord Gambier 1808–11, Sir Charles Cotton 1811–12, and then Admiral Lord Keith.

74 Mediterranean fleet commanders-in-chief: Vice Admiral Cuthbert, Lord Collingwood until his death in March 1810, then a succession of vice admirals – Sir Charles Cotton to 1811, Sir Edward Pellow to 1814 and Sir Charles Penrose until 1815.

Wellington considerable frustration. He believed it was the root of the problems he was experiencing but he was resigned to accept that 'it is a naval principle, which cannot be departed from'.[75]

The navy lent other assistance. It was not unusual for it to be called upon to haul heavy artillery ashore. Such help was requested during Sir John Murray's attack on Tarragona to support the capture of the fort of San Felipe on 7 June 1813. The fort, which was protected on two sides by almost inaccessible mountains, commanded the road between Barcelona and Valencia which the French reinforcements would use to come to the aid of the town.[76] Lieutenant Corbyn of the *Invincible* took a team of seamen to haul two 12-pounders and a howitzer up a very steep mountain to a position overlooking the fort at a distance of about 700 yards.[77] From this vantage point they began to pound the fort to divert attention from the Royal Engineers and Royal Artillery units who were preparing platforms for the main batteries that would breach the walls of the fort. Later that night the same seamen manhandled another five 24-pounders and all the powder and shot required for the battery into position, ready to be hauled onto the platforms. They toiled all night at this arduous task and at daybreak resumed their attack on the fort. The engagement was supplemented by a number of marines using artillery from the *Strombolo* bomb vessel including two 8-inch mortars and two 4-pounders. Before the preparations for the breaching battery were concluded the garrison in the fort ran out of ammunition and surrendered. As evidence of the close working relationship, the decision to accept the surrender was taken jointly by the army commander, Lieutenant Colonel William Provost, and the naval commander, Captain Adam. Provost later wrote in his despatches, 'The success of this expedition may in great measure be attributed to the zealous action and indefatigable exertions of Captain Adam and the men of the Royal Navy.'[78]

In February 1813 the French were planning to transport to France a large land convoy of colonial produce that had been assembled in Barcelona. Due to the British naval blockade it had to travel by way of a road which ran along the coast. A ship of the line, a frigate and a sloop were ordered to attack the convoy. These actions forced the French to abandon the road and resort to the more difficult mountain route where the convoy was at the mercy of Spanish guerrillas.[79]

During the battle for San Sebastian, Lieutenant-General Sir Thomas Graham calculated that the attack against the fort could be aided if a small battery could be established on the small island of Santa Clara just off the coast. This was achieved on the evening of 26 September 1813. The island was attacked, and the enemy troops in situ were taken prisoner by a detachment of the 9th Regiment led by Captain Cameron, supported by a detachment of Royal Engineers conveyed

75 Gurwood, *Dispatches* VII, 239, Wellington to Earl Bathurst, 1 Nov. 1813.

76 Hall, *Wellington's Navy*, 137, 187.

77 Gurwood, *Dispatches* VI, 566, Murray to Wellington, 9 Jun. 1813.

78 Gurwood, *Dispatches* VI, 484, Lieutenant-Colonel William Provost, commanding the 2nd Brigade, 1st Division of 67th Regiment, to Lieutenant-General Sir John Murray, 7 Jun. 1813.

79 Hall, *Wellington's Navy*, 186.

in boats commanded by Lieutenant James Arbuthnot. The naval contingent then manned the batteries which successfully breached the wall of the bastion, allowing the fort to be stormed and carried.[80] On 9 April 1814, before the cessation of hostilities and with the enemy on the run, in 36 hours a detachment of 600 seamen and marines performed a fatiguing march of more than 50 miles during which they successfully dismantled five forts, destroyed forty-seven pieces of cannon and seventeen mortars before re-embarking.

The progress of Wellington's army across the Peninsula was frequently inhibited by rivers which required bridging to avoid long detours. In some instances bridges had been destroyed by the enemy and had to be rebuilt; on other occasions new bridges were required. This work was generally done by the Royal Staff Corps which was under the command of the army, although it was not unusual for it to receive support from the navy to move pontoons and help with the heavy lifting that was required to manoeuvre spars and other materials into position.[81] This happened in February 1814 when Wellington planned to drive the French army from Bayonne in southern France. To achieve this he needed to get his forces across the mouth of the river Adour. The bridge that was designed to facilitate this has been described as 'one of the finest pieces of military engineering during the Peninsular War'. He sought the assistance of the commander of the naval fleet that was supporting the army, Rear Admiral Charles Penrose. It became clear that the best approach to the challenge was to anchor a number of small vessels across the river and to build a bridge across their decks. Wellington wrote to Penrose:

> I propose that our bridge should be constructed of vessels from 15 to 30 tons burthen, two masted and each well ballasted, and provided with anchors and cables to be anchored by head and stern … I propose to lay cables across these vessels from bank to bank, which we have reason to believe is an extent of 400 yards; and on the cables we shall tie the planks.[82]

Lieutenant John Debenham, an agent of the Transport Board, was ordered to find these ships. He procured forty-eight coastal trading boats, known as chasse-marées, in the local fishing ports. Each was of 40 to 53 feet with 15-foot beams and decks approximately 3–4 feet above the water. The point finally chosen for the bridge was about 1½ miles from the mouth of the river where it flows into the Bay of Biscay and where it narrows to about 270 yards. The boats were to be anchored some 30 feet apart across the river, anchored fore and aft. Cables would be laid across them from bank to bank. One end of each cable would be secured to an 18-pounder cannon which would be buried in the marsh on the

[80] Gurwood, *Dispatches* VI, 726, Wellington to Earl Bathurst, 2 Sept. 1813.
[81] Robert Burnham, 'British Bridging Operations in the Peninsula', in Rory Muir, Robert Burnham, Howie Muir, Ron McGuigan (eds), *Inside Wellington's Peninsular Army 1808–1814* (Barnsley, 2006), 226–74.
[82] Burnham, 'British Bridging Operations', 247–8.

bank. There was a 14-foot variation in tidal height so capstans were to be used on the other bank to allow the tension to be adjusted as the tide ebbed and flowed.[83]

Penrose was called to advise upon the defence of the bridge from attack by fire ships. He suggested that gunships should anchor above the proposed location for the bridge and lay a boom across the river of 50- to 60-foot spars attached to each other by chains. The navy was to take responsibility for building the boom. In fact two booms were subsequently laid. The bridge was completed in less than two weeks. On 22 February, under the command of Captain Dowell O'Reilly RN, the flotilla of chasse-marées put to sea from Socoa where they had been assembled and prepared. The army commenced an attack on Bayonne to divert attention from the river and to secure the banks.

The wind direction and strength created havoc for the plans to bring the flotilla into the river. In such conditions the mouth of the Adour is a very dangerous place because of the heavy surf that breaks over the sandbar. Even the local pilots were reluctant to attempt to cross.[84] Thus, Debenham, the Transport Agent, was the first to attempt the crossing but he beached the boat. He and the crew jumped out and hauled the boat over the bar into the calmer waters in the river. The next boat hit the bar and overturned, drowning five of the crew. The third boat succeeded but the tide was changing, falling quickly and starting to flow out of the river, so Penrose suspended the operation until the tide changed again. When the tide turned and sufficient depth of water started to flow into the river, Penrose was anxious to encourage the boats over the bar. The first boat, the *Lycra*, was wrecked, killing the master's mate and all the crew. The second boat suffered a similar fate, killing the captain, a ship's surgeon and the crew. No other crews were prepared to risk crossing the bar until Penrose offered 50 gold guineas to the men of the first chasse-marée to be successful. The crew of one vessel tied themselves to the masts and made an attempt. They were successful and this encouraged others to try. Of the original flotilla of 3 sloops, 2 brigs, 5 gunboats and 48 chasse-marées, only 34 chasse-marées, 2 sloops and 4 gunboats crossed the bar. Two chasse-marées and a sloop were destroyed at a cost of 30–40 lives. Twelve chasse-marées turned back to port. The four gunboats, under the command of Lieutenant Cheshire, managed to get into position. Each had an 18-pounder gun. Unusually, the guns were under the command of a Royal Artillery captain and were manned by ten Royal Artillery men.

The booms were built and positioned parallel to each other, 20–25 feet apart, by the navy. Twenty-six of the chasse-marées were directed into position within two days. Approximately 800 planks were laid over the cables to form the roadway. Some 40 hours after the boats crossed the bar the construction was complete. However, 'Despite the detailed planning and herculean efforts of the engineers, they were too late. Most of the attacking force had already been ferried across [on pontoon rafts] and had moved on.'[85] Nevertheless when the bridge was complete the remainder of the cavalry, the artillery, the officers'

[83] Burnham, 'British Bridging Operations', 249.

[84] Burnham, 'British Bridging Operations', 252–6; Hall, *Wellington's Navy*, 227.

[85] Hall, *Wellington's Navy*, 227; Burnham, 'British Bridging Operations', 261.

baggage and a Spanish division did cross over it.[86] Wellington later praised the naval contribution to Bathurst: 'It was effected with a degree of gallantry and skill seldom equaled.'[87] After the allied army had crossed the Ardour, Rear Admiral Penrose, at the request of Wellington, not only forced the entrance of the river Gironde, thereby opening important communications with the army in Bordeaux, but also destroyed all of the enemy armed ships, forts and batteries in the river. The fort at Blaye only avoided a similar fate by hoisting a white flag just as it was about to be stormed by a detachment of the army under the command of the Earl of Dalhousie in concert with naval forces under Penrose.

Given the extent of the support that the navy provided to Wellington's forces it was inevitable that there would be calls for sharing the prize money. After the war Admiral Keith petitioned Bathurst, claiming £150,000 as a share of the prize money in respect of artillery, ordnance and other captured stores through the 'perfect harmony, more ardent zeal and more cordial co-operation'. He also relied upon Wellington's stated position that he never could have maintained his position in the northern provinces of Spain if it had not been for the persevering exertions of the navy.[88] This petition received serious consideration but in end the secretary of state obfuscated. He refused the petition on the grounds that the sum of £800,000 granted by Parliament had been specifically and exclusively awarded to the land forces and could not be used for any other service.[89]

The War with America

From 1812 the Admiralty had been balancing Wellington's demands with the requirements for naval ships off the coast of North America where the navy was to play a major role in bringing America to the negotiating table. The navy established a blockade of the long eastern coast line. Among the blockading ships were numerous transports used as floating barracks for several thousand troops and supply ships for the warehousing of their ordnance, materials and supplies. From these floating barracks a number of amphibious attacks on coastal targets were launched and naval seamen and marines played an active role in them. In particular, seamen and marines participated in various attacks and played a prominent role in the action against Washington in 1814.[90] This was launched in retaliation for the burning of York in Ontario by the Americans. Vice Admiral Sir Alexander Cochrane led the British raid of 4,500 lightly armed, fast-moving troops and marines. They burnt ships in the navy yard and destroyed government buildings including the White House.[91]

86 Burnham, 'British Bridging Operations', 262.
87 Burnham, 'British Bridging Operations', 258.
88 HoCPP,1816 (449), XII.575, Papers relating to the claim of the navy under Admiral Lord Keith, to share the prize money on behalf of the navy co-operating with the Army under the Duke of Wellington, on the coast of Spain and France, May 1816.
89 HoCPP, 1816 (449), XII.575, Bathurst to Admiral Keith, 7 Aug. 1815.
90 Lavery, *Nelson's Navy*, 152.
91 Knight, *Britain Against Napoleon*, 447.

Plate 2. British troops burn the White House in 1814 during the War of 1812

Another raid in early September 1814, which became known as the Battle of Hampden, was described by Captain Robert Barrie of HMS *Sylph* in his report to his commander, Rear Admiral Edward Griffith Colpoys.[92] The latter, having learned that the USS *Adams* was undergoing repairs at Hampden, some 30 miles up the Penobscot River in Maine on the eastern seaboard, ordered Barrie to capture or destroy the enemy vessel. Barrie's force consisted of four vessels – the *Sylph*, the *Dragon*, the *Peruvian* and the transport *Harmony*. His ships received on board a detachment of twenty men of the Royal Artillery with one 5½-inch howitzer, a party of eighty Royal Marines, the flank companies of the 29th, 62nd and 98th Regiments, and a rifle company of the 7th battalion of the 60th Regiment. The whole force was under the command of Lieutenant-Colonel John of the 60th Regiment. They proceeded up the Penobscot in light, variable winds. They had no pilot familiar with this 'most intricate channel' and thick foggy weather slowed the passage. They received intelligence that the enemy had been strongly fortified with militia and that the *Adams*' commander, Captain Charles Morris, had placed nine ship's guns in a battery on a nearby hill and fourteen on the wharf by his

92 TNA, ADM, 1/508, To Rear Admiral Edward Griffith Colpoys, from Captain Robert Barrie of HMS *Sylph* off Bangor, in the Penobscot, 3 Sept. 1814.

crippled ship. On the way they had observed troops moving on the east side of the river in the direction of the nearby hamlet of Brewer. These troops were driven into the woods by a landing party and volleys from the guns on the boats. The enemy suffered one killed and some wounded. The British had no casualties.

At 5.00 pm they arrived off Bald Head Cove about 3 miles below Hampden. John and Barrie landed on the south side of the cove to reconnoitre the ground and obtain intelligence. Having climbed the hill they discovered that the enemy was advantageously posted near the highway leading to Hampden on the north side of the cove. They decided to land 150 men to establish and protect a landing site. The remaining troops were landed on the north side of the cove by 10.00 o'clock but it was found impossible to land the artillery at the same place. The troops bivouacked on the ground but it rained incessantly during the night. At daybreak the fog cleared for about a quarter of an hour, which enabled Barrie to reconnoitre the enemy position by water and identify a landing place for the artillery. The point selected was about two thirds of a mile from Bald Head. The artillery was landed there and was advanced with the troops towards Hampden. The boats under the immediate command of Lieutenant Peddlar of the dragoons advanced in line with the right flank of the army.

Intelligence indicated that the enemy force was 1,400 strong. It was situated in an excellent location on the hill, a good position from which to control both the road and the river. Meanwhile, the 18-pounders were mounted on a wharf and completely commanded the river, which was just less than three cables wide. The land on each side was high and well wooded. A rocket boat under Barrie's immediate direction was advanced a quarter of a mile ahead of the boats. As soon as the boats got within gunshot the enemy opened fire from the hill and the wharf. The British returned fire. The rockets were generally well directed and evidently threw the enemy on the wharf into confusion. In the meantime the troops stormed the hill with the utmost gallantry, and before the boats got within good grapeshot of the wharf battery, the enemy ran off. Morris, realising his force's position, successfully ordered that the guns be spiked and the *Adams* blown up before the British could capture it. John and Barrie left a force in possession of the fort and with the remainder pursued the enemy troops who were then fleeing down the Bangor road. The boats and ships moved up the river alongside the army in pursuit but the enemy was too nimble and most escaped into the woods. When they reached Bangor, the enemy soldiers threw off their military clothing and, as citizens and magistrates, declared an unconditional surrender of the town. About two hours later their commanding officer, Brigadier General Blake, came into the town to deliver himself up as a prisoner. He and 197 other prisoners were admitted on parole. During the operation more than fifteen vessels were captured or destroyed. Losses were minimal – four seamen of the *Dragon* killed and Captain Gill and seven privates slightly wounded. Colonel John praised the good co-operation between the services, the skilful navigation of the ships upriver by the captains and the actions of the marines and seamen attached to artillery.

Off the American coast the navy was also heavily involved in protecting merchant and transport shipping from American privateers. The British captured 278 privateers with 8,974 men and 1,248 American merchant vessels, in

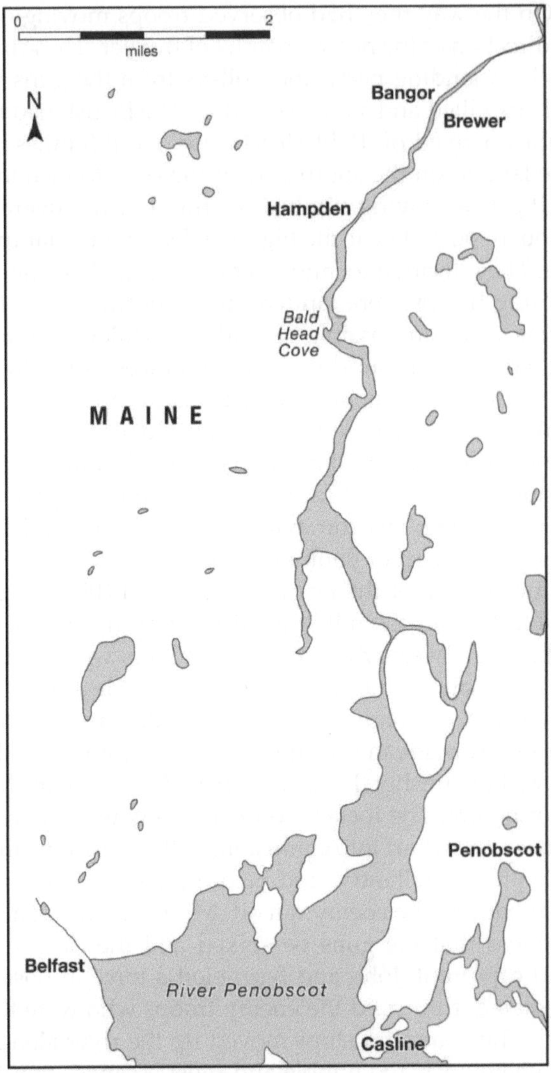

Map 6. USA east coast, Maine

addition to 34 US navy ships. In total 20,000 US seamen were captured. The US navy captured 172 British merchant ships but privateers took many more – 1,441. Fortunately 30 per cent of captured British ships were recaptured before they reached port.[93]

A significant number of American privateers lurked in the waters off the west coast of Britain and in the English Channel. Merchant shipping and transports were easy targets, particularly if they had lost their convoy protection, either deliberately or otherwise. The *Canada* armed transport suffered this fate on 28

[93] Arthur, *How Britain Won the War of 1812*, 216–26.

January 1812 when she was captured by the *Paul Jones*, an American privateer under Commander William Dobson. The *Canada* had 25 guns and a crew of 125; on board was a detachment of 18th Hussars, an infantry brigade and forty-four horses. The *Canada* had sailed from Spithead on 19 January under the convoy led by HMS *Aboukir* bound for Lisbon. She lost contact with the convoy the next day but proceeded on to Lisbon. On 28th at noon had she fell in with the *Paul Jones*, of 16 guns and 120 men, and received a broadside which was returned; the captain of the *Canada* surrendered. The privateers plundered arms and ammunition and threatened to burn the ship. Lieutenant Black, senior officer of the 48th Regiment, signed a bill of exchange for £3,000 in exchange for release of the ship, the crew and the troops on the understanding that 'if the bill is not honoured troops and crew should consider themselves prisoners of war'.[94]

Although there were few significant naval actions after Nelson's final battle at Trafalgar in 1805, the navy continued to play an indispensable role in the defeat of Napoleon. Its early successes against the French and their allied navies, combined with the rigorous blockading of enemy ports, severely restricted their freedom of movement and made it possible for the British government to implement a policy of launching major expeditions relatively free from enemy intervention at sea. The navy, under the system of blockade and convoy protection, secured the seas to allow British international trade to increase significantly. During this period the navy conducted a whole variety of tasks in the Peninsula. Its very presence off the coast forced the French to detach thousands of troops to defend against coastal attack. It manned gunboats and batteries on the Tagus and signal towers along the lines of Torres Vedras. It built bridges, hauled supplies inland, provided muscle-power to implement the landing of heavy artillery, secured communications with London, moved vast amounts of bullion and specie, and facilitated the supply of arms and ammunition to the numerous guerrilla groups that operated within reach of the coast.[95] It is clear to see why, despite his numerous criticisms of its effectiveness, Wellington realised that naval support was essential to keep his army fed and supplied. From 1812 the navy was carrying out this work whilst successfully holding the USA at bay by instituting a blockade along its 2,000-mile east coast in order to prevent it trading internationally, thus seriously impacting its financial stability. It was also attempting to protect British maritime trade and military convoys from the numerous American privateers. So whilst the nature of its operations changed during the course of the war, the navy nevertheless contributed significantly to the ultimate downfall of Napoleon. In a purely military sense the value to the Allied cause of the ability to move men and equipment safely by sea was truly immense.[96]

[94] TNA, ADM, 1/3764, Robert Charnside, Master of the *Canada* transport, to the TB, 13 Feb. 1813.

[95] Hall, *Wellington's Navy*, 233.

[96] Hall, *Wellington's Navy*, 113.

6

Troop Ships: King's Ships or Merchant Transports?

The notion of using naval rather than merchant vessels as troop ships was the subject of considerable debate throughout the wars. Though there were strong arguments that supported this in terms of flexibility, reliability and security, the Admiralty objected vociferously. It was adamant that there were not enough naval vessels that might be converted for such purposes. Furthermore, it claimed that to custom-build a fleet of troop ships would be cost-prohibitive and would take up valuable shipbuilding capacity. The professed insufficiency of ships was questionable. There appears to have been a considerable number of ships in ordinary which might well have been suitable. However, the more realistic obstacle was the limited availability of seamen. There were never enough experienced seamen to fulfil the operational demands of the navy.[1]

The Admiralty's overriding concern related to the precedence of authority when troops were on board naval vessels, and this issue compounded its reluctance to supply troop ships. Indeed there had been open dispute between naval and army officers about whose authority troops on board naval vessels should observe. When, in 1794, Henry Dundas, Secretary of State for War, applied to the Admiralty for three 44-gun ships to transport troops, Lord Chatham, the First Lord of the Admiralty, refused to comply unless it was clearly understood that the soldiers would be subject to naval command whilst on board: 'It was several weeks before this delicate issue of inter-service relations could be sorted out to the Admiralty's satisfaction.'[2]

Then, in 1795, when the navy court martialled an army lieutenant serving as a marine in the Mediterranean fleet for insubordination, the army commander in the Mediterranean refused to recognise the court martial and ordered the lieutenant to rejoin his regiment. The case was referred to London where government law officers eventually endorsed the court martial verdict. This worsened relations between the two services, and to avoid a reoccurrence the Duke of York, Commander-in-Chief of the army, established regulations empowering naval officers to arrest and detain soldiers only until they could be court martialled by the army. When these regulations were released there was great consternation

1 Robert Gardiner, *Frigates of the Napoleonic War* (London, 2000), 88.
2 Duffy, *Soldiers, Sugar and Seapower*, 51.

in the navy, leading to vigorous protests by a number of admirals and captains.[3] The issue was fudged to avoid delays to the Abercromby-Christian expedition which was then being prepared, but the matter continued to simmer and it does not appear to have been resolved in the twenty years that followed. Indeed, Henry Bunbury, then a junior army officer, later recalled that the management of naval troop ships was 'a badly arranged service and equally disagreeable to the officers of the army and of the navy who were always quarrelling'.[4] Even in 1810 the discipline question was raised as an objection to the proposal by Henry Dundas (now Viscount Melville) to use naval vessels as troop ships.[5] Although troops were subsequently conveyed on naval vessels this discipline issue was probably the principal stumbling block to the establishment of a dedicated fleet of naval troop ships.

The demand for troop shipments was not restricted to major campaigns. There were scores of smaller campaigns and almost daily demands to redeploy troops from one place to another, to ship out newly appointed officers and recruits, to relocate troops and to repatriate the sick and wounded. At home, troops were constantly being moved between depots. Where appropriate, sea transport was preferred for this because it was usually cheaper and speedier, and what is more, it considerably reduced the opportunities for desertion. There were constant troop movements to and from the Isle of Wight, Guernsey, Jersey and Ireland.[6] To illustrate this, on one day in April 1800 Henry Dundas, Secretary of State for War, instructed the Transport Board to facilitate the following troop movements: a regiment of 600 men to Halifax, Nova Scotia; two regiments, the Banffshire Fencibles (600 men) and the Second Argyleshire (600 men), from Portsmouth to Gibraltar; two regiments (the 69th and the 60th) plus an additional 100 recruits, in total 1,500 men, from Portsmouth to Jamaica; two battalions of the 24th and 16th Regiments (1,600 men) to be returned from Halifax, Nova Scotia to Portsmouth; horses for the 12th and 26th Light Dragoons to Lisbon from Portsmouth; and 1,722 militia from Jersey and Guernsey to Ireland.[7] By that day's instruction, which was by no means uncommon, the Transport Board was requested to provide transports for 6,622 men plus 500 horses, requiring over 17,000 tons, equivalent to 60 ships.

The Transport Board was frequently obliged to request the Admiralty's support to provide shipping to enable it to meet its obligations. In June 1796, when there were significant demands for tonnage from the Victualling Board to carry provisions to the Leeward Islands and the Mediterranean, Rupert George found it necessary to advise the Secretary of State for War, Henry Dundas, that

3 Duffy, *Soldiers, Sugar and Seapower*, 187. Duffy suggests that the significant extent of the protest by the officers may have encouraged the 'great mutinies of 1797'.

4 Henry Bunbury, *Narratives of Some Passages in the Great War with France from 1799–1810* (London, 1854), 93. Bunbury as Colonel became Military Undersecretary to the Secretary of State for War (Nov. 1809 to Jul. 1816) serving under Lord Liverpool and Earl Bathurst.

5 Melville, *Speech*.

6 Duffy, *Soldiers, Sugar and Seapower*, 345.

7 TNA, WO, 6/257, Henry Dundas to TB, 10 Apr. 1800.

'from enquiries made there is at present no adequate supply to be obtained from the several ports in the United Kingdom'. He asked, 'Can the Admiralty assist with ships of war?'[8] In April 1799 the Board again requested that Henry Dundas apply to the Admiralty requesting that some of its ships on the Irish station be called into service to convey the Irish Militia from Dublin to Guernsey because the Board was aware of the 'great expense and difficulty of procuring the necessary shipping at this time'.[9]

Dundas had long held the view that naval ships should be used to transport troops. In 1795 he had attempted to persuade Spencer's Admiralty of the virtues of this proposal when he wrote:

> I have no objection to a board for the transport service. I believe if proper men are appointed it is a most excellent institution, but I am decidedly of the opinion that if they are not provided with a set of shipping appropriated to the special purpose of transports and of a size to accommodate a considerable number of troops at a time, it is impossible that service can be carried on with any degree of propriety. Indeed, so much am I impressed with that conviction, no consideration on earth would induce me to take charge of any expedition, if the present system of providing transport in any chance way you can was to be continued. I am positive, exclusive of every other advantage, the saving to the public by such an arrangement would be immense. Twenty or twenty five ships of the size of India ships or 44 gun ships would be adequate to all the service of the country, and it would take a volume to point out all the advantages to the public service that would result.[10]

He continued to be strongly opposed to the use of merchant ships, commenting that 'I have long thought it radically erroneous.'[11] A limited number of naval ships were made available for den Helder campaign in 1799, but the lack of success, partially blamed on there being insufficient transports of suitable size to land battalions as a unit, may have strengthened Dundas's argument. Subsequently, forty naval troop ships were made available for the next big expedition to Egypt in 1801, which was also led by Sir Ralph Abercromby.[12]

Each year between 1798 and 1804 naval vessels were used as troop ships in limited numbers, whilst remaining under the command of the Admiralty. These ranged from 28-gun through to 64-gun vessels, manned by reduced naval crews – 64-guns by 250 officers and men (compared with the usual complement of 480–520), 36-guns to 44-guns by 155 (compared with 240–80) and the smaller ships

[8] TNA, ADM, 1/3731, TB to Henry Dundas, 21 Jun. 1796.

[9] TNA, ADM, 108/20/7, TB to Henry Dundas, 30 Apr. 1799.

[10] TNA, WO, 1/84, Henry Dundas to Spencer at the Admiralty, 24 Nov. 1795; Condon, 'Transport Service', 92. This letter was generated by Dundas's frustration with the delays in procurement and preparation of troop transports for the Abercromby-Christian convoy for the West Indies in 1795.

[11] Knight, *Britain Against Napoleon*, 191.

[12] Lloyd (ed.), *The Keith Papers*, vol. 2, 267–9.

by 121 men (compared with 160–220). However, these crews were considerably larger than would have been committed to equivalent-size merchant transports, although the naval vessels would have required gun crews which the merchant ships would not.[13] How the recommended naval crew sizes were determined is unknown. Could they have been pared back further whilst still operating safely, thus allowing more ships to be made available? Certainly Melville believed this to be so when he presented his case to the House of Lords.[14]

When John Schank, a commissioner of the Transport Board, appeared before the Committee on Finance in 1798 he was asked, 'Can any mode be devised for reducing expense and promoting dispatch in the business of transporting men, stores and provisions?' Schank suggested that this could be achieved by having some 40- or 50-gun ships, each capable of carrying 500 to 800 men, fitted as transports.[15] However, when asked to respond to the report, the Transport Board stated that 'we do not believe that a more speedy, cheap or efficacious transmission of stores can be made, than is already practiced'. Note there was no reference here to transmission of men or troops. The statement went on to suggest that the Transport Board did not believe that the navy could spare the required number of ships. The Board concluded, 'However if the ships could be supplied then the next objects of consideration would be their manning and discipline.' Merchant seamen's wages were higher than the equivalent navy pay. If the ships were to be manned by the former the cost would be higher. If, to reduce costs, the ships were manned at the same level as the armed transports which operated under the Transport Board, 'They could only repel privateers but would be no match for stout frigates.' Discipline was perceived to be an even greater problem because if the ships were under civil jurisdiction, military law would not apply. Crews would be able to leave the ship for a man-of-war at any time. Seamen would also be wary that should they be expected to join in hostilities with the enemy there would be no provision for the wounded or incurable in any of the hospitals or charitable provisions for seamen, nor relief for those who fell in action.[16]

Thus the Transport Board was reiterating the Admiralty's position, although this view seems to be contrary to the individual opinions of its commissioners. One of them, James Bowen, had at one stage proposed that the government should engage and man 100 collier brigs as troop transports.[17] Then, when he appeared before the Commission for Revision, he was asked, 'Do you consider that HM's troops would be conveyed with greater convenience in ships of war in commission, rather than in Transports?' He responded that 'Ships of war in commission cannot carry troops conveniently but ships of war of a certain

[13] Sutcliffe, 'Bringing Forward Merchant Shipping', 210–11; N.A.M. Rodger, *The Wooden World* (London, 1988), 348.

[14] Melville, *Speech*, 28.

[15] Committee on Finance (31st), 504.

[16] Committee on Finance, Further Proceedings (18th+31st).

[17] David Bonner-Smith (ed.), *Letters of the Admiral of the Fleet, the Earl of St Vincent*, vol. 2, Navy Records Society (1927), 402.

description properly fitted and manned as the troop ships as they were in the last war are the best ships for conveying troops.'[18]

The matter was regularly discussed in the press. A report in 1800 about a transport with 250 troops on board which had been captured and taken into a French port concluded that 'If any proof were necessary of the advantages of troops being, upon every occasion, conveyed in ships of war, this would furnish an additional one.'[19] Dundas claimed subsequently that he had discussed this proposal with Prime Minister William Pitt shortly before his death in 1806 and he remained convinced that 'If his valuable life had been spared, the suggestions I offered to him would not have been overlooked.'[20]

Army commanders and their troops were generally not enthusiastic about sailing in civilian vessels. Again the press regularly reflected these views. In 1800 the following report was filed:

> We learn today from Torbay, that the troops which sailed in the Armed troop ships from Deal last week expressed their highest satisfaction at the accommodation, convenience and safety, which they experienced in these ships in preference to the former mode of embarking them in small merchant vessels hired as transports.

Apparently they had sailed through a violent gale without detriment or inconvenience, whereas it was claimed that if they had been on merchant ships, 'dispersion, if not more serious disaster would have been the consequence'.[21] However to some, being confined in naval troop ships for any amount of time was pretty unpleasant. When the army was embarking in preparation to leave Marmaris Bay to sail for Egypt in 1801, the Earl of Dalhousie, who in future years would command a division under Wellington in the Peninsula, lamented, 'Again begins another life of misery, going on board is like going to prison.'[22]

By April 1807 when the Board was experiencing difficulty finding enough ships for the Copenhagen expedition, Castlereagh was the Secretary of State for War and Lord Mulgrave presided as First Lord at the Admiralty. The request for naval ships to be used 'in conformity with the custom on former occasions, particularly in the instance of conveying troops to Egypt', was not favourably received. Few ships were forthcoming; the navy claimed to have neither spare men nor ships available.[23] Later that month the Board was asked to ship a number of Royal Artillery troops to Halifax. Because there was insufficient accommodation on the transports that were about to sail, it was obliged to

[18] Commission for Revision (9th), 207. Bowen was referring to 44-gun ships which had the lower deck guns removed.
[19] *English Chronicle*, 15 Apr. 1800.
[20] Melville, *Speech*, 38.
[21] *English Chronicle*, 15 Apr 1800.
[22] Mackesy, *Egypt*, 49.
[23] TNA, ADM, 108/21/56, TB to Castlereagh, 20 Apr. 1807.

request that the Admiralty allow the troops to be carried on the convoy escorts.[24] This time the Admiralty acquiesced.

In 1808 Lieutenant-General Sir John Moore grumbled about the unwieldy nature of the convoys taking his troops to Sweden. He noted that it took seven or eight vessels to carry each battalion and that the resulting convoy numbered over 200 ships. He felt that much larger vessels of at least 1,000 tons were needed to make his command fully effective: 'What can a convoy of 200 small brigs do, commanded by north-country skippers, who will do nothing but what they like themselves?'[25] Moore's contempt for the captains of transports also highlights the antipathy between crews and their military passengers.

The Treasury was, of course, concerned about the considerable expenditure on transports. Spencer Perceval, the Chancellor of the Exchequer (March 1807–April 1810), was particularly exercised in November 1808 when the cost of transports was running at a rate equivalent to almost £3 million a year, far more than the £1,752,000 which had been agreed in that year's naval estimate. He asked the First Lord at the Admiralty, Lord Mulgrave, to consider any practicable means of reducing this figure, but the response was very defensive. Perceval was in favour of custom-building a fleet of troop ships, though he accepted that given the cost and the considerable time this would take, in reality the only way to avoid the use of merchant transports was to convert some existing warships with reduced complements.[26] The first lord's response was very defensive and this stance 'greatly disappointed' Perceval's expectations. He challenged it on several counts. Mulgrave had sought to prove that the cost of using naval ships was higher than the comparative cost of transports but the chancellor claimed that the cost comparison was flawed. The victualling of troops on board had been accounted for in the naval cost but not in the transport cost. Neither had the transport's fitting out costs been considered. If these costs were included, the result swung the argument in favour of the use of naval troop ships. Furthermore, Mulgrave had not acknowledged savings which were more difficult to quantify such as the ability of troop ships to carry regimental baggage in their holds, thus saving the cost of additional transports; the reduction in the cost of delays that invariably arose when a large number of transports had to be accumulated for an expedition; and the saving that would be made as a result of a reduction in the cost of convoy protection.

Perceval also believed that Mulgrave had not reflected the potential benefits that might follow as a result of the recent Spanish uprising, which ought to release some ships of the line that were then blockading the Spanish and Portuguese coast. This could easily free up the 7,000 seamen that would be required to crew a number of troop ships sufficient to convey 20,000–25,000 troops and reduce the transport expense by the equivalent of the cost of 60,000 tons (at 25s per ton per month this would represent a saving of almost £900,000 per annum). He was also of the view that reduced government demand for transports would reduce

[24] TNA, ADM, 1/3751, TB to Admiralty, 7 Apr. 1807.
[25] Hall, *British Strategy*, 43.
[26] Hall, *Wellington's Navy*, 99.

the charter rate of those transports retained, benefitting not only government expenditure but also the general public when these savings filtered through to the cost of all goods that travelled by sea. In addition he referred to the potential benefits of retaining the services of officers and seamen who might crew these troop ships which might also be used as cruisers when not actually transporting troops, and he envisaged that this might afford the possibility of providing a substitute for the Sea Fencibles, 'an establishment now being looked on with disgust and dissatisfaction', which would create further cost savings.[27] This exchange occurred only a few months before the Corunna evacuation in which naval ships and crews proved invaluable both in the effectiveness of the removal of troops and the provision of capacity to convey them out of harm's way. A few months later in July 1809 warships were again used, this time to convey troops to Walcheren. It should be noted that a considerable number of members of parliament were also ship-owners who were deriving financial rewards from the high level of government demand for their ships. This shipping interest was particularly powerful and possibly had some bearing on the rate of introduction of naval troop ships.

Perceval returned to the debate in 1810; by then he was Prime Minister. Like many politicians his perception of merchant transports ships was influenced by the army. He referred to the many delays and accidents that he unfairly perceived to be due to the 'desertion, indiscipline and want of seamanship' of both masters and men in transports. He also repeated the mantra of naval officers who 'complain of the want of skill or discipline from inadequate crews in obeying their orders and signals'.[28] He remained convinced that an Admiralty-controlled body of military troop ships, which, being larger vessels, would be able to carry guns, baggage and ammunition as well as troops and be under naval control their seaworthiness, would be better maintained and their crews would become more experienced in embarkation and disembarkation. He was still motivated by the potential to derive financial savings because armed naval troop ships would be able to protect themselves and would certainly reduce the number of escorts required.[29] It was argued that comparatively fewer naval troop ships would be required for the conveyance of an army – between a half and a third of the number of merchant ships; that the preparation and assembly of expeditions would be achieved faster and more discreetly; and that the superior sailing capabilities and enhanced armaments meant there would be fewer maritime losses and captures of troop ships. This would have increased the possibility of complete regiments arriving simultaneously, overcoming the problems created when troop ships were separated from the convoy.

[27] The Mulgrave Papers, Mulgrave Castle, VII, 19/29. Spencer Perceval, Chancellor of the Exchequer, to Lord Mulgrave, First Lord of the Admiralty, 18 Nov. 1808. Perceval also referred to the possibility that such ships might be required to assist the evacuation of some Spaniards to South America should the uprising fail.

[28] Hall, *British Strategy*, 43.

[29] Melville, *Speech*, 27.

Had it been possible to convert more warships to transports it might have been possible to reduce the considerable time and risk involved in disembarking troops, particularly on hostile shores. Landing troops from a large number of transports took a considerable time, even when the landing was unopposed. Merchant transports could not accommodate complete battalions which consequently had to be divided between a number of transports, each carrying 100 to 200 troops. This invariably meant that complete battalions rarely landed together, particularly if the transports were separated from the fleet due to bad weather. Fewer naval troop ships were required for the same force and they were more likely to arrive at their destination at the same time. The evacuation at Corunna demonstrated that naval ships also provided the most effective means of evacuating an army under fire very quickly; the greater number of boats from naval troop ships could re-embark troops quicker than merchantmen because the size of a warship's crew enabled it to change boat crews and avoid exhaustion. At Corunna it was calculated that the ratio of trips from beach to ship was 10:1 in favour of naval launches.[30] Whereas naval troop ships still had sufficient men on board to manoeuvre the ship if necessary, the absence of boat crews often left merchant ships short-handed.

Prior to the Walcheren campaign in 1809 Sir Home Popham told Castlereagh that transports were 'the greatest clog to every sort of expedition' and that dismantled line-of-battle ships should be used in the forthcoming expedition.[31] The minister had already reviewed a comparison of the costs of naval troop ships against that of hired transports and was aware that they marginally justified the use of naval vessels.[32] Anticipating that the Transport Board was likely to experience difficulty in raising transports for the expedition, he managed to persuade the Admiralty to supply naval vessels. The Admiralty proposed sufficient shipping for over 24,000 men in 129 vessels ranging from small brigs to 74-gun ships, whose lower gun decks would be removed and holds fitted out for horses, as demonstrated in table 6.1. This represented a significant revision of policy: never had so many ships been offered up. Was this an indication that the manning situation was easing? In fact only 17,000 troops were eventually shipped across the North Sea to the Scheldt in naval troop ships.

This development seems to indicate that the earlier opposition from the navy was beginning to evaporate. In March 1810, a few months before Melville's speech, Liverpool, the new Secretary of State for War, had written to the Admiralty:

> In order to obviate in some degree the difficulty and expense attending to the provision of sufficient quantity of Transport tonnage to meet the exigencies of public service I have received His Majesty's pleasure to desire that your lordships will take the necessary steps for immediately providing a number of armed ships well qualified for the transport of troops.[33]

[30] Melville, *Speech*, 10.

[31] Castlereagh, *Correspondence* I, 274, Sir Home Popham to Lord Castlereagh, 13 Jun. 1809.

[32] Castlereagh, *Correspondence* VI, 254, Memorandum respecting the Expense of Transports, 1 Apr. 1809. Author not identified.

[33] TNA, ADM, 1/4213, Liverpool to Admiralty, 23 Mar. 1810.

Table 6.1. Arrangement for conveying troops in men-of-war for the purpose of the Walcheren expedition

	No.	Troops in each	Total no. of troops
74-gun ships	18	550	9,900
Line of battle ships	8	250	2,000
Flutes	9	600	5,400
Frigates	14	250	3,500
Sloops	30	50	1,500
Gun brigs	30	30	900
Vessels from dockyards, victualling establishments, Transport Board etc.	20	50	1,000
Total	129		24,200

Source: Castlereagh, Correspondence *VI.278, 18 May 1809.*

As a result the Admiralty requested the Navy Board to identify which ships were available and appropriate for that service. In 1810 nine former frigates and small ships of the line were sent to replace merchant vessels in the Tagus; although few in number, they had the capacity to carry 3,000 men, twice as many as a similar number of merchant ships would have been able take.[34] The effectiveness of these troop ships was illustrated when two of them, the *Diadem* and the *Regulus,* shipped almost 1,000 troops of the 1st Battalion of the 23rd Foot Regiment from Halifax, Nova Scotia, to Lisbon in thirty-four days in October and November 1810. This compares to the fifty-six days that it had taken most of those troops to reach Halifax by conventional transports in 1808, and on that occasion one of the transports had got separated from the others and had taken even longer to arrive.[35]

Lord Melville's Proposal

In May 1810, Lord Melville finally forced a vote in the House of Lords on the subject of troop transports, although by that time he had been out of office for five years. This was after the Walcheren disaster. By then his proposal had developed well beyond the provision of naval shipping for troops. He proposed fitting out from the ordinary of the navy about 40,000 tons or 30–40 ships, sufficient for the accommodation of 24,000 men, to become a standing fleet permanently garrisoned by troops. He was effectively proposing the creation of a number of

[34] Hall, *British Strategy,* 45.

[35] Hall, *Wellington's Navy,* 100.

floating barracks. These were to be distributed along the enemy's coasts and to act separately or together, with the capability of annihilating the enemy coasting trade. He believed that their presence would cause constant alarm, obliging the enemy to deploy a very great part of its armies to defend the coast against attacks and predatory incursions. Melville's proposal was very similar to Lord Castlereagh's earlier scheme of 1805 to develop a 'disposable force' of about 40,000 men, probably better described as an expeditionary force, together with the appropriate number of transports.[36]

Melville's critics argued that this would be an unacceptable drain on the naval capacity, although he had made a very powerful argument that the navy could well afford to lose the 40,000 tons which he felt was necessary, given that the naval tonnage at the time was 800,000 tons. Indeed he reminded the House that ships had been used for this purpose previously in more difficult circumstances. He was referring to 1800 and 1801 when the number of naval troop ships and armed transports amounted to fifty sail and their tonnage to about 50,000 tons. This was a time when the fleets of France, Spain and Holland combined amounted to about 100 sail of the line. In his view that figure had reduced to seventy-one by 1810.[37]

Melville, with his wealth of experience, was firmly convinced that 'The naval establishment of this country is now upon a scale considerably exceeding that in wisdom, in sound policy and of every principle of economy, it ought to be.'[38] His justification was that Britain had about 105 ships of the line in actual service and another 103 sail of the line in ordinary. In fact he believed that only ninety ships of the line were required to provide effective security for Britain. He estimated that eighty-one would suffice in European waters with ten more for foreign service.[39] He was suggesting, in fact, that if fewer ships of the line remained in commission, there would be sufficient seamen to man a greater number of troop ships, which would have smaller complements. However, at that time he was

[36] Glover, *Peninsular Preparation*, 18.

[37] Melville, *Speech*, 31. His estimate was: 'The Russian fleet, in the Baltic, consists of thirteen sail of the line with several heavy frigates. The Danes have one ship of the line, and the Swedes eleven sail. The Dutch have agreed by a recent treaty, to furnish nine sail of the line to France. These are all the ships they have; and no doubt they will find great difficulty in fitting and manning them for active service. France has, in the Scheldt, ten sail of the line whose crews, though not nearly complete, are (as far as they can go) composed of various nations. She has also two sail of the line at Cherbourg, and, in the ports on the Bay of Biscay, about twelve sail of the line, most of which are at Rochefort. France has also in the Mediterranean, thirteen effective ships.'

[38] Melville, *Speech*, 32.

[39] Melville, *Speech*, 33. His estimate was: 'For the Baltic 20, To watch the motions of the Dutch fleet, And the French fleet in the Scheldt and at Cherbourg 18, for the ports in the Bay of Biscay 12, For the Mediterranean and Black Sea 21, for the service of Lisbon and Cadiz 10. The total for European Service 81. Regarding foreign service: France did not have any ports in the West Indies, so very few ships of the line were required for that station; and very few ships were required in the East Indies, Portuguese America, the Cape of Good Hope and North America. 10 ships should suffice.'

Table 6.2. Estimated annual cost of using a 64-gun naval ship as a troop ship

	Cost per man per month	*Total cost per annum*
Wages for 164 men	£1 17s 0d	
Victuals for 164 men	£2 19s 0d	
Total per man	£4 16s 0d	£10,233 12s 0d
Wear and tear 1,380 tons	£2 6s 0d per ton	£10,315 10s 0d
	Total cost per annum	£20,549 2s 0d
Cost per ton £15	**Number of troops 690**	**Cost per man £30**

Source: Melville Speech, 24.

not able to predict that there would be an increased demand for warships off the American coast in 1812–13. Nor does he seem to have considered the large number of naval ships that were involved in convoy duty, probably correctly because this responsibility would normally have been undertaken by smaller ships.

Melville constructed a persuasive argument to demonstrate that the costs were comparable to chartering the equivalent tonnage from the market. He compared the cost of hire with the cost of a 64-gun ship with an estimated tonnage of 1,380 tons and her usual complement of men of 491 reduced by two-thirds to 164. This was rather less than the 250 men that the Admiralty believed were required to handle a ship that size. He computed the annual cost to be £20,549, as shown in table 6.2. He compared this with the cost of hiring 1,380 tons of transport at 25s per ton per month for twelve months.[40] This would be £20,700. There was thus a marginal difference in favour of troop ships of £150.[41] Nevertheless, it should be noted that at any hire rate less than 25s a ton, which was not paid until April 1807 and had been temporarily reduced in 1810 to 21s per ton, the naval troop ship proposal would have been more expensive. Melville, however, had not estimated the savings highlighted by Perceval which would have strengthened his argument.[42]

Indeed the comparisons that Melville produced were very similar to those found in an unattributed memorandum dated 1 April 1809 in the Castlereagh papers. That document compared the cost of transports to the costs of naval vessels of various sizes including a 74-gun ship with a complement of 250 men, a 64-gun

[40] Melville, *Speech*, 25. The price paid during the previous year.

[41] Melville, *Speech*, 25.

[42] Melville, *Speech*, 27. Regarding the estimated cost, which his opponents said was too low, Melville used the example of the *Inflexible* 64-gun ship (1,386 tons), which ship was employed for 28 months as a troop vessel. His calculation would allow £796 19s per month for wear and tear and ordnance, or £10,360 7s per annum for a 64-gun troop ship. The actual cost appears to have been at the rate of only £771 6s 5d per month, for fitting, wear and tear and ordnance; or £10,027 3s 5d per annum.

ship with a complement of 200 men and a 36-gun frigate with a complement of 125 men (rather than 250 and 155 men respectively, as recommended by the Admiralty). Those calculations also resulted in a small annual saving per ship if the naval vessels had been used, suggesting that the annual cost of shipping 20,000 men in transports would have been £597,000 compared with £448,000 in converted men-of-war. Although it was not included in the calculation, there was recognition that there may well have been some savings of half-pay to sea officers who would have become employed again.[43] Both calculations ignored the flexibility of hiring and discharging transports in line with demand and the initial cost of refitting the ships for this service. When challenged on the basis of his cost estimates, Melville retorted, 'I reject all such calculations; for I hold the life of a British sailor or soldier to be inestimable: ... [I] cannot therefore consent to have this subject considered upon the ground of a trifling saving of expense.'[44]

In the debate that followed Lord Mulgrave, who up to a few months before had been the defensive First Lord of the Admiralty, expressed surprise that Melville had dwelt upon the advantages to be derived from the naval discipline on board ships of war employed for the conveyance of troops. This was despite Melville's having received letters from high-ranking naval officers which strongly expressed 'the great inconvenience that had arisen from it' and after what had happened on board some of the ships previously employed for this purpose. He also disagreed with Melville's cost estimate, citing the 1801 Egyptian expedition where 41,000 tons of troop ships had been employed for the conveyance of 17,000 troops. The apparent cost comparison – although no evidence for it seems to be available and it seems to be overstated – was that in troop ships the cost was £27 per ton and £55 per man; and in hired transports it was not more than £12 per ton and £24 per man.[45] Needless to say, the former first lord rejected the principle of reducing the navy, knowing how difficult it would be to restore it again if it became necessary. In addition, he believed that Melville had miscalculated the number of ships that could be mustered against Britain: it was actually 101 ships. For this reason he believed that it was quite impracticable to derive the tonnage necessary for the transport service from the 147,000 tons then employed in the navy. Mulgrave's logic appears sound if applied to the period up to 1805 when the naval fleet was comparatively well balanced with enemy fleets, except even in 1800 and 1801 between forty and forty-six ships had been converted to troop ships. However, post-Trafalgar there seem to be strong indications that support Melville's theory, given the government's success in preventing the ships of the line of Denmark, Portugal and Russia from falling under French control. Clearly

[43] Castlereagh, *Correspondence* I, 254–6. Unattributed Memorandum respecting the expense of Transports, 1 Apr. 1809.

[44] Melville, *Speech*, 28.

[45] Clearly Mulgrave was assuming the cost of transports at 20s per ton per month (whereas the rate was actually 21s) and the rate at 2 tons per man. The comparative cost of the 64-gun ship based on Dundas's model was £15 per ton and £30 per man per annum rather than the £27 and £55 that he quoted, and taking the actual freight rate his values would be £13 4s and £26 8s. Marginally in favour of using merchant transports.

irritated, Mulgrave described the motion to be an 'unnecessary interference with the executive government'.[46] Melville's proposal was, unsurprisingly, defeated.

Despite this, by June 1810 Liverpool's actions were bearing fruit. There were newspaper reports that 'At all dockyards the weakest ships of war are fitting as troop ships, it being intended to bring this description of force into service as extensively as circumstances will admit, instead of having transports.'[47] In 1814 the Admiralty was asked by the Transport Board whether it was the intention that troop ships able to accommodate 10,000 men were going to be maintained.[48] As preparations were being made for the end of the war, a list was drawn up of thirteen naval vessels to be retained as troop ships capable of carrying 5,000 troops. The cost of conversion of the nine that had not previously been used as troop ships was £28,300, between £2,000 and £5,000 per ship.[49]

East India Company Vessels used as Troop Ships

East India Company ships were occasionally used, their size being not dissimilar from that of some naval vessels, but the Transport Board viewed East India ships as 'not the most eligible in point of expense', only to be hired as an unavoidable necessity.[50] Indeed during the American war the Navy Board did occasionally use East India Company ships to convey troops to India. However, it was reluctant to use such ships because it could not gain complete operational control of them: the company insisted that the vessels 'be discharged from the King's service as the Company's service may render it necessary' and the Navy Board was reluctant to accept this condition.[51]

During his examination by the Committee on Finance in 1798, Commissioner John Schank was asked 'Does it increase or diminish the expense of the service to take up East India men?' He advised the committee that hiring such ships generally meant that the government carried responsibility for high demurrage 'in consequence of the disappointment of wind or the change of service in consequence of the operation of war' and that the liability for delays usually subjected the government to the risk of very significant additional costs.[52] He was no doubt referring to an incident in 1795 when the Board was requested to provide transports urgently to convey troops to the West Indies. Because of other demands there were not enough transports available. Rear Admiral Christian approached the managers of the recently arrived East India fleet but their owners could not be encouraged to accept the charter because of the risk of the vessels missing their next voyage. To persuade them Christian was forced to

46 Hansard, House of Lords, 21 May 1810.

47 *Aberdeen Journal*, 20 June 1810.

48 NMM, ADM, 359/346/195, TB to Admiralty, 18 Aug. 1814.

49 Sutcliffe, 'Bringing Forward Shipping for Government Service', 223, citing NMM, ADM, 359/37a/20, A list of ships to be reserved as troopships.

50 Committee on Finance, Further Proceedings (18th+31st).

51 Syrett, *Shipping and the American War*, 72.

52 Committee on Finance (31st), 503.

agree to pay demurrage at twice the normal rate and this subsequently proved to be very expensive.[53]

From time to time the East India Company was persuaded to pay for transports on behalf of the government but they were not always fit and suitable. In August 1803 the company selected twenty-four ships that it was proposing to charter for the government. When they were inspected by Captain Rains on behalf of the Board he found many of them unfit for the service; some were under the required tonnage whilst others did not have sufficient height between decks.[54] Only half were subsequently taken up. Even these had a further unhelpful restriction on their use beyond a fixed six-month charter. They were required to be discharged at Portsmouth or Deptford at the end of that period otherwise they would be subject to a high demurrage charge.[55] Clearly politics were at play. The Board did not welcome these ships but felt obliged to take them. In 1804 when offered further East India Company ships the Board wrote to the Admiralty that they should be rejected because they were not fitted as transports and were larger vessels than the Board normally hired. The Board explained that it could get ships at 19s per ton per month but these vessels, which had larger crews and were fitted as armed ships, would cost 21s per ton per month. It recommended that all those ships that were not coppered should be rejected.[56] It is not known how many were finally taken up, nor are the politics clear.

The East India Company had traditionally provided homeward cargoes from India or China for ships engaged by the government to take convicts to New South Wales or to be transports to India or Ceylon when homeward cargoes could not normally be guaranteed. However, the practice caused the company great inconvenience, so in October 1806 the Court of Directors declined to continue to guarantee homeward cargoes, stating that it believed that cargoes could, by then, be procured in New South Wales.[57] Thus in May 1807 when Castlereagh requested transports to take 600 men to Ceylon, the Board was forced to respond that it could not engage vessels to convey these troops and the 540 tons of ordnance stores bound for the same island without paying a very high rate unless the Court of Directors could guarantee a return freight for the transport.[58]

[53] TNA, ADM, 108/19, Rupert George to Huskisson, 12 Apr. 1796. This arrangement was made by Admiral Christian. When subsequently advising Dundas of the facts, Rupert George felt obliged to make it clear where the responsibility lay but did stress that 'It may be necessary to further observe that this arrangement was not made by a man or set of men just entered office with improper experience as has been erroneously stated', implying that appropriate judgement had been exercised.

[54] TNA, WO, 1/801, TB to Lord Hobart, 24 Aug. 1803.

[55] TNA, WO, 1/801, TB to Lord Hobart, 24 Aug. 1803.

[56] TNA, ADM, 1/3774/221, TB to William Marsden, 16 Feb. 1804.

[57] TNA, WO, 1/803/195, W. Ramsey, East India Company to TB, 21 Oct. 1806.

[58] TNA, ADM, 108/21/53, TB to Castlereagh, 15 May 1807.

Merchant Transports as Troop Ships

Troops invariably found that long stretches at sea were unpleasant: the ships were often damp and generally overcrowded, there was little to occupy them and the poor conditions encouraged ill health. It has been suggested that 'On board ship private soldiers suffer ten times as much, almost, as before an enemy.'[59]

In 1794 a number of troops conveyed to Barbados from Ireland were reported to be 'very fickle [sickly] on their arrival'. Between 800 and 900 were sick and sent ashore without any provision for their reception in terms of stores, utensils nor medicines.[60] This was not an unusual situation. In September 1794 Sir Jeremiah Fitzpatrick MD was appointed by Dundas as Inspector of Health for the Land Forces, 'with a particular view to their situation when on board Transports'.[61] Fitzpatrick demonstrated early successes but was ultimately side-lined by the vested interests of regimental commanders and the army medical board.[62]

The main specification for troop ships was that there were to be 5 feet between decks and the carrying capacity for each ship for extended voyages was 2 tons per man, but in European waters this was frequently reduced to 1½ tons per man. The Board was very strict about observing these regulations. In 1795 it reacted strongly to a complaint from the army, following sickness on board troop transports, by publishing a chart which showed the actual tonnage appropriated for the reception of troops to the West Indies compared with the tonnage required at the rate of 2 tons per man. This demonstrated clearly that the mandated tonnage was exceeded, although it was marginal in some instances.[63]

In December 1799 Lieutenant Ebenezer Fisher, Agent at Liverpool, hired ships to convey troops to the West Indies but did not follow the rules: 'It is the constant rule of this Board not to appropriate any ships for troops unless they exceed five feet between decks, and to apportion on West Indies voyages two tons per man. The agent at Liverpool has fallen into great error.' However, the Board did recommend that, because of the urgency, it was expedient to proceed with the

59 Mackesy, *Egypt*, 226.

60 IHR *House of Commons Papers 1795–96*, vol. 100, 226, Mr Mallet, Director General of Hospitals, Barbados to Mr Keate, 21 Jan. 1794.

61 IHR *House of Commons Papers 1795–96*, vol. 100, 257, Letter confirming appointment Instructions for conduct of duties. Duffy, *Sugar, Soldiers and Seapower*, 353: 'Dundas found however that the appointment of a trouble shooter, to expedite reform and improvements, seldom broke through the traditional procedures of the ramshackle British administrative system. Fitzpatrick's energetic start soon bogged down in the face of the vested interests of regimental commanders and the army medical board. His tactlessness undermined the support that Dundas was prepared to give him and his powers were soon reduced to those of advice and exhortation though many of his ideas found their way into the Southampton Board's report.'

62 Duffy, *Sugar, Soldiers and Seapower*, 353; O. Macdonagh, *The Inspector General: Sir Jerimiah Fitzpatrick and the Politics of Social Reform 1783–1802* (London, 1981).

63 Sutcliffe, 'Bringing Forward Shipping for Government Service', 227 citing TNA, WO, 1/790/157, TB to Henry Dundas, 23 Jan. 1795.

arrangements and allow the troops to sail in those vessels on that occasion only as the 2 tons per man rule had been observed.[64]

The encouragement of good health and the welfare of troops on passage were taken very seriously. The 2-tons-per-man rule went some way to preventing overcrowding but if disease broke out on board then the limited confines of the ship invariably encouraged its spread. In 1795 William Windham, Secretary at War, issued a set of instructions regulating the conduct of troops on board troop ships which had been recommended by the Board of Military and Medical Gentlemen held on 28 September.[65] The principal objectives of these instructions were to guard against fire and to preserve the health of troops by 'cleanliness, disciplines, regularity of messing and general attention to the treatment and conduct of the soldiers while on board'. To prevent fire a sentry was constantly posted 'at the caboose, or one on each side if the number of soldiers on board were sufficient'. There was to be no fire of any kind without permission, and when fumigating, which was deemed 'highly material', every precaution was to be taken against accidents caused by fire. Materials used for fumigation included 'brimstone with sawdust or brimstone thrown over hot coals, nitre to which a little vitriolic acid is added; common salt, with the same addition as vitriolic acid or gun powder wetted or the heated loggerhead in the pitch pot'.[66] Fumigating between decks was to be undertaken immediately after the cleaning of the berths and again if practicable after each meal once all the dirt or fragments of victuals had been removed by sweeping.

To preserve cleanliness the men were to wash their feet every morning; for this purpose there had to be two tubs of salt water upon the forecastle by 06.00 every day and again in the evening. They had to comb their hair every morning. Twice a week they had to shave, wash thoroughly and put on clean shirts. There was also to be provision for changing their clothing when wet. Other measures included great attention being paid to the cleanliness of the privies; buckets of water had to be thrown down frequently during the day to prevent the soil from sticking to the sides of the ships. Bedding was to be brought up on deck every morning, weather permitting, by 07.00 and to be well aired. Once the bedding was taken up on deck, the men were to proceed in sweeping, scrubbing and scraping the berths and decks. The decks were not to be washed more often than once a week and then only when the weather was dry.

Breakfast would be at 08.00 after the cleaning, followed by a general parade at 09.00 when every man should appear as clean as his situation would allow. At dinner time the officer on duty was to see that the men were at their messes and that their rum was mixed with at least three parts water to one of spirits. He was also to observe any 'neglect in victualling of the troops' and report such events to the commanding officer who would report it to the transport agent.

64 TNA, ADM, 108/20/29, TB to W. Huskisson, 11 Dec. 1799.

65 BL, c.194.a456, *Regulations to be observed by Troops embarked in Transports for service abroad particularly those destined for the West Indies* (War Office, 10 Oct. 1795), 10.

66 BL, c.194.a456. *Regulations to be observed by Troops*, 11.

The greatest care had to be taken to ensure that the copper cooking pots were cleaned before and after use. The troops then had to parade again half an hour before sunset. At sunset the bedding was to be taken below and at 20.00 every man was to be in his berth except the men on watch. There were instructions regarding married couples, who were not allowed to make separate berths all over the ship by hanging blankets which obstructed the circulation of air. They were to have adjoining berths in one part of the ship if possible. The women were obliged to rise at 07.00 hours so that all the partitions could be cleared away for the day.

All the troops aboard were to be divided into three watches. The whole duty watch was to be always on deck except when it was raining heavily. In fine weather every man was to be on deck the whole day and drilled as often as weather permitted to give regular 'bodily exercise'. There were also instructions for the surgeon to examine the men's hands at morning parade when he was to identify the sick and decide whether he needed to isolate an individual or transfer him to the hospital ship to minimise contagion. There were also to be regular divine services to boost morale.[67]

Returning troop ships were stripped of beds and mattresses. These were usually destroyed if the ship had been in the West Indies. Otherwise if it was practical they were washed and repaired.[68] The ships were then fumigated and thoroughly dried out.[69] Occasionally Fitzpatrick ordered that even the ballast was removed and replaced. Additional air holes were cut into the decks to allow a better flow of air. Whilst naval troop ships carried a surgeon, merchant transports did not. For major expeditions troops were embarked and then the ship moved into the roads to allow other vessels into port to take on troops and supplies. They could often be waiting for several weeks while the fleet was assembled before commencing the voyage. This was not pleasant: boredom set in and the damp crowded conditions were ideal for the fermentation of disease. Also, men were frequently detained on board at the end of a voyage because land-based camps were full. This must have been particularly damaging to morale.

Although the health issues of troops on transports improved during the Revolutionary war as the focus moved away from the West Indies, this was still a very important consideration for the Transport Board. In 1803 Captain Charles Patton, Agent at Portsmouth, wrote to the Transport Board that the *Ceres* (455 tons) 'from continuing at Spithead in this damp and rainy weather, has become sickly'. Even though the *Ceres* was a very high and roomy ship, and no more crowded than transports commonly were for short trips, sickness broke out and

[67] BL, c.194.a456, *Regulations to be observed by Troops*, 12.

[68] *Hampshire Telegraph*, 30 Mar. 1807, request for tenders to 'wash and repair bedding … The materials for mending the same to be supplied by the contractor; and the foul bedding to be taken away', placed by Charles Patton, Resident Agent, Portsmouth.

[69] TNA, ADM, 108/32, TB Minute, 12 Nov. 1794. Sir Jeremiah Fitzpatrick condemns the airing machines that are supposed to be used in ships and recommends those of his own invention, which Mr Winlow of Margaret Street will make for £6 6s each. He recommends an improvement to his own invention which Mr Winlow will add without charge to some machines now lying unused at Portsmouth.

two men died. The inspector advised that any newly infected person should be removed to the military hospital. The appointed escort, the frigate *Apollo*, had been delayed at Spithead, so Patton requested an additional transport and the appointment of a new escort. In the covering letter to the Admiralty the Transport Board said that it had ordered the agent to find another vessel for the reception of the sick and requested that a new convoy escort be urgently appointed.[70]

The owners of merchant vessels that were to be used as troop ships were expected to provide, at their own cost and in sufficient quantities, the coppers or furnaces for boiling and dressing the provisions for the troops, together with cans and pumps for serving them with beer and water as well as platters, spoons, candles and lanterns.[71] Troops were to be victualled in accordance with the standing instructions, which were equivalent to two-thirds of the seaman's allowance. Women were to be victualled at three-quarters of a soldier's allowance and children at half a woman's allowance. Masters were obliged to fix to the mainmast an explanatory notice of the victualling arrangements for the troops, as illustrated in table 6.3, so that the troops were well informed of the rations to which they were entitled and to ensure that masters complied with the rules. There were also specific instructions about the daily allowance for horses. In addition masters were required to keep a good record of when any of the soldiers deserted, died or were absent from the ship.

Accommodation arrangements on board troop transports were a frequent source of friction between the senior army officer on board and the transport agent when they were both allocated to the same ship. The agents afloat took priority. The Articles of General Instructions to Agents of 1803 were revised with the provision that in all cases where there was no agent embarked on board a transport the starboard stateroom should be allotted to the officer of the troops on board if he was a commissioned officer, but when a transport agent was on board the stateroom was to be allotted to the agent. There is much evidence in correspondence about the ill-will that was generated by this.

Quite often soldiers' cabins on transports were broken or destroyed during the passage, to the extent that they required considerable repairs when ships returned to port. Recognising this, the Board amended the Articles of General Instructions to Agents, directing that when any troops were embarked on transports the agent should apply to the officer commanding them to appoint a sergeant to take responsibility for the cabins during the time the troops remained on board. This procedure had previously been adopted in shore barracks in order minimise damage.[72] One of the arguments in favour of the use of troop ships was the degree of comfort in terms of accommodation which officers and men enjoyed on the voyage compared with that which they experienced in merchant transports.[73]

70 TNA, ADM, 1/3744, Captain Patton, Agent at Portsmouth, to TB, 18 Dec. 1803.
71 Committee on Finance (18th), 202, Form of General Charter Party.
72 TNA, ADM, 106/3096, handwritten update to printed *Articles of General Instructions to Agents*, dated 1803.
73 Melville, *Speech*, 7.

Table 6.3. Rules to be observed in victualling the troops when embarked: six soldiers' allowance for every day of the week

Day	*Bread pounds*	*Beer gallons: ½ pints of spirits or pints of wine*	*Beef pieces of 8 pounds*	*Pork pieces of 4 pounds*	*Pease pints*	*Oatmeal pints*	*Butter pounds*	*Cheese pounds*	*Vinegar*
Sunday	4	4		1	2				One quart in a week
Monday	4	4				4	½	1	
Tuesday	4	4	1 or 6 pounds of flour, ½ pound of suet and 1 pound of raisins						
Wednesday	4	4			2	4	½	1	
Thursday	4	4		1	2				
Friday	4	4			2	4	½	1	
Saturday	4	4	1 or as above						

In cases of substitution of one article for another the following proportions are to be observed:
When flour, suet and raisins are put on board, they are to be served in equal proportions with beef, viz. one half in beef, the other half in flour, suet and raisins, on each beef day.
Four pounds of Flour, or three Pounds of Flour, with half a pound of raisins or a quarter pound of currents, and a quarter pound of Suet are equal to four pounds of Beef or two pounds of Pork with Pease.
Half a pound of Rice is equal to a pint of Oatmeal, half a pound of sugar is equal to half a pound of butter, or two pounds of cheese; that is, a pint of Oil is equal to the allowance of Butter and Cheese.
One pound of fresh Beef is equal to one pound of Salt Beef, and one pound and a half of fresh Beef is equal to One pound of Pork.
A pint of Wine or half a pint of Brandy, Rum or Arrack is equal to a gallon of Beer

Source: Commission for Revision (13th), 220.

Agents were frequently called upon to provide accommodation for females on board. Their instructions specified six women to every 100 men. These were sometimes overruled by specific instructions like these issued by Dundas in 1801: 'No women should be allowed to embark with those troops belonging to the regiments actually serving in the field with Sir Ralph Abercromby, but those intended to join any of the regiments actually serving in any of the garrisons in the Mediterranean may take with them the proportion of women allowed under regulations.'[74]

Generally relationships between merchant ship masters and their crew and the infantry on board were workmanlike or even good. However, given the number of troops conveyed, often in crowded, cramped, uncomfortable, frequently damp circumstances, it is quite natural to expect that there were many expressions of discontent. Most of these were resolved by the agents accompanying the transports. Similarly the agents were able to work amicably with the army. Occasionally, however, there were local outbursts of frustration which escalated unnecessarily. In one instance in 1813 Bathurst received a complaint from Colonel Ferreno (serving under Sir Thomas Graham) about stores, particularly blankets and camping equipment, which were being shipped for troops in Holland. On receipt of the application of the storekeeper-general on 23 November, the Board had immediately ordered its agent at Ramsgate to arrange for these stores to be shipped with the troops who were under orders for embarkation from there to Holland. The transports allotted for that service had been detained at Portsmouth by contrary winds. Therefore seven small transports which were lying at Dover ready to receive troops were ordered to proceed to Ramsgate to transport the supplies. After arriving they were diverted, on Graham's instructions, to convey the troops to the Downs where they were transferred to the ships of war. This prevented the stores from being loaded until the 13th. The store-ships finally sailed from Ramsgate on 17 December.[75] Graham's intervention rather undermined Ferreno's complaint.

Another instance arose in May 1810 when seven officers and 134 men under orders for Canada were left behind when the convoy sailed from Cowes. The agent was accused of not allowing sufficient time for embarkation of the troops on the *Sanderson* transport. The Board defended the actions of Lieutenant Parker, the Transport Agent, who had advised Brigadier General Taylor, the complainant, on 15 May that the *Sanderson* had been engaged to ship the troops, giving him twelve days' notice to prepare for embarkation. However, the troops were not ready to be embarked on the *Sanderson* until 5 June, by which time the convoy had sailed. The Board pointed out that the timing of the departure of convoys was very much in the hands of the navy.[76]

[74] TNA, WO, 1/801, Dundas to TB, 16 Mar. 1801.

[75] TNA, ADM, 108/ 24/100, TB to Earl Bathurst, 12 Jan. 1814.

[76] TNA, ADM, 108/21/240, TB to Liverpool, 7 Jun. 1810.

Plate 3. Hussars Embarking at Deptford, 1793, by William Anderson.

Horses at Sea

Horses were usually loaded and unloaded by hoisting them in slings; with practice, about 30–40 horses could be loaded in an hour.[77] As with troop ships, horse ships were loaded and then moved out into holding areas until the remainder of the fleet was loaded and the weather was suitable for sailing. In 1800 the Transport Board advised Dundas:

> The artillery and other horses for the expedition under command of Gen Sir Charles Stuart having now been some time embarked on board the cavalry ships at Portsmouth; we request that you will be pleased to inform us, if it is intended that they shall be landed or remain on board, as we apprehend some inconvenience from their being suffered to remain long embarked, previous to their departure from Spithead.[78]

It was well known that such delays had a detrimental effect on the horses. It is not clear that this issue was ever fully addressed: such delays occurred frequently throughout the wars.

[77] Lieutenant-Colonel George Armand Furse, *Military Transport* (London, 1882), 172.
[78] TNA, ADM, 108/20/40, TB to Henry Dundas, 28 Apr. 1800.

During passage the daily allowance for each horse was 10 pounds of hay, 8 pounds of oats, a small quantity of bran and 6 gallons of water.[79] Under the terms of the charter party, owners were obliged to provide the proper number of pumps and buckets for drawing and serving water for the horses.[80] Horse ships required a considerable volume of forage. Masters of horse ships were expected to sign receipts for quantities taken on board and to keep an exact daily record of the usage. Following a stock-take at the end of the voyage, any deficiency was charged against the ships.[81]

There appears to be little surviving information about the construction of horse stalls in the period, but in an 1868 publication there are good illustrations of constructions which are likely to have been very similar to those of the day.[82] There were numerous factors to be taken into consideration. Horses varied in size; those that hauled wagon trains were much bulkier than cavalry horses. Horses needed space but not so much that they could fall over in rough weather. Some of the bars in the construction had to be rounded to avoid the horses being injured during sudden movement. In certain strong weather conditions the horses would be suspended in slings to avoid dramatic movements. There also had to be strict guidelines to ensure that the horses had sufficient ventilation and that the stalls were cleaned out regularly.

Merchant ships, supported from time to time by naval vessels, transported hundreds of thousands of troops and thousands of horses to join major expeditions and subsequently repatriated the survivors. In addition, they supported numerous minor operations and general troop movements which occurred with high frequency. In the debate about whether naval vessels should be used to a greater extent for this purpose, the political and professional view was that there were not enough to meet the need. There was still great uncertainty about how long the war might last, about the changing allegiances of the nation states and about the development of the French navy.

Commissioner Bowen's assessment that naval vessels were the preferred method of shipping troops suggests that although some naval ships were used, an opportunity was missed by not providing more to be used as troop ships.[83] That forty-six ships were available in 1800 at a time when the navy's capacity was lower than in any subsequent year of the war (105 ships of the line in commission out of a total of 469 vessels) possibly undermined the navy's argument that there were not enough ships. This was particularly true post-Trafalgar when significant numbers of suitable ships were placed in ordinary, and even the question of the availability of sufficient experienced seamen was

[79] Commission for Revision (13th), *Regulations to be observed by all Masters of Ships and Vessels employed and hired in HM's Transport Service.*

[80] Committee on Finance (18th), Form of General Charter Party.

[81] Commission for Revision (13th), *Regulations to be observed by all Masters of Ships and Vessels employed and hired in HM's Transport Service.*

[82] Furse, *Military Transport*, 159–77.

[83] Commission for Revision (9th), 207.

challenged from time to time by politicians including Spencer Perceval and Lord Melville. The latter proposed that what was required was a change in the profile of the fleet so there were fewer large ships of the line, which would have released seaman to man troop ships that could also have become cruisers.[84] However, the question of the accountability of troops on board naval vessels was never resolved and consequent disputes may have negated all the other benefits which might have arisen otherwise from this. It is difficult to anticipate whether the use of more naval vessels as troop ships would have had a material impact on the length of the war. This seems unlikely, but maybe more merchant ships were diverted from trade than was really necessary. Yet it is important to reflect on the high number of troop passages and recognise that, despite the discomforts and the close proximity of troops and crews, the majority of passages were successful and passed with surprisingly low levels of disaffection.

[84] James, *Naval History*, vol. 2, 402. Melville, *Speech*, 27. Mulgrave Papers, VII, 19/22, Perceval to Mulgrave, 18 Nov. 1808.

7

Castlereagh's European Expeditions, 1805–08

Viscount Castlereagh was Secretary of State for War between July 1805 and September 1809, except for the fourteen months of the Ministry of All the Talents administration, between January 1806 and March 1807, during which time the office was occupied by William Windham. Shortly after Castlereagh's first appointment in 1805, and prior to Trafalgar, the French military camps at Boulogne were broken up and the troops marched eastwards, removing the immediate threat of invasion. This was a period of great uncertainty regarding the loyalties of the states of Europe, but on 9 October the Third Coalition was established, uniting Britain, Austria and Russia. Prussia refused to join, preferring instead to form an alliance with France in exchange for the possession of Hanover. Napoleon reneged on the deal so no alliance was concluded, but France did transfer some of its force stationed in Hanover to Bavaria. In light of this, Britain prepared to send a light corps of about 10,000 men, by way of the Elbe, to reoccupy Hanover.[1]

Whilst planning this campaign, Castlereagh had to resolve a conundrum that immensely frustrated army commanders: the consistent shortage of horses for their cavalry battalions, for the Commissariat, which required the animals for hauling supply wagons, and for the artillery divisions, to haul the heavy guns. This was a perennial challenge because of the volume of tonnage required for transporting each horse and rider, for forage and supplies, and for artillery equipment and wagons. There was a shortage of suitable ships and the costs were significant: £8–£10 per horse per month compared to troop ships at just under £2 per man per month.[2]

The Expedition to Northern Germany, 1805

Lieutenant-General George Don was appointed to lead the vanguard of 10,000 men, including 5,000 troops of the German Hanoverian legions who were then

1 Castlereagh, *Correspondence* I, 3–5. For further explanation of the politics behind this strategy see Hall, *British Strategy*, 118–24; Harding, 'Expeditionary Armies and Naval Power'.

2 This was the case when the charter rate was 19s per ton. Then later in 1807–09 and again in 1813–14 when the rate increased to 25s per ton per month these costs increased to £12 to £15 per horse per month and £2 10s per man.

in Ireland. Admiral Lord Keith was the naval commander while Lieutenant-General Lord Cathcart was appointed Commander-in-Chief of the army. The scope of the operation matured into a quest to drive the French from Holland; the 10,000 men from England would be joined by a further 10,000 and also by 20,000 Russians, 10,000 Swedes and 26,000 Danes to create a force of 70,000–80,000, possibly even 100,000 men.[3] A vast fleet of transports of over 305,000 tons was required. This was more than the total tonnage chartered at any other time during the wars. Almost 200,000 tons of this was for horses.[4]

When Castlereagh assumed office there were 322 chartered transports of 88,835 tons, of which 102 were in the Mediterranean, 47 in the West Indies and another 49 at Cork preparing to embark more troops for the West Indies. Twenty-six transports were provisioning the fleet while the remainder, just fewer than 100 transports, were at home, unallocated to any specific service. Early in August 1805 the Transport Board was instructed to procure 18,000 tons fit for distant service; half was to be sent to Cork and the other half to Portsmouth. In addition transports for 10,000 men, provisioned for six months, were ordered to be assembled in the Downs. Towards the end of August Castlereagh increased the tonnage requirement by 15,000 tons for infantry, cavalry ships for 700 horses, and 3,000 tons of store ships. He also demanded that the Board report to him regularly on the progress of engaging and fitting these ships.[5] The total demand was almost 60,000 tons, to be sent to Cork, to Portsmouth and to the River Thames in equal proportions. They were to embark 2,000 cavalry men and horses, 1,500 artillery and associated horses and equipment, and 20,000 infantry men. Castlereagh wanted the total tonnage to be brought forward before the end of December 1805, which allowed a realistic sixteen weeks to prepare. Fifty-thousand tons were to be provisioned for two months and 10,000 tons for six months.[6] To achieve this, three-month ships were chartered by the score. Captain Stephen Rains, the Resident Agent at Deptford, was responsible for organising the surveying of the ships, preparing them for service and then transferring them to the Nore where another agent, Lieutenant William Braithwaite, monitored their arrival, reviewed their preparedness, remedied any shortfalls and hastened them on to the Downs. There the transports were assembled to await the embarkation of troops and supplies, the formation of the convoy and the arrival of the convoy escort. Captain John Culverhouse was the agent at the Downs but, due to the level of activity, he was soon joined by three more. Meanwhile, hammocks and beds were in great demand: an order for 3,000 to be delivered to Deptford was placed on 3 September, followed by another order for 5,000 two days later; similar quantities were sent to Portsmouth.

3 Castlereagh, *Correspondence* VI, 8, Memorandum for the Consideration of the Cabinet, Sept. 1805.

4 Castlereagh, *Correspondence* VI, 86, Memorandum, 21 Dec. 1805.

5 TNA, WO, 6/156/321, Castlereagh to TB, 29 Aug. 1805.

6 Castlereagh, *Correspondence* I, 17, Memorandum relative to the Provision of Transports, 16 Oct. 1805.

Table 7.1. Scheme of embarkation for 10,000 troops at 1½ tons per man in transports victualled to six weeks (Transport Office, 9 September 1805)

	No. of ships	Tonnage	
In the Downs	38	9,310	
On passage to the Downs	14	2,926	
Fitting at Deptford	7	1,663	
Ordered from Portsmouth to the Downs	4	1,118	
Total	**63**	**15,017**	
Cavalry ships foraged to 8 weeks	Ships	Tonnage	Horses
In the Downs	19	4,899	596
Fitting at Deptford	3	899	102
Ordered from Plymouth and Cork to the Downs	2	562	66
Total	**24**	**6,360**	**764**
3,000 tons for ordnance and victualling stores	Ships	Tonnage	
Ordered to Deptford	8	1,876	
Cleaning the bottoms at Portsmouth afterwards to proceed to Deptford	8	1,299	
Total	**16**	**3,175**	

Source: TNA, WO, 1/802.

By 9 September the Board was able to advise Castlereagh that preparation of transports for 10,000 troops which were required in the Downs was well advanced. These transports comprised 103 vessels, including 63 infantry ships, 24 horse ships for the cavalry and 16 for ordnance and provisions, as demonstrated in table 7.1. Anticipating that the Transport Board would have some difficulty procuring all the ships required, Castlereagh asked Lord Keith 'What number of infantry could you transport to the coast of Holland in the King's ships under your command, should that course of action be necessary?' Keith estimated that his ships would be able to receive about 14,000 men.[7] By December the proposed British and German force had increased to 65,090 troops, 1,268 Commissariat staff and 1,000 hospital staff corps (in total 67,358 men), 17,000 horses and all the attendant stores, supplies and forage. By 23 December almost 74,000 tons were ready and provisioned; a further 101,000 tons had been hired and were awaiting provisions. Nearly all were three-month ships hired since early August. This was a significant achievement but a task made easier when availability improved as the summer sailing season drew to a close. The remaining 126,000 tons of horse ships were still to be located.[8]

In recognition of this difficulty Castlereagh proposed that 3,000 draught horses should be procured on the Continent and 4,000 horses for the light dragoons,

[7] Castlereagh, *Correspondence* I, 26, Memorandum relative to the Projected Expedition, for the Consideration of the Cabinet, 21 Oct. 1805.

[8] Castlereagh, *Correspondence* I, 85, Memorandum by Castlereagh, 23 Dec. 1805.

the Commissariat horses and some of the artillery horses should be transported ahead of the main contingent so that the transports which delivered them could return and prepare for a second voyage. That would leave 10,000 horses to be ferried to the river Weser with the main body of troops. The transports should be able to make two trips over a five-week period, thus reducing the requirement for horse ships to about 50,000 tons, equivalent to approximately 150–200 ships.[9] This would reduce the cost by £190,000, but despite this the bill for the transports would still be £422,000.[10] The relatively short distance made such plans practical, assuming the weather conditions were conducive, but as the expedition would sail between November and January, this could not be guaranteed. The British government was once again reacting to events in Europe. The timing was far from ideal and consequently there was significant risk of bad weather in the North Sea, which could seriously jeopardise the safe passage of the expeditionary force. Castlereagh was obliged to take this risk, one that he had limited opportunity to mitigate. However, the far greater challenge to the success of the venture was in maintaining the goodwill and co-operation of the allies, particularly the Russians.[11]

The vanguard of 10,883 troops embarked in the Downs in mid-October where they were joined by the 5,000 Hanoverian troops from Cork; contrary winds prevented them from sailing until 5 November and they arrived at the Elbe on the 17th, having been on board the transports for more than a month. In this convoy – in addition to the 83 troop transports with 11,272 men, 465 women, 228 children and 676 horses – were two ships with arms and hay for the King's German Legion and three army victuallers.[12] Captain Culverhouse, the Principal Agent, sailed with Agents Lieutenant Mears and Lieutenant Ross who were responsible for the ships carrying the British contingent, whilst Agents Captain Thomas Withers and Lieutenant John Curtis had responsibility for the ships carrying the German troops.

Ships were still being brought forward at Deptford until 23 October 1805 when the Board instructed the agent, Captain Stephen Rains, that 'No more ships for either troops or stores are required for Germany but you should continue to take up such ships as are offered giving cavalry the preference.'[13] On 30 October transports loaded with 6-pounder guns left Deptford for the Downs. At the

9 Castlereagh, *Correspondence* I, 86–7, Memorandum by Castlereagh, 23 Dec. 1805.

10 The horses originally required 170,000 tons at 19s to 20s per ton per month for up to two months at a cost of £340,000, whereas this plan only required 50,000 tons, making three voyages possibly over three months at a cost of £150,000, saving £340,000 less £150,000 = £190,000. Thus the total cost: original tonnage of 306,000 tons less horse ships at 170,000 tons = 136,000 tons; for two months this tonnage cost £272,000 and the horse ships cost £150,000. Total cost £422,000. However, many of these ships were still in service when William Windham took over as secretary of state for war in February 1806, so the actual cost of these vessels may well have been 50 per cent more.

11 Harding, 'Expeditionary Armies and Naval Power', 70–1.

12 Sutcliffe, 'Bringing Forward Shipping for Government Service', 244.

13 TNA, ADM, 1/80, TB Minute, TB to Rains at Deptford, 23 Oct. 1805.

same time ammunition, camping equipment and supplies were being loaded in the River and at Portsmouth. All of these ships would be sent to the Downs as soon as possible for onward passage to join Lieutenant-General Don's combined force.[14] The frenetic activity at Deptford was being supervised by Rains, who was a very experienced agent. Unfortunately, however, he became embroiled in a controversy fomented by a disgruntled contractor, who accused him of corruption. He felt obliged to resign his post until he had cleared his name. After a review the Board dismissed the claim and appointed him to be the agent at Ramsgate. He was succeeded at Deptford without delay by Captain James Young. The pace of activity at Deptford was barely disrupted.[15]

To minimise congestion in Ramsgate and speed up the embarkation process transports were ordered to the Downs where troops were then ferried out to them. Cavalry and store ships, however, were ordered into the harbour to embark horses and heavy equipment from the quay. By 12 November there were 32 infantry ships, cavalry ships for 2,018 horses and 48 army victuallers and store ships in the Downs. Even these were insufficient; more were required, particularly horse ships, obliging the Duke of York to insist that priority be given to transports for heavy horses for artillery and wagon trains. The cavalry would have to await the return of horse ships that had already sailed.[16] On 25 November, eleven cavalry transports for 302 horses arrived from the River; six of them were ordered to Ramsgate where another twelve cavalry ships for 414 horses had just arrived from Spithead.[17] The transport commissioners were updated about the preparations and movement of transports by the daily returns prepared by the agents at the various ports. This allowed them to monitor the process remotely and to react appropriately; for instance, they ordered ships to be diverted to Northfleet to receive 700 heavy horses of the German legion and two troops of wagon train to relieve the congestion at Ramsgate.[18] A second fleet of transports with cavalry and artillery sailed for Germany on 27 November carrying 978 men, 19 women, 14 children and 989 horses, accompanied by 25 store ships carrying 4,915 tons of ordnance and camping equipment. Three transport agents sailed with the fleet.

As might have been predicted, fate then played its hand. A third force of nine infantry battalions and one cavalry regiment sailed on 10 December. Tragedy struck this force. Severe gales caused the loss of eight transports at the cost of the lives of 664 seamen and troops, although 1,552 were rescued.[19] The surviving vessels were forced back to English ports from where they sailed again on 22 December.[20] Culverhouse sent regular reports of arrivals: 'every effort being

14 Castlereagh, *Correspondence* I, 35.

15 TNA, ADM, 1/80, TB Minute, 23 Oct. 1805.

16 Castlereagh, *Correspondence* I, 43.

17 TNA, WO, 1/802.

18 TNA, ADM, 1/80, TB Minute, 13 Nov. 1805.

19 Castlereagh, *Correspondence* I, 125, List of Eight Transports employed on the Continental Expedition which have been wrecked, 15 Jan. 1806.

20 Castlereagh, *Correspondence* I, 37, Chronological Memorandum, undated.

made to send them back but the wind is very adverse'.[21] However, many of the vessels had been badly damaged.[22]

By early January 1806 there were roughly 25,000 troops in Hanover. All the troop transports had been provisioned for two months. In addition, more supplies were shipped from England: salt provisions and biscuits for 15,000 men for five months, 12,000 bags of meal to supply 50,000 men for three months and 12,000 quarters of oats, equal to the supplies for 12,000 horses for one month. It was estimated that these provisions would supply the whole of the 25,000-strong force for four months, in addition to such supplies as might be procured locally. In anticipation of winter conditions, supplies of flannel waistcoats, drawers, blankets, 30,000 pairs of shoes and additional hospital and camping equipment were loaded into transports and dispatched to the Downs. Some 3,214 horses had been transported; more were waiting for transports to become available. However, Lieutenant-General Don was not experiencing any great difficulty in procuring wagons and draft horses locally for the transporting of supplies from the place of debarkation to the magazines and then onwards in support of the army.[23] Meanwhile transports with supplies had to remain fully laden at anchor where, despite naval protection, they were harassed and attacked by small Dutch gunships

Whilst the military preparations proceeded, the political landscape changed; the Austro-Russian army was defeated at Austerlitz on 2 December. The Austrians sought an armistice with France and dropped out of the coalition and the Russians were in full retreat.[24] The expedition was doomed without allied support, which was rapidly diminishing. There was a growing clamour for the British troops to be brought home. Further troop shipments were deferred and planning of the withdrawal commenced at the end of December. Empty transports had been retained for this purpose; only additional horse ships were required to complete the withdrawal. Meanwhile the weather conditions deteriorated. In early January Captain Withers, the Agent in the Elbe, told the Board he had been advised by the harbourmaster that it would be advisable to haul the transports ashore so they would be out of danger from ice. He said he would ensure they were not hauled out so far that they were prevented from getting off again when the ice broke up.[25]

The politicians prevaricated, delaying the departure until early February 1806. Eventually the troops returned home safely without firing a shot.[26] Of course, whilst this dramatic build-up of ships for the expedition was in progress, there were many other more routine demands for transports. To facilitate these demands plus those of the expedition the fleet was doubled in size between 1

[21] TNA, ADM, 1/80, TB Minute, 18 Dec. 1805.
[22] TNA, ADM, 1/80, TB Minute, 26 Nov. 1805.
[23] Castlereagh, *Correspondence* I, 73–4.
[24] Hall, *British Strategy*, 121.
[25] TNA, WO, 1/802, Lieutenant Withers, Agent in the Elbe, to TB, 3 Jan. 1806.
[26] Hall, *British Strategy*, 122.

September and 1 November, rising from 506 ships of 130,655 tons to 1,028 ships of 261,240 tons; the average tonnage per ship was 254 tons.[27]

Preparations for an expedition of this magnitude demanded a great effort from contractors and manufacturers to increase production of munitions, clothing, camping equipment and other essential supplies and by the agents of the Transport, Victualling and Ordnance Boards to co-ordinate these activities to ensure that the transports were prepared and loaded quickly, thus minimising the turnaround time and reducing congestion in harbours and at the rendezvous points. The size of this particular expedition grew to the point where it was too large to be effectively co-ordinated in terms of procuring supplies and preparing sufficient transports and troops.[28] Every service was operating at its capacity. In addition Castlereagh was taking risks, relying on good weather in the depth of winter to allow repeat voyages to deliver troops and horses without incurring delays. Duffy reached a similar conclusion in relation to the 1795–96 West Indies expedition which was 'too logistically ambitious' and suffered from 'a succession of unkindly weather accidents'.[29] Glover suggests that, despite some demonstrable successes, the Anglo-Russian den Helder campaign of 1799 also suffered from over-ambition and lack of planning; the landings from the 296 transports were slow and chaotic and the anticipated Dutch resistance against the French failed to materialise.[30] The Walcheren campaign of 1809, the largest expedition of the wars, was also a disaster. Its objective was to eliminate the continuing threat to Britain of an invasion launched from Holland by capturing and destroying the heavily fortified dockyard at Antwerp, 60 miles up the river Scheldt, in the process destroying the warships under construction there. Weak leadership, poor inter-service co-operation, the speed of the arrival of French reinforcements and then rampant marsh fever all contributed to the failure.

The Collapse of the Ministry of All the Talents

In March 1807, the ageing Duke of Portland was appointed to lead a new Tory administration.[31] The Ministry of All the Talents led by Lord Grenville had collapsed over the matter of Catholic emancipation, the thorny issue that had brought down the Pitt administration in 1801. Grenville's administration had become embroiled in South America and had not been particularly active in Europe. However, in reaction to the shifts in the European political situation,

[27] Castlereagh, *Correspondence* I, 97.

[28] Harding, 'Expeditionary Armies and Naval Power', 75. Harding's assessment is that the operation was smoothly organised but had transport problems.

[29] Duffy, *Soldiers, Sugar and Seapower*, 374.

[30] Glover, *Peninsular Preparation*, 24–6.

[31] The new government included: Canning as Foreign Secretary and his rival Castlereagh as Secretary of State for War for the second time. The new First Lord at the Admiralty was Lord Mulgrave and the Home Secretary was the 2nd Earl of Liverpool (the former Lord Hawkesbury). The Master-General of the Ordnance was the Earl of Chatham.

the new government brought the European dimension back into focus. Within months the emphasis of the war shifted from South America back to mainland Europe, in particular to Denmark, the Spanish Peninsula and Holland.

When Castlereagh left office for the first time, in January 1806, the number of transports was inflated by the ships hired for the German expedition. In total there were 630 transports of 157,222 tons. However, by the time he was reappointed as secretary of state for war in 1807 the fleet had been reduced by the previous secretary, William Windham, to 340 ships of 100,000 tons, and as recently as 24 January 1807 the Board had been ordered to make further reductions.[32] As can be seen in table 7.2, in March 1807 the main concentrations of transports were in the Mediterranean, where there were 164 ships, and to a lesser extent in South America, where there were 91 ships. The landing of troops in Egypt to capture Alexandria occurred on 17 March, ten days after the preparation of the report on which table 7.2 is based, and in South America in the previous month Sir Ralph Auchmuchty had captured Montevideo. In addition there were 72 ships serving as navy victuallers, of which 38 were in the Mediterranean, along with 41 store ships and 11 army victuallers, and 26 servicing the home fleet off Ushant. There were only 38 unallocated transports at home and those would have been discharged, in accordance with the instruction from the Treasury, but for the fact that terminating 17 of them whose charter parties were still within the guaranteed six-months-certain clause would have incurred a significant financial penalty and the remainder were being held in quarantine after returning from the Mediterranean.[33]

The previous administration's desire to reduce the number and cost of transports had denuded Castlereagh's transport reserve, which was intended to support a 'disposable force' of the army.[34] Within weeks of taking office for the first time, in July 1805 Castlereagh had proposed to the Duke of York that a force of some 30,000–35,000 infantry and 8,000–10,000 cavalry should be established to create a 'disposable force' which would be permanently ready to 'menace or attack the enemy on their maritime coast and particularly in Holland'. He had promised that a fleet of transports 'perfectly equipped and victualled for foreign service, capable of carrying about 10,000 men in one lift, would be constantly stationed between Cork and Portsmouth'.[35] On his return to office, Castlereagh recognised that because he required large numbers of transports to support existing operations, there were not enough to support a disposable force as well, nor indeed were there available troops to form that force.

[32] HoCPP, 1807 (117) IV.109, 1, Treasury to TB, 22 Jan. 1807.

[33] HoCPP, 1807 (115) IV.105, 1, Return of Tonnage of Transports at Home, 25 Mar. 1807.

[34] The Treasury's last instruction to further reduce numbers had been issued in January, just two months earlier: HoCPP, 1807 (117) IV.109, 1, Treasury to TB, 22 Jan. 1807. This was issued under the previous administration and was in line with William Windham's determination to reduce the transport fleet.

[35] Castlereagh, *Correspondence* II, 6; Glover, *Peninsular Preparations*, 33; Aspinall, *Later Correspondence of George III*, Castlereagh to HRH Duke of York 25 Jul. 1805. To date no trace of any response has been located.

Office)

	Troop ships		*Navy victuallers*		*Army victuallers*		*Store ships*		*All ships*	
Expedition	*Ships*	*Tons*	*Ships*	*Tons*	*Ships*	*Tons*	*Ships*	*Tons*	*Ships*	*Tons*
Sailed from Falmouth Oct. 19 1806 with General Auchmuchty for Montevideo	20	6,533					1	220	21	6,753
Ditto Nov. 12 to join ditto	4	1,537					1	149	5	1,686
Sailed from ditto ditto with General Crawford	26	9,289			3	861	3	541	32	10,691
At Buenos Aires	12	4,170	5	1,451	2	883			19	6,504
Under orders under General Whitelock for Buenos Aires	13	4,441					1	332	14	4,773
West Indies – In the West Indies	1	389							1	389
West Indies – On passage, sailed Jan. 4 & Mar. 2	4	1,574					4	1,272	8	2,846
West Indies – Under orders for.	2	614					9	2,926	11	3,540
Mediterranean – In the Mediterranean	68	23,575	24	6,824	4	1,023	22	6,336	118	37,758
Mediterranean – On passage, sailed Jan. 4					6	1,097	1	364	7	1,461
Under orders for Mediterranean.	6	1,850	14	3,626	1	120	18	4,636	39	10,232
Cape of Good Hope – At Cape	2	748	2	664	2	580			6	1,992
On passage, sailed Oct. 4 & Feb. 26							2	938	2	938
Under orders for the Cape							2	575	2	575

	Troop ships		*Navy victuallers*		*Army victuallers*		*Store ships*		*All ships*	
Ceylon, sailed from Portsmouth Oct. 4							1	272	1	272
New South Wales. On passage, sailed Jan. 4 & Feb. 26	2	805			1	327			3	1,132
North America – Under orders for							4	933	4	933
Particular service – On passage, sailed Feb. 1							1	165	1	165
Particular service – Under orders for.							4	724	4	724
Home service – Victuallers to the fleet off Ushant			26	3,346					26	3,346
Home service – Miscellaneous services	7	1,706	1	290			4	509	12	2,505
Home service – Defence ships.	2	843							2	843
Unappropriated at home	38	10,844							38	10,844
Total tonnage employed	207	68,918	72	16,201	19	4,891	78	20,892	376	110,902
Deduct freight tonnage	5	2,024			8	1,544	23	6,840	36	10,408
Total tonnage on monthly pay	202	66,894	72	16,201	11	3,347	55	14,052	340	100,494

Source: TNA, ADM, 1/3751, Transport Office, 7 Mar. 1807.

However, he certainly acted quickly to increase the number of transports. On 8 April 1807, shortly after his return to office, he instructed the Transport Board to hire 36,000 tons of shipping urgently – 16,000 tons for infantry, 10,000 tons fitted out for horses and 10,000 tons for ordnance transports and victuallers – in anticipation of an attack on Denmark. The Board was authorised to increase the hire rates, if necessary, to achieve this; thus the price of coppered transports would be raised from 19s to 25s per ton per month and that of wood-sheathed ships from 17s to 21s per ton per month, both for six months certain, after which time the latter rate would be reduced to 19s. Single-bottom ships would be increased from 15s to 20s for three months certain; this rate would then to be reduced to 15s. Despite this considerable increase, no coppered or sheathed ships were tendered, only a few single-bottoms. The Board learned that the prevailing high demand for transports was inducing owners to withhold any available ships in expectation of further increases in the rate of hire.[36] So two commissioners of the Board were dispatched to visit the River and the docks but they saw very few ships fit for the transport service and as a result the Board strongly recommended an application to the Admiralty for use of its ships, as had been done before in similar circumstances. The Admiralty offered up eight ships.[37]

An important contributing factor to the scarcity of transports was the number of victuallers and store ships that had been retained in the Mediterranean where, despite numerous requests that they be returned, there were still eighty-two ships of 23,079 tons. The Board requested the Admiralty to instruct Vice Admiral Lord Collingwood to return them with as little delay as possible, retaining on station only those which were absolutely necessary. The Admiralty subsequently issued instructions to Collingwood to give 'most positive orders to the flag officers under his command' to return of transports, but this had little effect and the Admiralty refused to apply further pressure.[38]

Nevertheless Castlereagh demanded that procurement of additional transports 'should be pushed forward with the utmost possible despatch' and requested a weekly return of the progress of the preparations.[39] The Board reported that the difficulties in procuring transport were arising 'owing to the unusual scarcity of unemployed shipping at this season of the year, as well as the very great amount of tonnage already in the service of this department'.[40] Such was the sensitivity about the rate of progress that the Board sought instruction from Castlereagh in matters where it might have been expected to use its own judgment, such as attributing priorities to troop movements from Jersey to Cork, Portsmouth to Jersey, and Portsmouth to Guernsey.[41]

36 TNA, ADM, 1/3751/245, TB to Admiralty, 24 Apr. 1807.
37 TNA, ADM, 108/21/56, TB to Castlereagh, 20 Apr. 1807.
38 TNA, ADM, 1/3751/277, TB to the Admiralty, 7 May 1807.
39 TNA, ADM, 108/21/53, TB to Castlereagh, 15 May 1807.
40 TNA, ADM, 108/21/57, TB to W. Faulkner, 8 Jun. 1807.
41 TNA, ADM, 108/21/58, TB to Castlereagh, 10 Jun. 1807.

The higher charter rates did have an impact eventually as owners who had been holding out for further increases realised that they had received the best offer that the Board was prepared to make. During the following four months, April to August 1807, the number of troop transports and store ships at home was increased by 32 regular ships and 146 three-month ships, whilst the number of cavalry ships increased by 149, mainly three-month charters, raising the capacity for horses by almost 4,000.[42]

The Bombardment of Copenhagen, September 1807

The Treaty of Tilsit, which was ratified on 7 July 1807 following Russia's decisive defeat at Friedland, provided Napoleon with the opportunity to build an extensive combined naval fleet, incorporating ships of Russia, Denmark, Spain and Portugal.[43] The news of the treaty ratification jolted the new British administration into action. A naval force of about twenty ships was ordered to the Baltic to observe the Danish navy and to be prepared to react to events as they materialised.[44] Meanwhile Castlereagh was refining plans to attack Copenhagen and capture the Danish fleet. This reactive response transpired to be one of the most successful amphibious attacks of the wars. Admiral Lord Gambier was appointed naval commander, the military commander was to be Lieutenant-General the Right Honourable Lord Cathcart. In a preliminary move on 20 July, Castlereagh ordered that all transports at Portsmouth, including some recently returned from the Mediterranean, were to proceed with all possible speed to the Downs to await the arrival of troops and supplies.[45] Captain James Young was appointed senior transport agent to oversee the final preparations. Imagine the noise and confusion as thousands of men were marched into Portsmouth and Ramsgate to be embarked on ships whilst tons of supplies were delivered to the dockside and loaded on board. Nevertheless the first contingent of troops sailed ten days later on 30 July 1807. This force was joined by 10,000 troops under Lord Cathcart that had sailed in transports from Stralsund.

It was intended that the army would land on the morning of 14 August but strong offshore winds delayed the disembarkation until early on the 16th when it proceeded with little opposition. By 1 September the army had surrounded Copenhagen and the naval force was anchored just off the town. The commander-in-chief of the Danish land forces, General Peyman, was invited to surrender the Danish fleet. He refused, so a naval bombardment commenced the following day. The Danes finally opted for a truce on the 5th and the capitulation was signed on 7 September 1807. In consequence, the Danish naval fleet, comprising 16 ships of the line, 9 frigates, 14 sloops and brigs together with 31 smaller merchant ships,

[42] HoCPP, 1807 (115) IV.105, 1.
[43] Hall, *British Strategy*, 158–62.
[44] Wendy Hinde, *George Canning* (London, 1973), 47.
[45] TNA, WO, 6/156, Castlereagh to TB, 20 Jul. 1807.

was spirited away to England.[46] To man these vessels, additional crews were recruited of seamen off whale fishery ships from Whitby and other seamen from Hull, Grimsby and Newcastle. However, before agreeing to travel to Holland, these men demanded assurances from the Admiralty that the navy would not impress them but would release them on their return.[47]

The embarkation of the army commenced on the 13 October.[48] On that day, the eight battalions of the line of the King's German Legion were embarked and on the 14th the two more light battalions of the legion, together with the 7th and 8th Regiments of British troops. These corps, the depot and garrison company of the legion, and the sick and wounded of the army completely occupied all the troop ships which had not been appropriated to the conveyance of naval stores. By Sunday afternoon, 18 October, only the rear-guard remained. Its departure was delayed for a few days due to gales which prevented embarkation.[49] This campaign was a welcome success. Foreign Secretary George Canning's assessment was that 'a Northern Confederacy, an invasion of Ireland and the shutting of Russian ports had been prevented'.[50] After an examination of the Danish naval fleet, Gambier reported that 'it is certain that the fleet could have been completely ready for sea in three weeks, or a month at the most'.[51]

The campaign involved 374 transports of 78,420 tons of which 15 (4 per cent) were wrecked, captured or missing, representing the tragic loss of 402 troops and 14 seamen, 60 horses and a significant volume of naval stores. However, 495 troops, 125 seamen and 18 horses were rescued from those vessels, and 362 transports completed their missions, returning safely.[52] Of the fifteen losses, five were captured and nine were wrecked on the return voyage and one was blown up off Copenhagen.[53]

Several ships owned by Michael Henley & Son were involved in the action. *Friendship* sailed from Deptford to Harwich where it embarked troops for Copenhagen, returning to Yarmouth with captured naval stores and thence to Portsmouth. *Pitt* sailed from London to Sheerness and then Copenhagen, returning to Sheerness then Deptford. *Norfolk* carried coal to London, then ballast to Copenhagen where it was loaded with captured naval stores which it

[46] T. Munch-Petersen, *Defying Napoleon: How Britain Bombarded Copenhagen and Seized the Danish Fleet in 1807* (Stroud, Gloucestershire, 2007), 215. The author does not mention the stores and supplies.

[47] *Hull Packet*, 29 Sept. 1807.

[48] HoCPP, 1808 (10) IX.53; 1808 (236) IX.95, Papers presented to the House of Commons relating to the expedition to Copenhagen, Feb. 1808.

[49] HoCPP, 1808 (10) IX.53; 1808 (236) IX.95, Papers Presented to the House of Commons relating to the Expedition to Copenhagen, Feb. 1808.

[50] Hall, *British Strategy*, 160.

[51] HoCPP, 1808 (10) IX.53; 1808 (236) IX.95, Papers presented to the House of Commons relating to the Surrender of Copenhagen, 28 Feb. 1808, Gambier to W.W. Pole, 28 Jan. 1808.

[52] TNA, WO, 6/156, Castlereagh to TB, 29 Jan. 1808.

[53] HoCPP, 1808 (10) IX.53; 1808 (236) IX.95, Account of Transports employed against Copenhagen, 2 Feb. 1808.

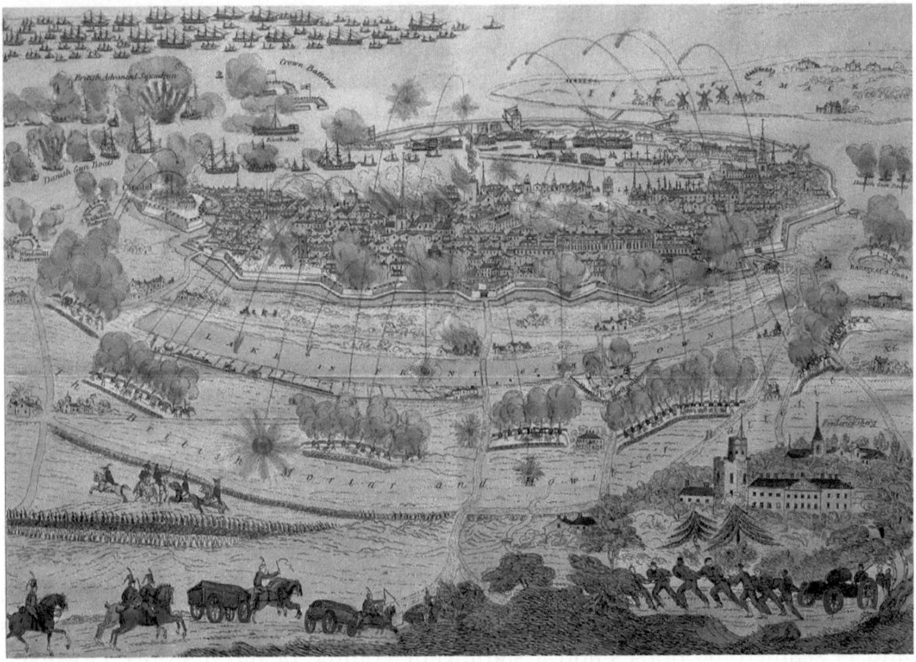

Plate 4. The bombardment of Copenhagen and the surrender of the Danish fleet, 7 September 1807

delivered to Deptford. Similarly, the *Salus* was used to bring back stores captured from the naval arsenal.[54]

The captured victualling stores comprised casks and staves with an estimated value of £4,611.[55] The confiscated naval stores included hemp, cordage, timber, masts, canvas, sails, iron, anchors, copper bolts and copper sheets; the value attributed to these stores was £270,240.[56] Planning the removal of the stores and ships, the re-embarkation of troops and the subsequent return to England was a complex process. The stores and the German cavalry were re-embarked onto

54 NMM, The Henley Papers, HNL 61/53, HNL 106/16, HNL 99/35.

55 HoCPP, 1808 (236) IX.95, 7, Accounts presented to the House of Commons of Stores taken at Copenhagen, 13 May 1808.

56 HoCPP, 1808 (236) IX.95, 11, Accounts presented to the House of Commons of Stores taken at Copenhagen, 13 May 1808, TNA, WO, 1/803, 18 Dec. 1807. Seven transport agents presented a petition to the naval commander-in-chief requesting that they should receive part of the Copenhagen prize money. Led by Captain John Halstead and Lieutenants John Grieg and Richard Chessman, they claimed that although as naval officers working for the Transport Board they did not normally qualify for prize money, their work had been unremitting, landing and re-embarking troops and shipping the naval stores from the arsenals and they had experienced the dangers attached to those services. In their view they should be treated as other officers of the navy present on that occasion. The outcome of this petition has not been established.

transports for the return to Yarmouth, whilst the whole of the British infantry returned on board British naval ships and the Danish ships of war.[57] It is reported that on one day, 21 October, some 460 ships and vessels sailed through the sound between Denmark and Sweden (Øresund), of which 370 were transports.[58]

Subsequently the government came under attack in Parliament for over-estimating the capacity of the Danish forces. It had thought that there were 35,000 troops garrisoned at Copenhagen whereas it was subsequently reported that there were only 6,000 men.[59] The suggestion was that more troops, and consequently more transports, were committed to the campaign when they might have been better used elsewhere. However the report did not mention the 7,500 troops stationed nearby at Köge, nor the fact that Gambier claimed he needed 12,000 troops to help navigate the Danish fleet to Britain.[60] This is a good illustration of the weaknesses of the intelligence that was available when such operations were being planned.

By Christmas 1807 a significant number of transports had returned from the Mediterranean, the Baltic and South America. Castlereagh advised the Transport Board to 'proceed to discharge all horse ships that are employed over and above the tonnage that may be required for 1,000 horses'.[61] He again recommended that the disposable force be reintroduced, proposing that 30,000 men be positioned at the main embarkation points – Cork, Portsmouth, Harwich and close to the Downs; these were to include 10,000 'selected with a view to more distant operations'. To provide transport for these troops he proposed retaining all the infantry transports, knowing 'the delay and difficulty attendant on taking up a large supply of transports when immediately wanted for service'.[62]

On 24 March 1808 the Board received orders to prepare the fitting out of the transports for 1,000 horses which it had previously been instructed to retain.[63] However, by 21 April, the Board was probably wishing that it had retained more horse ships: it received a secret instruction from Castlereagh to bring forward 34,807 tons for service in the Baltic under Lieutenant-General Sir John Moore. Of this tonnage, 17,276 was for 11,000 infantry, 10,793 for a regiment of cavalry with a proportion of artillery, 2,046 for army victuallers and 4,692 for store ships [64] Moore was to sail with Vice Admiral Saumarez to support Sweden

[57] HoCPP, 1808 (9) IX.113, 5, Papers presented to the House of Commons relating to the Expedition to Copenhagen, 2 Feb. 1808, Admiral James Gambier to convoy commanders, 20 Oct. 1807.

[58] HoCPP, 1808 (9) IX.113, 7, Papers presented to the House of Commons relating to the Expedition to Copenhagen, 2 Feb. 1808, Note to letter number 8.

[59] Cobbett, *Weekly Political Register*, 22 Oct. 1808.

[60] Munch-Petersen, *Defying Napoleon*, 213.

[61] TNA, WO, 6/156/387, Castlereagh to TB, 25 Dec. 1807. He clearly had forgotten his own statement in the House of Commons on 31 Jul. 1807 regarding the difficulty of raising horse ships. This instruction is contrary to that statement.

[62] Castlereagh, *Correspondence* III, 161, undated, State of the Army.

[63] TNA, WO, 6/156/387, Castlereagh to TB, 24 Mar. 1808.

[64] TNA, WO, 6/156, Castlereagh to the TB, 21 Apr. 1807.

against Russian aggression. The expedition arrived at Gothenburg on 13 May 1808. Almost immediately the relationship between Moore and Gustavus IV, the Swedish monarch, failed. This was principally attributable to the restrictive scope of Moore's orders, which fell well short of Gustavus's expectations. The Swedish king had Moore arrested. He escaped and responded to the humiliation by bringing his whole force back to Britain without the troops having disembarked.

Wellington's 1808 Expedition to the Peninsula

Prior to 1808, the British had determined not to invade the Peninsula for fear of driving the Portuguese and Spanish populations into the French camp. Fortuitously, Napoleon inadvertently created a nationalist backlash in both countries by replacing the Spanish king with his own brother, Joseph, and by attempting to subdue Portugal. In May 1808 the Spanish rebelled against French rule. In the summer Spanish and Portuguese delegations visited Britain to persuade the foreign secretary to engage in a joint campaign against the French. Britain decided to support the revolt by landing troops in Portugal.[65] Wellington was initially appointed to lead this force. At that time he was waiting with troops at Cork for favourable weather to sail for South America where he was to support General Miranda in Venezuela. His orders were urgently countermanded and he sailed for Portugal ahead of his troops to establish the situation in northern Spain and Portugal and to select a landing point. His force of 9,196 infantry men, 384 cavalrymen (but only 238 horses) and 693 artillerymen (but only 306 horses) sailed from Cork on 12 July 1808 in 40 troop transports, 18 cavalry ships, 7 ordnance vessels, 9 victuallers (4 of which carried oats and hay) and 2 store ships with camping equipment – 76 transports in total of 25,257 tons.[66] Oman attributes the shortage of horses to the authorities' (mistaken) belief that both draft and saddle horses could be readily procured in Portugal.[67]

In consultation with Vice Admiral Sir Charles Cotton, the naval commander of the Portuguese station, Mondego Bay situated about 100 miles north of Lisbon was selected as the most suitable landing point. The plans for the landing specified that the horse ships and the ordnance ships would be unloaded in the river; troop ships would be unloaded in the bay and the troops rowed into the river. In the meantime the following arrangements were to be made. The haversacks and canteens in the regimental stores were to be issued to the men and tin camp kettles were to be issued from the quartermaster-general's stores to the regiments. The commissary should issue to each regiment's paymaster the sum of £1,000. A month's pay should also be issued on the same account to the officers of the staff. The officer commanding the 20th Light Dragoons should receive a sufficient number of horses to mount all his men; he should therefore

65 Fedorak, 'Royal Navy and British Amphibious Operations', 141–6.
66 HoCPP, 1809 (17) XII.1, 224, Proceedings of the Inquiry relative to the Armistice and Convention in Portugal.
67 Oman, *Peninsular War*, vol. 1: *1807–09*, 225.

be prepared to land the horse appointments (equipment) of all the men who at present had no horse. Each man was to land with a spare shirt, a pair of shoes, a comb, razor and brush, which were to be packed in his greatcoat. Knapsacks were to be left in the transports, under the control of the baggage master. The men were to land with three days' bread and two days' cooked meat. The commanding officer of artillery was to land the three brigades of artillery, each with half the usual amount of ammunition, the forge cart etc. He was also to land 500,000 rounds of musket ammunition, for which carriage would be provided. Each soldier would have with him three good flints. Besides the bread carried by each man, three further days' bread for all the troops on the transport was to consolidated and landed in bags containing 100 pounds. A commissary and a number of clerks were attached to each brigade. The commissary took charge of the bread and made subsequent arrangements for victualling the troops. Three days' supply of oats was to be landed for each of the horses. The officers commanding companies were instructed to make arrangements to purchase mules for the carriage of camping equipment, for which they had received an allowance.[68]

The expedition landed unopposed. It took five days to get the whole of the force ashore even when the transports' ships boats were supplemented by those from the accompanying naval vessels. Later landings were supported by hiring local small boats.[69] Shortly after the landing the force was joined by 7,204 troops under Major-General Spencer. These troops, who had sailed from Gibraltar on 14 May, 6 and 9 June, were embarked on 46 infantry ships, 2 horse ships with artillery regiments and staff horses, 2 hospital ships, 2 victuallers and a store ship with camping equipment and provisions. They had previously sailed from England for the Tagus and Sicily in January 1806. Meanwhile another force of 4,817 infantrymen under Brigadier-General Anstruther sailed from the Downs on 22 July 1808 in 28 troop ships, 2 horse ships and 13 provision ships, arriving ten days after Spencer's force. They were landed further south at Maceira. They were followed by a convoy of a further 181 transports which sailed from Portsmouth on 31 July with a mixed British and German force under Lieutenant-General Sir John Moore consisting of 11,774 infantry, 1,557 artillery, 646 cavalry, 241 staff and 1,519 horses. The fleet included 57 horse ships and 34 store and provision ships. This force arrived at Mondego Bay but was ordered to land further to the south, which it did between 20 and 24 August.[70]

These arrivals are summarised in table 7.3. Within a period of three weeks, 32,620 infantry troops, 1,815 cavalry with 1,653 horses, 2,537 artillerymen with 984 horses, and 349 staff with 294 horses – in all 37,311 men and 2,931 horses and accoutrements – had been shipped, mainly from Ireland, England, Cadiz and Gibraltar. This combined force had been transported in 204 troop ships,

[68] Gurwood, *Dispatches* III, 41.

[69] Hall, *Wellington's Navy*, 30.

[70] HoCPP, 1809 (17) XII.1, 224, Proceedings of the Inquiry relative to the Armistice and Convention in Portugal.

Table 7.3. Return of Officers and Men embarked and sailed for Portugal and Spain assembled un the command of Sir Hew Dalrymple and also of the Transports employed on that Service (Trans Office, 16 November 1808)

Force commander		Lieut.-Gen. Sir Arthur Wellesley	Major-Gen. Spencer	Brig.-Gens Anstruther & Aceland	Lieut.-Gen. Sir John Moore	Brig.-Gen. Stewart	
Sailed in 1808		Cork 12 Jul.	Gibraltar 14 May–14 Aug.	Harwich 20 Jul.	Portsmouth 31 Jul.	Portsmouth 9 Aug.	**Total**
Infantry	Men	9,196	6,879	4,771	11,714		32,620
Cavalry	Men	384			646	775	1,805
	Horses	238			642	773	1,653
Artillery	Men	693	287		1,557		2,537
	Horses	306			678		984
Staff	Men	24	38	46	241		349
	Horses	34	21	40	199		294
Total	Men	10,297	7,204	4,817	14,218	775	37,311
	Horses	578	21	40	1,519	773	2,931
Troop-ships	Number	40	46	28	90		204
	Tons	15,565	14,137	5,784	19,616		55,102
Cavalry ships	Number	18	2	2	57	29	108
	Tons	5,065	429	510	13,243	6,411	25,658
Army victuallers	Number	4	2	3	18		27
	Tons	1,389	452	434	4,711		6,986
Forage-ships	Number	4			2	3	9
	Tons	859			285	642	1,786
Camp stores	Number	2	1		4		7
	Tons	453	334		876		1,663
Ordnance stores	Number	7			10		17
	Tons	1,926			2,122		4,048
Total	Number	75	51	33	181	32	372
	Tons	25,257	15,352	6,728	40,853	7,053	95,243

Source: HoCPP, 1809 (17) XII.1, 224, Proceedings upon the Inquiry relative to the Armistice and Conventio Portugal.

108 horse ships, 27 army victuallers, 9 forage ships, 7 store ships with camping equipment and 17 ordnance store ships – in all 372 transports of 95,243 tons.

The landings preceded Wellington's victory at the Battle of Vimeiro. After it, Junot, Napoleon's defeated commander, sought terms of surrender which resulted in the contentious Convention of Cintra. One of the terms of this convention established that the French troops would be removed in British vessels.[71] It was decided that the French army would be taken to Rochefort and L'Orient because those were the French ports furthest from Spain. The convention also addressed the security of the transports that would repatriate the troops. This was felt necessary because fifty British transports which had returned the French army of Egypt had been detained by French authorities. Subsequently 24,735 men, 213 women, 116 children and 759 horses were returned in October 1808 in 151 transports at the rate of one man per ton, almost 30,000 tons in all. The French were permitted to leave commissaries in Spain to sell the horses that could not be shipped in British transports or to arrange alternative transport.[72] However, Captain Halstead reported on 17 October that nine of the vessels had been caught in a storm and returned to their anchorage.[73] Later, during the debates on the Corunna debacle in Parliament, there was a suggestion that the use of transports for this purpose had forced Sir John Moore to march his troops into Spain rather than convey them by sea and that this had contributed to the unfortunate outcome in Spain. Castlereagh strongly denied this and claimed that there were sufficient transports available had it been deemed expedient: 'The movement by land was not adopted by necessity but because it was deemed most advantageous.'[74] In September 1808 there were 451 transports of 113,273 tons off the coast of Spain and Portugal. This was surely enough transports to repatriate the French army and ferry Moore's army around the coast of Spain if he had opted for that.

For Want of Horse Ships

The British army overseas never had enough horses. This led to some cavalry brigades fighting as infantrymen and to a shortage of heavy artillery and of Commissariat carts for moving supplies to the front line.[75] The Duke of York regularly complained that the proportion of cavalry in continental armies was never less than a sixth of their total force, often as much as a quarter, whereas in the British army the cavalry represented little more than a tenth.[76] There may

[71] Hall, *Wellington's Navy*, 35.

[72] HoCPP, 1809 (17) XII.1, 243, Proceedings of the enquiry relative to the armistice and convention made and concluded in Portugal in August 1808.

[73] TNA, WO 1/804.

[74] Hansard, House of Lords, 7 Feb. and 9 May 1809.

[75] See Hall, *British Strategy*, 46; Flayhart, *Counterpoint to Trafalgar*, 107.

[76] Richard Glover, *Britain at Bay: Defence against Bonaparte 1803–1814* (New York, 1973), 224, Duke of York to Windham, 18 Mar. 1806.

have been other reasons for this but, as demonstrated above, the shortage and cost of horse transports were certainly significant contributory factors.

The average horse ship was between 250 and 300 tons. At 8 tons per horse, each ship of that tonnage could only carry 31–37 horses. However, horses were frequently transported in smaller vessels which could get over bars in river estuaries or manoeuvre close in shore so that the animals could be lowered in slings into shallow water and then swim ashore. This resulted in far fewer injuries than when they were hoisted from a heaving deck and lowered into flat-bottomed boats which were themselves rising and falling in the swell of the tides.

Although 8–10 tons per horse was the standard allowance, the size of ships used varied considerably. In 1809, replacement horses for the 20th Light Dragoons were sent out to Portugal in four ships of various sizes: the *Elizabeth and Ann* of 286 tons with 30 horses at 7½ tons per horse; the *Frederick* of 131 tons with 18 horses at just under 7½ tons per horse; the *Liberty* of 123 tons with 20 horses at 6 tons per horse; and the *Rodney* of 312 tons with 32 horses at almost 10 tons per horse.[77]

The most commonly used vessels for this service were colliers which normally plied the coastal route between the coalfields in the north-east of England and London. Taking large numbers of these vessels out of that service at one time had a significant impact on the availability and price of coal. For instance in August 1799 it was reported that the price of coal was 'daily advancing' owing to the great number of the largest colliers being employed as transports for den Helder campaign and that the colliers remaining were insufficient to supply enough coals for the London market.[78] Again in 1809 there was a report that 'the people of Plymouth and the whole kingdom are paying extra price in consequence of the scarcity of colliers due to so many being used on the Walcheren expedition'.[79] Even Castlereagh was forced to admit that the high number of colliers taken for Walcheren was having a major impact on the price of coal in London.[80] There was even press suggestion that during the preparation for that campaign colliers were being taken up as they arrived in the River, 'some say impressed for service'.[81] This is likely to have been press exaggeration, but it does possibly give an indication of the pressure that the Board was under to find horse ships. In 1805 when Castlereagh was planning the northern Germany expedition he needed to ship out 17,000 horses. This would have required 140,000–170,000 tons of horse ships, which were unlikely to be available. In recognition of this he had to make alternative plans. The relatively short voyage time meant that he was able to plan for repeat sailings of a smaller number of vessels besides some provision for local procurement. In 1808, when preparing to send additional infantry and cavalry brigades to Portugal, it was necessary to await the return of horse ships

77 TNA, WO, 1/804.
78 *New Lloyds Evening Post*, 9 Aug. 1799.
79 *Bury and Norwich Post*, 20 Dec. 1809.
80 Hall, *British Strategy*, 43.
81 *Trewman's Exeter Flying Post*, 8 Jun. 1809.

from earlier convoys, sufficient for 2,300 horses, before the additional horses could be shipped. These vessels took some time to return, causing some delay.[82]

One of the most critical events of the wars occurred in 1809 when the preparations for the Walcheren expedition were underway. At that time most of the Transport Board's fleet of horse ships was in Portugal where there was shipping capacity for over 5,600 horses. On 22 May Castlereagh made an urgent request to Sir Arthur Wellesley to send home shipping capacity for 3,000 horses because 'so much depends on their arrival being accelerated'.[83] For various reasons it was seven weeks before the first convoy arrived at Portsmouth on 11 July, only two weeks before the Walcheren fleet sailed. These ships and infantry vessels returned from Portugal and elsewhere were prepared in a remarkably short time. They sailed with the fleet on 28 July.[84]

Colliers intended to carry horses took some time to prepare. The holds had to be cleaned of coal dust and stalls had to be built. After 1803 this cost was the responsibility of the owner. Naturally, for short charters there was some reluctance on the part of owners to pay these expenses, particularly when there was alternative employment for their ships. Then at the end of the charter the stalls were dismantled, again at the owner's cost; the materials were often sold at auction, as demonstrated by an advertisement in September 1811 for an auction for the sale of 'pressed hay, oats, bran, horse mangers, fitting for troop and horse transports all landed from transports'.[85] This meant that newly taken up ships had to be fully fitted again, even if they had been used as horse ships on previous occasions. There was no apparent planning for the retention of horse ships during periods when they were not required so that there would be a reserve for when they were needed. Given the difficulty of bringing forward such vessels, this probably ought to have been a serious consideration; however, it was outweighed by the cost and the shortage of vessels and the impact on coal prices.

Though they may have survived the journey and escaped injury during the landing, horses frequently suffered from debilitating sea sickness for several days after reaching shore.[86] Then the challenge was to obtain an adequate supply of oats and forage. The standard allowance for each cavalry horse was 11 pounds of oats and 9 pounds of hay per day, for draft horses 13 pounds of oats and 9 pounds of hay.[87] There was rarely enough, as was reported in the diary of a cavalry officer during the Peninsular war in 1812:

82 Castlereagh, *Correspondence* I, 401.
83 Castlereagh, *Correspondence* II, 68, Castlereagh to Sir Arthur Wellesley, 22 May 1809.
84 Castlereagh, *Correspondence* II, 95, Castlereagh to Sir Arthur Wellesley, 11 Jul. 1809.
85 *Hampshire Telegraph*, 23 Sept. 1811.
86 John Pimlott, *British Light Cavalry* (London, 1977), 43.
87 Glover, *Peninsula Preparation*, 295.

> There was no straw in the country and from the horses being so starved they eat the weathered grass with so much avidity that they swallowed many of the stones at the roots and some died in consequence.[88]

and:

> The weather was extremely severe; the horses got nothing by way of long forage but the long grass that the men cut themselves from out of the woods. We were frequently 8 to 9 days without corn and in consequence lost many horses.[89]

Even if there were local supplies of forage the local population took advantage of the opportunity to inflate prices dramatically. For this reason forage and oats were constantly being shipped from England.

There was a high attrition rate of horses. During the Battle of Talavera on 27 and 28 July 1809 it was reported that 211 horses were killed, 71 wounded and 159 missing.[90] There were constant demands for replacement horses. In 1809 Wellesley frequently wrote to Castlereagh requesting more: 'It is inconceivable how fast the horses of both the cavalry and artillery fall off. When horses as well as men are new to war, I believe that the former are generally the sacrifice of their mutual inexperience.'[91] There were no horses to be procured in Portugal and in Spain the demands of the French had invariably reduced local availability. During one short period in 1812 the following remounts for various cavalry regiments were shipped to Spain: 21 horses sailed on 7 June; 32 on 28 June; 100 on 6 August; 729 on 27 August; 125 on 4 September; and 535 on 2 and 9 November – in total 1,542 replacement horses, representing 23 per cent of the total number of cavalry horses in Wellington's army, which at that time was 6,546.[92] The situation was so extreme that in 1812 Wellington was to declare, 'The Peninsula is the grave of horses; I have lost no less than 12 for my own riding.'[93] A local supply of horses could not be guaranteed.

During retreat, horses were often sacrificed because of lack of transport or in expedience. During the Corunna retreat William Werner, an officer in the 7th Hussars, noted that his regiment had shipped out 640 horses but was only

[88] James Tomkinson (ed.), *The Diary of a Cavalry Officer in the Peninsular War and Waterloo Campaign, 1809–1815, Lieutenant-Colonel Tomkinson, 16th Light Dragoons* (London, 1971), 121.

[89] Tomkinson (ed.), *The Diary of a Cavalry Officer in the Peninsular War*, 121.

[90] Robert O'Byrne, *The Victories of the British Army in the Peninsula and the South of France, from 1808 to 1814* (London, 1889), 53. Also in Michael Glover, *Wellington as Military Commander* (London, 1968), 211. It was reported in August 1809 that a cavalry brigade which had landed in Spain four to five months earlier with 1,300 horses was reduced to 800 horses, but 'not above 20 taken or killed by the enemy'.

[91] Gurwood, *Dispatches* III, 215, 318.

[92] Castlereagh, *Correspondence* III, 248.

[93] Glover, *Wellington as a Military Commander*, 212.

taking 60 back.[94] Oman reports that at Corunna 'the horses were in a terrible state, very few were worth reshipping ... more than 2,000 were shot, stabbed or flung into the sea'.[95] Even during the return voyage horses could be lost. In 1795 the *Freedom*, a ship owned by Michael Henley, was hired to bring back some of the remnants of the Duke of York's army that had evacuated from Holland and Germany. She embarked some Irish troops and forty-six horses but was subsequently delayed in the river Weser for a month whilst awaiting an escort. The convoy eventually arrived at Shields but by that time ten of the forty-six horses had died.[96] In September 1809 when Castlereagh was planning to increase the tonnage in the Tagus to provide accommodation for the whole army should an evacuation from Portugal become essential, he pointed out that the cost of the 40,000 tons required would be £50,000 per month. He said:

> it is quite impossible to think of retaining in the service tonnage for horses, unless with a view to an immediate evacuation, the monthly expense for each horse being £10.00. Government will have to make up their mind, that in the event of the army continuing indefinitely in Portugal, to leave the horses to their fate, that is, to be put to death should you be forced to retire.[97]

When Sir Rupert George was interviewed by the members of the Scheldt commission much of his testimony was in response to questions about the availability of cavalry shipping. He acknowledged that 'there was much more difficulty in providing cavalry transports than in providing ships for infantry or stores'.[98]

In 1810 Liverpool recognised, just as Castlereagh had done in 1805, the advisability of keeping troop and horse transports in constant readiness for sea at Cork, Plymouth and Portsmouth, 'for the most expeditious performance of any service that may arise'. He directed that at Cork 3,500 tons of coppered troop ships and cavalry vessels for 70–80 horses should be held in readiness; at Portsmouth 3,500 tons of coppered troop ships; and at Plymouth a large amount of cavalry tonnage. Aware that all the available horse ships were fully employed, he ordered that 'no horse ships should be discharged'.[99]

Castlereagh was Secretary of State for War at a crucial time. He instigated several major campaigns – northern Germany in 1805, Copenhagen in 1807, the Peninsula in 1808, and then in 1809 the Peninsula again and Walcheren. Each required

94 Pimlott, *British Light Cavalry*, 44.

95 Oman, *Peninsula War*, vol. 1: *1807–09*, 582.

96 John Barney, 'North Sea and Baltic Convoy 1793–1814: As Experienced by Merchant Masters Employed by Michael Henley & Son', *Mariner's Mirror* 95.4 (November 2009), 429–40.

97 Castlereagh, *Correspondence* II, Castlereagh to Sir Arthur Wellesley, 12 Sept. 1809.

98 William Cobbett (ed.), *The Parliamentary History of England* (London 1806–20), vol. 15, Papers relating to the Scheldt expedition.

99 TNA, WO, 6/158/11, Liverpool to TB, 6 May 1810.

the movement of a large number of men and huge consignments of materials and supplies. A large number of transports were indispensable for this service. However, this was also a period when the demand for shipping from trade was at its peak and so it took time to build up a fleet to ship thousands of men, horses and equipment. The hire rate had to be increased by 20 per cent to 25s per ton per month to persuade ship-owners to commit their vessels to government service when there were many other opportunities to generate returns.

To transport a mixed force of 10,000 required 103 ships of 25,000 tons, representing a significant element of the available suitable transport fleet. The organisation of such expeditions required significant planning and co-ordination. So when Castlereagh allowed the plans for the British contingent of the expedition to northern Germany to expand from 10,000 men to over 67,000 men and 19,000 horses, he repeated Dundas's error of creating an expedition that was too logistically ambitious.[100]

Despite this, Castlereagh appears to have been a man of detail and of foresight, described by Fortescue as 'the ablest man that ever controlled the War Office'.[101] He certainly consulted widely on the best approach before committing to an expedition but he failed to bring the Transport Board into the planning at an early stage. The reactionary nature of many of the campaigns invariably meant that the timing of expeditions was imposed by events which forced Castlereagh to take risks by restricting the planning process, by failing to anticipate the length of the preparation time required or by failing to take into account the dramatic impact of winter weather.

Through this period the Transport Board acquired significant status. It had developed a reputation for competence. Whilst there was acceptance that its role was essential, the cost of the service was under constant scrutiny. Castlereagh's vision, later reaffirmed by Liverpool and Dundas, of maintaining a disposable force of the army together with sufficient transports to convey it quickly proved to be impracticable because there were not enough troops or transports that were not required elsewhere. The cost of horse transports was immense: the bill for 6,000 horses for six months would have been £360,000. This excluded the cost of procuring and shipping forage and oats and the increased price of coal due to the shortages created when colliers were removed from their normal service for prolonged periods.

100 Castlereagh, *Correspondence* I, 85.

101 Glover, *Peninsular Preparation*, 32, described by Glover as the right sentiment but wrong department. Castlereagh was Secretary of State for War; he was never secretary at war, the minister who controlled the War Office.

8

1809: A Year of Military Disappointments[1]

1809 began with the enforced retreat of Lieutenant-General Sir John Moore's army in Spain to Corunna. Moore himself was killed defending the evacuation of his men. The year's nadir was the ill-fated Walcheren campaign, which was a military disaster.[2] There were, however, some military successes: Martinique surrendered to Lieutenant-General Beckworth's 10,000-strong force on 24 February, St Domingo fell to Spanish and British troops in July, and Wellington had initial success on his return to Portugal, defeating the French at Talavera on 28 July. However, these successes were overshadowed by Walcheren, and the year ended with Wellington having to make a tactical withdrawal behind the defences of Torres Vedras, following the success of the French over the Austrians in the Battle of Wagram and the consequent threat of a renewed French push into Portugal. The Walcheren debacle contributed to the demise of the Portland administration, culminating in its replacement by another Tory administration led by Spencer Perceval, with Lord Liverpool becoming Secretary of State for War.

Ironically, 1809 was a good year for the Transport Board. In January Commissioner Captain James Bowen was highly praised for his strenuous efforts in managing the Corunna evacuation under extraordinarily difficult circumstances. In April a number of transports converted into fire ships were the vanguard of Captain Thomas Cochrane's success at Aix Roads; at the same time, despite the severe competition from trade, the Transport Board managed to procure a large tonnage for shipping troops returning to the Peninsula. Then in May, June and July, when trade use of merchant shipping was at an unprecedented high, the Board managed to organise the transport for 46,000 men and their equipment to Walcheren. A few months later, transports played a highly important role in the evacuation of the sick troops. The success of the Board and its agents in procuring and preparing sufficient tonnages of shipping in very difficult circumstances on these occasions demonstrated its important position in the government war machine. The significance of this was reinforced in May when Rupert George, the Chairman of the Board, was elevated to the baronetcy. The reputation of the Board was further enhanced when, in April, the 9th Report of the Commission for Revision on the Transport Service acknowledged that

1 Rodger, *Command of the Ocean*, 556.
2 Hill, *British Strategy*, 178.

the benefits anticipated from merging the procurement of transports under one Board had been achieved, even exceeded. More importantly, the report was free of the censure that the Commission had attributed to other Boards in earlier reports.[3]

Arrival at Corunna

Captain James Bowen, a Transport Board commissioner, had been appointed to lead the transport service off the coast of Spain. He sailed from Falmouth in September 1808 with the first division consisting of 12,691 rank-and-file troops under Sir David Baird. They were to join up with those under Sir John Moore who were marching from Portugal to meet them. The combined force was to support the Spanish army's bid to drive the French out of Spain. Bowen sailed with almost 200 infantry and horse transports. This brought the total number of transports used in 1808 to support the Peninsular campaign to 676, totalling 166,906 tons. These transports conveyed 65,627 men (including over 6,000 King's German Legion troops), 7,155 horses and tons of stores and supplies from England to Portugal and Spain. The infantry were shipped at 1½ tons per man, cavalry at 8 tons per horse, and artillery at 10 tons per horse. The tonnage to ship supplies of ordnance, rockets, camping equipment, forage and victualling for such a significant force added almost 25 per cent to the requirement over that needed for troops and horses.[4] Thirty-six of the transports conveyed the 10,000 Spanish troops who had originally been part of the French army of occupation in Denmark under the Marquis de la Romana and their equipment back to Spain to support the uprising of their fellow countrymen. They had been repatriated from a small island off the Danish coast by the British to join the uprising in Spain.[5]

Bowen arrived off Corunna on 14 October 1808. Soon after his arrival, in accordance with the standing instructions for agents afloat, he began to send regular reports back to the Transport Board in London.[6] These reports serve as an interesting study of the role of transport agents involved in overseas amphibious expeditions. In his report dated 25 October, he enclosed a log of his daily activities and a return of transports under his charge, including a note of those who were short of complement. He also sent a disembarkation list of horses and men and he recorded in detail the arrival and departure of transports.

3 Commission for Revision (9th). The commission was critical of the dockyard management, the Victualling Office and naval hospitals and it rewrote the regulations governing these organisations.

4 Sutcliffe, 'Bringing Forward Shipping for Government Service', 275.

5 James Davey, 'The Repatriation of Spanish Troops from Denmark, 1808: The British Government, Logistics, and Maritime Supremacy', *Journal of Military History* 74 (July 2010), 689–707 and Hall, *Wellington's Navy*, 65.

6 Sutcliffe, 'Bringing Forward Shipping for Government Service', 317–18: a summarised version of the 'Articles of General Instructions to Agents relative to the Service of Transports'.

He reported that Sir David Baird wanted to save the salted provisions, so Bowen had entered into an agreement with the local American consul for the supply of fresh beef. Bowen also reported the misconduct of James Harris, master of the *Yare*, for repeated neglect and inattention to signals and for having only one boat – and that one unfit for service. He also reported the masters of the *George & Mary* and the *Ocean* for disobeying orders, which had slowed down the rate of disembarkation of the horses. The Board immediately advised the brokers of those ships that they would be severely mulcted.

The arrival at Corunna was not warmly received. Initially the Galician Junta refused to let the troops land. This issue was only resolved when the Central Junta in Madrid intervened.[7] Even then the disembarkation of the infantry was delayed because of the difficulty of feeding and accommodating the army ashore.[8] However, by 30 October Bowen was able to report that seven regiments and two companies of artillery had been landed but progress was still being hampered by the limited facilities in the town, where no more than 2,000 men could be accommodated. Further troops could only be disembarked as those that had landed earlier left the accommodation and marched into the interior. He predicted that the whole force would be landed by 3 November.

On 31 October a further thirty-four ships sailed from Portsmouth for Corunna, including nineteen ships with artillery horses and the Royal Wagon Train.[9] Bowen subsequently reported their arrival, although some of the troops were 'very sickly' and many had been removed to ships appropriated as temporary hospital vessels.[10] In the first two weeks in November Baird's command was increased by the arrival of another 5,414 troops. Bowen had received no instructions regarding the return of empty transports so he sought the Board's view. It in turn had no insight into the plans for the operation and so had to pass the enquiry on to the secretary of state.[11] In the meantime Bowen had ordered some vessels to return to England under the convoy of the *Racoon* and the *Sparrowhawk*.

Bowen had also sought the views of Rear Admiral Michael De Courcy on the subject of returning empty transports for which he did not have any current use. He pointed out that the increasing number of ships was overcrowding the small harbour and that returning ships would reduce the likelihood of accidents. It would also reduce the expense: the ships that he proposed to return were costing the government £15,000 per month. The admiral declined to give his consent but he must have subsequently acquiesced because on 20 November Lieutenant Debenham reported his arrival at Weymouth from Corunna with a number of empty transports, though some of his fleet had been dispersed by a heavy gale. A few days later nine more ships of the convoy arrived at Portsmouth.[12]

7 Hall, *Wellington's Navy*, 66.
8 TNA, ADM, 108/81, TB Minute, 16 Nov. 1808.
9 TNA, ADM, 108/185, TB Office at Deal to TB, 31 Oct. 1808.
10 TNA, ADM, 108/81, TB Minute, 16 Nov. 1808.
11 TNA, ADM, 108/81, TB Minute, 16 Nov. 1808.
12 TNA, ADM, 108/81, TB Minute, 18 and 22 Nov. 1808.

Bowen was able to report on 10 November that all the infantry was disembarked. He confirmed that the agents were then engaged in disembarking the 7th, 10th and 15th Light Dragoons, artillery horses and the Royal Wagon Train.[13] Meanwhile transports were still arriving. Five vessels, including one with hay, arrived on 5 November, as did the *Sally* with Lord Paget's horses from Lisbon. Bowen explained that he was accumulating numerous troop ships, victuallers and store ships at Santander and intended to send three agents to manage the transport service there.[14]

At about the same time Transport Board Commissioner Towry was sent to Portsmouth to resolve delays that had been reported in the preparation of more horse transports bound for Spain.[15] By 29 November Towry reported that there was capacity for 892 horses but there had been a communications breakdown because the army commander in Portsmouth had not been advised of the arrangements. The cavalry had not arrived and the artillery horses were only just about to commence the march from their depots. In addition, there was some confusion about the number of horses expected. At that time Portsmouth harbour was exceptionally crowded with shipping and this was causing Towry some anxiety, for he was keen that embarkation should start without delay because he judged that 'a gale of wind would probably cause much mischief'. Not only that, the presence of a large number of transports in the harbour caused severe friction with the navy in relation to the anchoring arrangements. This was despite the considerable time devoted by the agents to making the best arrangements for anchoring the transports out of the way of naval ships going in and out of the harbour.[16] To add to Towry's problems, the ropes of the horse slings were of poor quality; several had given way, injuring a number of horses. The Board instructed the contractor that if any horse was injured or killed he would be accountable for the loss.[17] Finally the transports, under the charge of the agent, Captain James Anderson, with 1,230 horses and several cavalry brigades sailed from Portsmouth for Vigo on 15 December.[18]

Meanwhile, ashore in Spain, Moore's army had run into problems. His allies, the Spanish, had been routed at Durango on 31 October, at Espinosa on 11 November and at Tudela on 23 November. So in order to join up with Baird's force, Moore was obliged to break through Marshal Soult's corps in northern Spain. His success allowed the two British forces to unite on 20 December. Napoleon, who had arrived in Madrid on 4 December, responded by ordering a much larger army to the north-west of Spain. When Moore learned of this he

13 TNA, ADM, 108/81, Captain Bowen, Corunna to TB, 10 Nov. 1808, TB Minute, 21 Nov. 1808.

14 TNA, ADM, 108/81, TB Minute, 21 Nov. 1808. There is a gap in the Transport Board Minutes between 22 Nov. and 28 Dec. 1808 which impacts on the reporting of Bowen's communications.

15 TNA, ADM, 108/81, TB Minute, 25 Nov. 1808.

16 TNA, ADM, 108/81, TB Minute, 10 Dec. 1808.

17 TNA, ADM, 108/81, Commissioner Towry at Portsmouth to TB, 7 Dec. 1808.

18 TNA, ADM, 108/81. TB Minute, 15 Dec. 1808.

concluded that his best course of action was to withdraw to the coast to ship his army back to the Tagus by sea.[19] Events were moving quickly. Suddenly all of Castlereagh's plans to send more infantry, cavalry and materials to Spain were thrust into reverse: on 16 December, the day after the fleet sailed, he received early indications from Moore that he was proposing to retreat to the coast. On the same day he received a letter from Sir David Baird informing him that he was withdrawing to Vigo where he would need transports for 2,000 horses.[20] Castlereagh immediately ordered the fleet that had just sailed to return to Portsmouth, where the horses and cavalry were to be disembarked and the empty vessels sent to Vigo along with 2,400 tons of empty troop ships, 2,000 tons of store ships laden with provisions, 3,000 tons loaded with hay and 1,400 tons loaded with oats.[21] These ships were prepared quickly, but shortly after they sailed adverse weather forced part of the fleet of back to Portsmouth.[22]

Meanwhile, in accordance with earlier instructions, Major-General John Brodrick, the army commander of the forces at Corunna, was sending most of the transport ships from Corunna (which was 'ill adapted for re-embarkation') around the coast to Vigo. It had been identified as the port of departure, being 'the most eligible place for re-embarkation', by Sir David Baird and Admiral De Courcy.[23] He ordered 143 transports there on 2 December, retaining about 32 at Corunna.[24] Conditions at sea were not good. Strong easterly winds were making coastal passages hazardous. Bowen reported the arrival at Corunna of the *Perseverance* on 13 December with army provisions from Santander, having been 'beating about nearly three weeks'. By then there were 25 transports at Corunna and 156 at Vigo.[25]

On 31 December, well after the decision to withdraw had been made but presumably without knowing what was happening ashore, Brodrick ordered back to England empty twenty-seven light horse transports that had recently arrived. Three weeks later these ships would have been a welcome addition to the evacuation fleet.

In England there was a period of uncertainty. Plans in place had to be reconsidered quickly to fit the new circumstances, troops had to be disembarked and some provisions and supplies had to be hastily prepared to feed and shelter

19 Hall, *British Strategy*, 173.

20 HoCPP, 1809 (69) XI.97, 146–54, Correspondence relating to the expedition to Spain and Portugal, Jun. to Dec. 1808, Lieutenant-General Sir John Moore to Castlereagh, 24 Nov. 1808 and Lieutenant-General Sir David Baird to Castlereagh, 24 Nov. 1808.

21 HoCPP, 1809 (66) XI.1, 77, Correspondence relating to the expedition to Spain and Portugal, Jun. to Dec. 1808, Castlereagh to Lieutenant-General Sir John Moore 16 Dec 1808; TNA, WO, 6/156, Castlereagh to TB, 10 Dec. 1808.

22 TNA, ADM, 108/21/147, TB to Castlereagh, 27 Dec. 1808.

23 HoCPP, 1809 (112) XI.277, 2, Correspondence relating to the expedition to Spain and Portugal, 1 Mar. 1809.

24 HoCPP, 1809 (112) XI.277, 1, Correspondence relating to the expedition to Spain and Portugal, 1 Mar. 1809, Major-General Brodrick at Corunna to Castlereagh, 2 Dec. 1808.

25 TNA, ADM, 108/81, TB Minute, 28 Dec. 1808.

the troops and horses at Vigo and provide further ships for the evacuation. By 5 January another 36 transports were ready to sail to Vigo, including 23 empty troop ships, 4 army victuallers, 2 transports loaded with hay, 1 with clothing, 3 with camp equipment and 3 with ordnance stores.[26]

On 11 January Castlereagh advised Moore that he was sending even more transports to support the evacuation; a further 5,000 tons had been ordered around from Lisbon accompanied by eleven ships of the line which had also been directed to receive troops. A further 10,000 tons which were being assembled at Portsmouth were expected to be dispatched within two weeks. Castlereagh envisaged that with these there ought to be sufficient tonnage to embark at the rate of one man per ton – more crowded than normal, but this was an emergency measure.[27] Castlereagh directed that if the Marquis de la Romana's Spanish troops were to be evacuated then an additional 40–50 men should be evacuated in each horse transport. As a temporary measure the Spanish troops were to be taken, with camping equipment and stores, to the Bayonne Islands until sufficient transports could be allocated to move them on. At Corunna and Vigo there were transports sufficient for 2,670 horses, and transports for a further 1,230 horses were proceeding to Vigo from England. In the Tagus there were transports for 2,155 horses; Lieutenant-General Sir John Craddock, who was at the Tagus, had been directed to send some of these direct to Vigo. If these vessels were insufficient then the Bayonne Islands were to be used as temporary depots for horses. The hope was that not only the cavalry and artillery horses could be saved but also the draft horses and mules. To sustain these beasts there were large stocks of forage at Vigo and more was on the way.[28] However, it was later recorded that 2,000 horses were slaughtered rather than saved.[29]

Early in January, Moore realised that he was not going to be able to get the army to Vigo where most of the transports were waiting because it was too great a distance to march. He wrote to Sir Samuel Hood advising of his change of circumstances; he reported that only one corps of about 3,500 men would reach Vigo for embarkation and that some transports should remain for them, but the remaining transports should be urgently brought around the coast from Vigo to Corunna. He was exceptionally disturbed to discover on the next day that 'my letter was lost through the negligence of a drunken Dragoon'.[30] Moore immediately wrote again to Hood, who in turn wrote to the governor of Vigo advising him that he would be moving most of his men-of-war and the transports. He was wary of the Vigo citizens' response when they saw the fleet leaving. He felt obliged to advise the governor that there was a corps of troops on the way to Vigo and that as they were likely to be pursued by French troops, they would

26 TNA, ADM, 1/3757, TB to Admiralty, 5 Jan. 1809.

27 HoCPP, 1809 (66) XI.1, 92, Correspondence relating to the expedition to Spain and Portugal, Jun. to Dec. 1808, Castlereagh to Sir John Moore, 11 Jan. 1809.

28 HoCPP, 1809 (66) XI.1, 54, Correspondence relating to the expeditions to Spain and Portugal, Castlereagh to Sir John Moore, 30 Sept. 1808.

29 Oman, *Peninsula War*, vol. 1: *1807–09*, 582.

30 NMM, MKH, 111, Sir John Moore to Rear Admiral Sir Samuel Hood, 4 Jan. 1809.

immediately embark. However, if the winds were against them, preventing the vessels' escape from the bay, then the warships would have to defend them. As a preventative measure the guns in the town that were pointing seaward would be rendered unserviceable. He also advised that for their defence the warships might be compelled to fire on the town, 'which to friends who have such strong ties would grieve me to the heart'.[31]

Later, in Parliament, the government was accused of incompetence by allowing this catastrophe to develop. Castlereagh defended his record: 'It was entirely as a consequence of an arrangement between Sir David Baird and Admiral De Courcy that the transports were sent from Corunna to Vigo.'[32]

Evacuation from Corunna

Unfortunately, the majority of transports from Vigo had not reached Corunna by 11 January when Sir John Moore arrived with the main body of the army. On the 13th he wrote, 'Had I found them here on my arrival on the 11th the embarkation would easily have been effected, for I had gained several marches on the French.'[33] On the 13th the decision was taken to destroy the gunpowder magazines, which had recently been replenished, to avoid them falling into French hands. In two huge explosions 4,000 barrels were destroyed; two people were killed by falling debris.

The main body of transports did not arrive back at Corunna until 14 January. Subsequently, Commissioner Bowen, working closely with Rear Admiral Sir Samuel Hood, played a principal role in the evacuation, which began immediately the transports were available.[34] There were over 30,000 British soldiers plus several thousand Spanish troops, horses, equipment and stores to be evacuated. This was a huge challenge to be accomplished in a very short time. Two evacuation points were established, one at the quay near the Citadel in the town and the other at the nearby beach at St Lucia. Effective organisation at the evacuation points was paramount to the success of the evacuation. Hundreds of vessels, including ships of the line, were at anchor in heaving seas waiting to embark the thousands of wet and weary troops, many of whom were injured. The smaller vessels with shallower drafts tried to maneouvre closer inshore. The beach-masters organised the progress of the ships' boats, mostly from the warships, filled them quickly and dispatched them to the waiting ships. There was a great shortage of small boats to embark such a large numbers of troops. Some Spanish boats had been requisitioned but they were not in good condition and most had to be abandoned. Meanwhile the enemy was getting closer. It was not long before the troops were under direct fire from artillery on the surrounding hills. Despite this, before daylight on 16 January many of the

[31] NMM, MKH, 111, Hood to His Excellency the Governor of Vigo, 12 Jan. 1809.

[32] Hansard, House of Lords, 9 May 1809.

[33] HoCPP, 1809 (112) XI.277, 2, Extract of a letter from Lieutenant-General Sir John Moore to Viscount Castlereagh, 13 Jan. 1809; Hall, *Wellington's Navy*, 71.

[34] TNA, WO, 1/805, Commissioner Bowen, Corunna to TB, 28 Jan. 1809.

wounded, some prisoners, and horses and artillery of the Royal Wagon Train had been embarked. Thirty-four loaded ships were sent off under convoy of HMS *Plantagenet*. These ships included 3 with French prisoners and guards, 6 with ordnance stores, 8 hospital ships, 3 with forage, 1 baggage ship, 1 with artillery staff and forage, 4 with the 4th Hussars cavalry, 2 army victuallers, 2 with staff horses, 1 with engineers and victualling, 1 with 10th Hussars and 2 with artillery and horses.

From the beach at St Lucia the troops were brought off entirely by boats from the transports. Captain Bowen himself was brought off on the last boat with fourteen men of 26th Regiment at 08.00 on 17th. By that time the French were in sight of St Lucia and had begun to kill stragglers. Bowen then went to the Citadel where the men-of-war's boats were embarking the wounded. Major-General William Beresford was still defending his lines. By 13.00 on 17th the enemy fire had 'became very brisk'. At the same time the wind was blowing 'with great violence'. Despite these hardships the embarkation continued from the Citadel until after midnight when General Beresford left to complete the embarkation.

Bowen was to describe the embarkation as a 'complete success in unfavourable circumstances': 'The tide being out when it commenced, the troops were obliged to wade into the boats up to their necks in water, the night was excessively dark, the transports were obliged to lay out at a considerable distance, a gale of wind whipped up the surf greatly endangering the safety of the boats.'[35] The boat crews from the men-of-war were able to take breaks for rest and food whilst colleagues manned the boats and continued shuttling men and materials. The merchant crews, being much smaller in number, had no one to spell them and had to keep toiling away; they became so exhausted that they found it difficult to row to windward in the strong squalls. Similarly, the agents had to work without breaks for three days and two nights, 'almost without refreshment, without rest'. The saving grace was that the wind eventually swung round to the south, a great stroke of luck. As a result of these efforts 3,000 troops were evacuated from Vigo and 25,097 from Corunna. Bowen was pleased to report that the artillery attacks had not given the French the satisfaction of 'sinking a single boat or destroying a single man'. However, five or six transports had been lost in the melee due to the 'violence of the winds and occasional confusion in the minds of masters'. Bowen pronounced that 'the general conduct of the Transports has been highly meritorious, notwithstanding the actions of a few, some masters had remained to get their allotment of troops long after the signal to weigh had been made'.[36] However, congestion caused by so many ships meant that the travails of the transport masters continued even after they received the troops and horses. The master of the *Belle*, a horse ship with twenty-six horses on board, tried to leave the bay but there were so many ships at anchor that despite tacking from side to side he could not make progress. He anchored again and for some unknown reason left the vessel. Lieutenant Fisher of the *Resolution* was sent by his captain

35 TNA, WO, 1/805, Commissioner Bowen to TB, 28 Jun. 1809.

36 HL, STG Collection, STG 136 (21), Captain James Bowen to TB, 28 Jan. 1809.

to ensure that the *Belle* did not fall into enemy hands. After refusing to allow the master to re-board, he managed to get the ship and all twenty-six horses safely back to Portsmouth.[37]

This successful evacuation was probably one of the Transport Board's finest moments. On 31 January the Board sent Bowen's report to Castlereagh, commenting, 'We flatter ourselves that it will appear to your lordship, notwithstanding rumours which have gone abroad, that Commissioner Bowen and the several agents under his direction, performed their duty, in the embarkation of the army at Corunna, with judgment, activity and unwarranted perseverance.'[38] Bowen declared that he was much indebted to Rear Admiral Sir Samuel Hood 'for his countenance and support he gave me and for his ready compliance with all my wishes'. The two men clearly got on well; their professional relationship was the foundation of the successful evacuation. In the days prior to the evacuation, Bowen had moved from his quarters on the *Montreal* transport to the flagship *Barfleur*, to strengthen the communications between the two men.[39] However, it was Sir Samuel Hood who received the approbation of the House of Commons for his 'effectual and able disposition of the ships and transports in the embarkation of HM's troops at Corunna'.[40]

Some army officers were very critical of the evacuation. No doubt some were justified in their comments but others were ungracious in their exaggeration as they sought to promote demands to use naval ships as troop transports. Indeed Lord Melville (Henry Dundas) used some of these views to reinforce his motion that naval ships be used as troop ships, declaring that 'The naval and military officers who superintended that midnight embarkation, endured by far more anxiety than they had experienced in the hour of battle, owing to the want of order and discipline amongst the transports.' He claimed that 'This want of discipline produced the utmost confusion and embarrassment and created in the minds of those present the greatest alarm for the fate of the army.' His view was that serious consequences were only prevented by the zeal and active exertions of the officers and seamen of the navy and of the superintending commissioners of the Transport Board.[41]

It has been suggested that Moore's apparent failure overshadowed the fact that his campaign had irretrievably shattered Napoleon's plans for the conquest of the Peninsula, thus enabling Wellington to go back to a 'free' Portugal to return to the fray.[42]

37 TNA, ADM, 1/1546, Captain's in-letters.
38 TNA, ADM, 108/21/156, TB to Castlereagh, 31 Jan. 1809.
39 HL, STG Collection, STG 136 (21), Captain James Bowen to TB, 28 Jan. 1809.
40 Hansard, House of Commons, 1 Feb. 1809.
41 Melville, *Speech*, 9.
42 Gates, *The Spanish Ulcer*, 115.

The Cost of the Board's Peninsular Service

Following Corunna, Parliament was keen to have an understanding of the cost of transports employed in the service of conveying the army, ordnance, stores and provisions to Portugal and Spain in 1808.[43] The Board reported that it had used 803 transports of 194,670 tons to ship troops, horses, artillery, wagon trains, supplies and provisions at a cost of £1,178,112 and in addition had spent £114,669 fitting out transports with bedding, cabins, horse stalls and various other stores including forage for horses. The Victualling Board produced detailed listings of the supplies and provisions shipped to Spain and Portugal. These included beef (4,346,192 pounds), pork (2,498,358 pounds), flour (3,886,296 pounds), bread (10,175,403 pounds) and spirits (374,123 gallons).[44] The Ordnance Board produced a schedule of arms and ammunition sent out to Spain and Portugal as shown in table 8.1.

Table 8.1. A return of the arms and ammunition sent to Spain and Portugal between 1 May 1808 and 20 March 1809

Guns	98	Pistols	316
Howitzers	38	Swords	61,391
Carronades	20	Pikes	79,000
Mortars	6	Infantry accoutrement sets	39,000
Ammunition for guns	31,600	Ball cartridges	23,477,955
Ammunition for howitzers	7,200	Lead balls	6,060,000
Ammunition for carronades	4,000	Lead (cwt)	10,544
Musquets	200,177	Powder barrels	15,428
Rifles	120	Flints	2,442,080
The Royal Artillery contingent consisted of 2,255 officers, non-commissioned officers, gunners and drummers and 1,684 drivers and 2,469 horses.			

Source: HoCPP, 1809 (82–85) XI.193,12, Supplies embarked for Portugal and Spain, 20 Mar. 1809.

Wellington's Return to Portugal in 1809

In March and April 1809, 20,000 infantrymen and several regiments of cavalry were embarked for Portugal. Following the return of some of these ships, the Board received an instruction from Brigadier-General Charles William Stewart, military undersecretary (and Castlereagh's half-brother), on 5 April 1809 specifically directing that 8,000 tons of three-month ships were to be discharged

[43] TNA, ADM, 108/21/160, TB to Brigadier-General Stewart, 27 Feb. 1809.
[44] Sutcliffe, 'Bringing Forward Shipping for Government Service', 287.

and paid off.[45] By that time freight rates were reducing. The Transport Board responded quickly to this development: on 20 April 1809 the Board issued a notice to owners via the brokers 'that any owners not accepting the new rate of 20 shillings per ton to find that their vessels will be returned to this country for discharge and are unlikely to be hired again'.[46]

Wellington returned to Portugal in April 1809 to assume command of all British and Portuguese troops. He drove Marshal Soult and his forces out of Oporto on 12 May and then crossed into Spain and achieved a further, though costly, victory at Talavera. Faced with the threat of a renewed onslaught from a strengthened French army he retreated into Portugal where he built the Torres Vedras defensive system to protect Lisbon. Wellington's correspondence clearly demonstrates both his awareness of the importance of the transport service and his significant involvement in the transport arrangements. On 5 April 1809 Castlereagh had forwarded to Wellington a list of eighty-three transports which were in Portugal, requesting that after careful survey he send home as many as he felt able, retaining only those that he considered 'indispensably necessary to the security of your army, in the event of its being obliged to re-embark'. He was advised that a regiment of cavalry was under orders to join him but as there was a great shortage of horse transports he should send sufficient for 800 horses from the Tagus to Cork, 'as the expense of horse transports presses severely on the public finances at the present moment'.[47] By 24 April this had been arranged, plus sufficient tonnage for an additional 300 horses for the Commissariat.[48]

On 7 May Wellington confirmed that before he left Lisbon he had made arrangements for the return of 5,000 tons of infantry ships plus tonnage of every description. However, he was not sure if these instructions had been executed because 'The Agent of transports is the worst hand of that description, and as you know transports are not solely under my directions.' This was pretty strong condemnation of the transport agent and also possibly an indication of Wellington's frustration that he shared control of the transports with the naval commander. He requested that the current agent be replaced by Lieutenant Fleetwood, an agent with whom he had worked before. This commitment to return transports was agreeable not only to Castlereagh but also to the Treasury to whom Wellington wrote, 'You may depend upon it that I shall keep the expenses as low as possible. I have already made arrangements for sending home a quantity of transports and victualling tonnage.'[49] Vice Admiral George Berkeley, the naval officer commanding the ships off the Peninsular coast, was also aware of the costs of retaining transports. He proposed to Wellington that the cargoes of some of the store ships containing provisions, hay and oats should be unloaded and stored ashore in depots. Wellington agreed: 'The hire of store

[45] TNA, WO, 6/156, Castlereagh to TB, 5 Apr. 1809.

[46] TNA, ADM, 108/81, TB Minutes, 20 Apr. 1809.

[47] Castlereagh, *Correspondence*, II, 50, Castlereagh to Sir Arthur Wellington, 5 Apr. 1809.

[48] Gurwood, *Dispatches* III, 190, Wellington to Castlereagh, 24 Apr. 1809.

[49] Gurwood, *Dispatches* III, 213. Wellington to W. Huskisson, Secretary to the Treasury, 5 May 1809.

houses will be a much less heavy expense upon the public than that of victuallers, which may, of course, be sent back to England.'[50]

The transports on the Portuguese station were in constant demand to withdraw troops and to bring in reinforcements, particularly to and from Gibraltar whence troops were sent for periods of relief from battle duties: on 14 June, 'The 23rd Light Dragoons having arrived, the 20th will proceed to Sicily as soon as transports shall be ready for them'; on the 15th, 'Please arrange transports for the 2nd Battalion of the 9th Regiment to Gibraltar to relieve a garrison so that those troops might be returned to Portugal.' Such movements were originated or sanctioned by Wellington. When he wanted transports he always communicated with Admiral Berkeley rather than directly with the transport agent, since that relationship had broken down completely.

Wellington was continually issuing requests to Castlereagh for replacement horses and for further supplies to be transported to Portugal. For instance, on 31 May he requested the supply of 30,000 pairs of shoes of the best quality and 1,500,000 pounds of biscuit, 3,000,000 pounds of hay and 3,000,000 pounds of oats; and on 11 June 1809 he sent a request for '40,000 knapsacks, 5,000 saddles and bridles, 6,000 carbines, 6,000 swords, 4,000 pairs of pistols, 2,000 rifles and 5,000 saddlebags'. These were needed for the Portuguese allies.[51]

Preparations for Walcheren

Back in England, George Canning, the Foreign Secretary, had wanted to attack Flushing for some time as he believed that there were five enemy ships of the line there. He also hoped that a continental expedition might encourage Austria, which was vacillating, to move against France.[52] A successful military venture would also help to eradicate memories of the failure of Corunna. He was supported by Lord Mulgrave, the First Lord of the Admiralty, who was naturally in favour of an attack on France's naval capacity. When the Earl of Chatham was appointed to command the operation he was told that it was to be nothing else but a 'coup de main'. It was intended that once the naval forces had been destroyed, the troops, excluding a small garrison on Walcheren Island, were to return home immediately.[53] Unfortunately many of the transports that would be needed for this expedition were in Portugal.

On 22 May Castlereagh wrote again to Wellington explaining that there was tonnage for 5,654 horses in Portugal and directing him to return shipping for 3,000 horses to Portsmouth where they were urgently required to support the Walcheren campaign. Castlereagh explained, 'I do not make this order peremptory but unless a crisis is anticipated send them immediately. Much depends upon their arrival being accelerated.'[54] Although he had not received

50 Gurwood, *Dispatches* III, 263.
51 Gurwood, *Dispatches* III, 265, 289.
52 Hall, *British Strategy*, 177.
53 Hall, *British Strategy*, 179.
54 Castlereagh, *Correspondence* VII, 69, Castlereagh to Wellington, 22 May 1809.

that letter, on 31 May Wellington complained that he was not being kept informed of the situation regarding the transports because 'the Agent of transports keeps so much aloof from us'; he had already requested a replacement. Despite this, he had begged Vice Admiral George Berkeley to send home all the three-month infantry ships and a great proportion of the cavalry ships.[55]

Wellington did not receive Castlereagh's letter of 22 May until the night of 6/7 June – it had taken over two weeks to catch up with him.[56] No doubt this was because, as he said in another letter to the admiral of the same date, he had 'been constantly moving'. Immediately on 7 June he wrote to Berkeley: 'I enclose the copy of a dispatch, which I received in the night from the Secretary of State. It is my opinion that the transports for 3,000 horses may, with safety, be sent to England as soon as you think proper to dispatch them.'[57] However, some of those cavalry ships had already been dispatched to Oporto to collect French prisoners of war, captured there when Marshal Soult and his force were driven from the city on 13 May, to be delivered to England. It was to be another four weeks before the first batch of horse ships arrived at Portsmouth on 11 July. Castlereagh had been under the impression that these transports were under sailing orders at Lisbon on 5 June but this was clearly not the case.[58] The main body of the fleet arrived just seventeen days before the Walcheren fleet finally sailed.

The Board had been advised of the need to raise significant tonnage on 20 May but there was no clarification at that time about the mix of tonnage required, nor was the objective defined. The lack of clarification about the required tonnage was probably a result of Castlereagh's only just starting to formulate the plan of the operation; at that time he had not determined the nature of the force that would be sent. On 20 May there were 23,000 tons of unallocated transports at home. Only 24,000 tons were forthcoming over the next few months despite the rate being 25s per ton. The 48,000 tons of cavalry and infantry ships returned from the Peninsula were absolutely vital to make up a complement for the fleet of over 100,000 tons.[59] Later Rupert George was to explain that although the Board had given serious consideration to recommending an increase in the hire rate to 30s per month per ton, it had finally agreed that the benefit of this in terms of additional tonnage raised would be materially offset by the 'great inconvenience' to the service generated by the 'grave discontent' of existing transport owners. 'I believe', he told the Scheldt Inquiry, 'that my opinion in some measure decided the case.'[60] However, the Board did take up some ships for six months certain, and in some cases twelve months, instead of the usual three. This encouraged

55 Gurwood, *Dispatches* III, 266–74, Wellington to Castlereagh, 31 May 1809. At this time he had not received Castlereagh's letter dated 22 May. He did not receive it until the night of 6 June, thus his earliest letter on the subject to Admiral Berkeley is dated 7 Jun. 1809.

56 Gurwood, *Dispatches* III, 273. Wellington to Vice Admiral Berkeley, 7 Jun. 1809 and Wellington to Castlereagh, 7 Jun. 1809.

57 Gurwood, *Dispatches* III, 274, Wellington to Vice Admiral Berkeley, 7 Jun. 1809.

58 Castlereagh, *Correspondence* I, 281,Castlereagh to HM King George III, 21 Jun. 1809.

59 HoCPP, 1810 (12) 37, Scheldt Inquiry, Sir Rupert George's evidence.

60 HoCPP, 1810 (12) 37, Scheldt Inquiry, Sir Rupert George's evidence.

some ship-owners to come forward that might not have done so otherwise, and because the price was at the current rate this arrangement did not antagonise existing transport owners.

During May and June Lieutenant-Colonel James Willoughby Gordon, military secretary to the Duke of York, proposed that naval ships be used to convey troops.[61] This proposal was supported by Captain Home Popham, who had written to Castlereagh that 'Transports are the greatest clog to every sort of expedition, particularly those in which promptness and celerity are so essential to success.' He argued that embarking whole brigades into fewer ships would enable them, their horses and their equipment to arrive together and allow them to go into action 'on the first moment of their landing'.[62] The Admiralty proposed that the operation should consist of nine sail of the line and 87 gunboats unencumbered by troops, but that up to 21,400 troops could be conveyed in other naval vessels including eighteen 74-gun ships, which could carry 550 troops each, and fourteen frigates, which could each carry 250 men, together with a number of flutes, sloops and gun brigs. It was finally agreed that 17,000 men would be conveyed in naval ships.[63] This naval support was crucial given the Board's difficulty in procuring enough transports. It had even considered procuring some of the great number of foreign ships that were in the Thames at that time but, in a discussion between Rupert George, the minister and the chancellor of the Exchequer, it was agreed that it would not be proper to take them.[64]

Given the large body of troops allocated to this campaign it was necessary to embark them from numerous ports, so they were marched to Portsmouth, Dover, Ramsgate, Chatham and Harwich. On 17 July Castlereagh advised the king that 'The whole force, with the exception of the cavalry, will be on board tonight.'[65] It was expected that of the 616 sail, 352 would be transports and 264 ships of war.[66] The fleet, with over 44,000 men and 4,500 horses on board, finally sailed on 28 July 1809, only ten weeks after the Transport Board had received the first notification of the requirement. This was very speedy preparation, achieved mainly because almost three-quarters of the transports used were already in service and a number of naval troop ships were quickly made available.

The men-of-war were provisioned for four months for their complement and the transports carrying troops had six weeks' full allowance together with an additional fifteen days' supply of meat and thirty days of bread for 40,000 men. In addition the commissary-general was funded and ordered to establish supplies of provisions locally in an attempt to avoid the anticipated exploitation

61 Castlereagh, *Correspondence* I, 257, Lieutenant-Colonel Gordon to Castlereagh, 31 May 1809.

62 Castlereagh, *Correspondence* I, 273, Sir Home Popham to Castlereagh, 13 Jun. 1809.

63 HoCPP, 1810 (12) 37, Scheldt Inquiry, Sir Richard Strachan to Rear Admiral Otway, 27 Jun.1809.

64 HoCPP, 1810 (12) 37, Scheldt Inquiry, Sir Rupert George's evidence.

65 Castlereagh, *Correspondence* I, 286, 17 Jul. 1809.

66 Castlereagh, *Correspondence* I, 283, 15 Jul. 1809

and inflated prices that the local population might try to exact.[67] However, in mid-August further provisions were sent for 20,000 men for a month plus oats for 3,000 horses together with a supply of coals. It was expected that even though the local population would no doubt inflate prices, fresh meat would be purchased from them at the best price that could be negotiated.[68] Once again the transport agents received criticism, this time from Rear Admiral Sir Richard Strachan, the naval commander. He was frustrated with his inability to communicate with them: 'I can never get an agent of transports near me and from the neglect of these people the light [empty] transports are gone into Veer Gat with the loaded ones.'[69]

Once it landed, a combination of bad weather, inadequate planning and incompetent leadership meant that the expedition made little progress. Valuable time was lost. Flushing finally surrendered on 16 August but by that time the French had taken the opportunity to reinforce Antwerp's garrison and strengthen its defences, and by 27 August Chatham acknowledged that there was little or no chance of capturing the port. By 2 September the health of the troops was deteriorating rapidly. On Chatham's advice that the expedition be aborted Castlereagh confirmed the order to commence the evacuation of Walcheren down to the minimum troop level need to defend the island.[70] Meanwhile further medical men were sent to care for the sick soldiers.[71]

Once again, however, no consideration was given in the planning to the disposition of the transports when they had disembarked troops and horses. On 26 September the Board sought guidance on the future plans for the 122 cavalry ships of 27,576 tons fitted for 3,456 horses which had recently arrived from Walcheren: 'Are they to be completed for further service or discharged?'[72] Castlereagh confirmed that any unappropriated cavalry ships should be discharged.[73] The subsequent reduction caused some delay when Liverpool instructed the Transport Board to prepare to ship 1,817 horses to Lisbon only three months later.[74]

On 28 September Captain Daniel Woodriff, Principal Transport Agent at Flushing, requested directions for action in relation to the many ordnance and other store ships stated to be now lying idle at the Scheldt.[75] Two weeks later it

[67] Castlereagh, *Correspondence* I, 305, Mr Huskisson to Commissary-General Robinson, 17 Aug. 1809.
[68] Castlereagh, *Correspondence* I, 306, Mr Huskisson to Commissary-General Robinson, 17 Aug. 1809.
[69] HoCPP, 1810 (12) 37, Scheldt Inquiry, Sir Richard Strachan to the Earl of Chatham, 2 Aug. 1809.
[70] Castlereagh, *Correspondence* I, 320, Castlereagh to Chatham, 2 Sept. 1809.
[71] TNA, ADM, 108/21/189, TB to J.F. Robinson, newly appointed military undersecretary, 9 Sept. 1809.
[72] TNA, ADM, 108/21/192, TB to Castlereagh, 26 Sept. 1809.
[73] TNA, WO, 6/157, Castlereagh to TB, 28 Sept. 1809.
[74] TNA, WO, 6/157, Liverpool to TB, 27 Dec. 1809.
[75] TNA, ADM, 108/21/193, Captain Woodriff, Principal Agent at Flushing to TB, 28 Sept. 1809.

was reported that the transports there were suffering considerably from loss of anchors and cables because of the 'foul ground' and that it would be impossible for them to remain there in safety. Many of them contained ordnance stores which the Board was anxious to land or return to Sheerness.[76] It had sought advice from the Ordnance Board which in response denied further responsibility for the supplies: 'The stores in question were no longer an ordnance charge as they had been supplied on the signification of the King's pleasure by Mr Canning, then Secretary of State for the Foreign Department.' The Ordnance Board advised the Transport Board to seek directions from the foreign secretary himself.[77] The Board also sought guidance regarding the two transports which had been appropriated for over a month to receive recruits from the prisoners recently arrived from Walcheren, at an annual cost of £14,922: 'Can these men be landed?'[78]

Although Chatham had ordered the bulk of his command home on 2 September, Walcheren was not evacuated completely until November due to the vague hope of stimulating Austria to renew hostilities. A statement of transports at the Scheldt or preparing to go there to remove the troops dated 11 November shows that there were still eighty-five transports involved as well as thirty-one others preparing for the Peninsula which might be diverted to the Scheldt if necessary. By 28 November 1809 the majority of the troops, both sick and healthy, had been removed, leaving only the force that was to remain for some time, together with enough transports to accommodate them all.[79] Meanwhile horse transports were anxiously awaited so the surviving horses could be repatriated. The military fiasco was exacerbated by the high levels of sickness due to insanitary conditions and bad weather: fever was rife, 4,000 men died and 11,000 were classified as sick.

The transport cost was assessed to be £280,966 but this figure excludes the cost of the 48,000 tons of transports that Wellington had arranged to ship back urgently to England to be used on the expedition. It was claimed that these ships would probably have continued to be used off Spain and Portugal had they not been wanted for this service. The gross cost before this deduction was £501,101. The main expenses were not only the shipping of troops, horses and ordnance stores but also the cost of building cabins for officers and stalls for horses and the provision of forage.[80] As Castlereagh stated in Parliament when justifying the cost, three-quarters of the Walcheren force had been carried in ships already in government service that could not have been paid off even if the service not been undertaken.[81] Given the experience of Corunna and the recent evacuation from the Scheldt, Castlereagh was forced to consider the implications of an urgent

76 TNA, ADM, 108/21/194, TB to Castlereagh, 10 Oct. 1809.

77 TNA, ADM, 21/195, TB to the Earl Bathurst, 17 Oct. 1809.

78 TNA, ADM, 108/21/194, TB to Castlereagh, 10 Oct. 1809, the ships were the *Norfolk* (642 tons) and the *Royalist* (420 tons).

79 TNA, WO, 6/157.

80 Sutcliffe, 'Bringing Forward Shipping for Government Service', 298.

81 Hansard, House of Commons, 26 Mar. 1810.

evacuation from Portugal should the need arise. He promised Lord Wellington that he would ensure that there were enough ships for the reception of the whole army in the Tagus as soon as the transports returned from the Scheldt. However, this would be at a significant price. Indeed the 40,000 tons that would be required would cost £50,000 per month. In Castlereagh's view it would be too expensive to retain horse ships indefinitely. At £10 per horse per month, it would not be viable: over 5,000 horses would add another £50,000 per month to the bill. He recommended that if the army was forced to withdraw then the majority of the horses would unfortunately have to be put to death.[82] One of his last instructions as secretary of state for war was to order that the transport tonnage at Lisbon, where there was 20,143 tons, should be increased to 40,000 tons as speedily as possible.[83] This fleet was intended to lie in the Tagus to facilitate an emergency evacuation of the entire British army from Portugal should that prove to be necessary.

In addition to satisfying the demands for shipping to support the major European campaigns, the transports conducted a multifarious assortment of routine tasks. The Board committed 130 transports to the Mediterranean station and over thirty to the Baltic station. It was called upon to supply transports to support the lesser campaigns in 1808–09 to achieve the surrender of Martinique and support the conquest of St Domingo by Spanish and British troops in July 1809. Transports also facilitated the subsequent 1809–10 campaign to capture Guadeloupe, the last remaining French colony in the region. This was finally achieved on 6 February 1810, bringing to an end the British West Indies campaigns. In all more than fifty transports of over 15,000 tons had been committed to the North American and West Indies station to support these campaigns. In addition the Board sent transports to South America and the Cape of Good Hope. There were also some thirty transports involved in victualling the English Channel and North Sea fleets, and twenty-two transports were required to ship Russian troops back to Russia.

The Aftermath of Walcheren

The Portland administration fell in October 1809 to be replaced by another Tory administration, led by Spencer Percival, in which Lord Liverpool became the new secretary of state for war. Following the debacle of the Walcheren expedition the new government established a major inquiry to investigate what had gone wrong. Sir Rupert George appeared before the commission. His attendance was preceded by that of General Sir Thomas Trigge, Lieutenant-General of the Ordnance, who was asked: 'Were not transports supplied for the embarkation of the ordnance equipment, in the proportion that they were required, within the time that was necessary to complete the armament, at the period it was required to assemble in the Downs?' His response was very positive: 'The transports were

[82] Castlereagh, *Correspondence* II, 118, Castlereagh to Lord Wellington, 12 Sept. 1809.
[83] TNA, WO, 6/157.

very quickly supplied, and I have no recollection of any transports being waited for, for any considerable time.'

During his cross-examination Sir Rupert George was questioned in particular about the timing of the first instruction to prepare transports and the nature of the instruction, the timing of the readiness of the transports and whether, if the hire rate had been increased sooner or set higher, more ships might have been available more quickly. He was also challenged about the provision of hospital ships and the arrangements established once the scale of the ill-health of the troops became known. He was asked, 'Do you imagine that there would have been any difficulty in procuring a considerable amount of transports earlier in the spring, if timely notice had been given?' and 'Do you imagine that it would have been impossible to have procured transports in the spring if a sufficient price had been offered?' George did not consider that more ships could have been obtained earlier in the year at a price that was favourable to the transport service.[84] In his evidence he answered questions but he did not refer to the instruction that the Board had received on 5 April 1809 from Brigadier-General Stewart specifically directing that 8,000 tons of three-month ships be paid off and discharged.[85] Had these ships been retained then the additional cost would have been £12,000 per month for, say, two months, but the task of preparing the fleet would have been completed more speedily, assuming that the troops had been ready to embark. The Scheldt Inquiry was exercised about the difficulty of finding cavalry ships and when precisely they had been ready to embark troops. George explained that they had been ready to receive cavalry on 17, 18 and 19 July. Understandably he did not mention the significance of the transports that were sent back from Portugal which only arrived in the nick of time. The inquiry found no fault with the performance of the Transport Board.

The unintended consequence of this expedition was to alert Napoleon of the potential of the Scheldt. He subsequently refortified its naval establishments during 1810–11 by restoring Flushing's docks and arsenal, enlarging the dockyards at Antwerp and increasing the size of his fleet there. By 1812 the Scheldt formed a significant part of a revived French naval capability. Bond even suggests that but for events elsewhere this might have enabled him to seriously threaten Britain with invasion again.[86]

The forced evacuation of Moore's army following the dismal outcome of the Battle of Corunna was a national disaster but it was possibly the Transport Board's finest hour.[87] The Walcheren expedition has been described as the largest amphibious operation in the war and the greatest disaster suffered by

[84] Cobbett, *Parliamentary History*, vol. 15, 23 Jan.–1 Mar. 1810, exlii.

[85] TNA, WO, 6/156, Castlereagh to TB, 5 Apr. 1809.

[86] Gordon C. Bond, *The Grand Expedition: The British Invasion of Holland in 1809* (Athens, Georgia, 1979), 164.

[87] Mackesy, *War in the Mediterranean*, 357, citing TNA, WO, 6/56. Castlereagh to Stuart, 5 Oct. 1809.

British arms during it.[88] However, the Transport Board again performed well, responding quickly to government demands and in the evacuation of the sick. Throughout the year it used the price mechanism skilfully to attract ships, refusing to pay a higher rate which would have brought forward more ships but not in numbers sufficient to make it worth generating dissatisfaction among the existing transport owners. Where necessary it extended the usual three-months-certain commitment to six or even twelve months to attract ships.

Despite the Board's successes, both Wellington in Spain and Strachan at Walcheren were critical of the transport agents on station with them. The circumstances are not clear but Wellington, who had already requested a replacement, accused the agent of being aloof, whilst Strachan said the agents were elusive. Whether these situations were due to the workloads placed on the agents or were dereliction of duty is not clear. Nevertheless, it is clear that Sir Rupert George and his fellow commissioners were well regarded – and his elevation to the baronetcy is clear evidence of that. It is fair to view 1809 as one of the best years in the Transport Board's 23-year history.

[88] Tracy, *The Naval Chronicle*, vol. 4, 263.

9

The Turning of the Tide

From the nadir in 1809–11, Britain's prospects in the war began to improve. The navy's dominance of the supply routes allowed merchant transports to provide a relatively unimpeded, regular flow of troop reinforcements, horse replacements, arms and munitions, materials and provisions to the army based overseas. The main British forces overseas during the period 1808–15 were in: Canada, Newfoundland and Nova Scotia; the Mediterranean on Malta, Sicily and the Ionian Islands; the Peninsula, not only Spain and Portugal but also Gibraltar; India and Ceylon; and the Caribbean Islands, including Jamaica and the Leeward and Windward Islands, the Bahamas and Bermuda. However, by 1813 there were also many small forces in less immediately obvious locations such as Anholt (a Danish island in the Kattegat), the Cape of Good Hope, Honduras in South America, Madeira, Mauritius, New South Wales, Stralsund and the west coast of Africa in Sierra Leone, Senegal and Gorée.[1]

The principal overseas dispositions of the regular forces are shown in table 9.1. This analysis demonstrates that the regular army grew from the July 1808 total of 165,284 men to peak in 1814 at a strength of 238,730.[2] The line in the table for northern Europe illustrates the manpower commitment in 1809 to the Walcheren expedition; Sir Thomas Graham's attempt on Heligoland in late 1813–14 is shown under 1814 and is followed by the build-up for the Waterloo campaign in 1815. The Mediterranean line illustrates the extent of the forces committed to that theatre throughout the period, mainly to support the garrisons at Gibraltar, Malta and Sicily. The Peninsula entries show the build-up of the force under the Duke of Wellington which peaked at 63,121 in 1812. At the same time there was a constant commitment of approximately 20,000 troops in the East Indies and 30,000 in North America and the West Indies, rising further to 38,524 in 1813 to support the American war and increasing to 43,803 towards the culmination of that conflict. Table 9.1 includes foreign troops who served as pseudo-British, such as the King's German Legion, but excludes artillerymen and wagon-train drivers. In

1 Hall, *British Strategy*, Appendix 2.

2 Andrew Bamford, *Sickness, Suffering and the Sword: The British Regiment on Campaign, 1808–1815* (Norman, OK, 2013), 41. Whilst the total number of rank and file for the army as whole and the total overseas do not fully accord with those in Appendix 2 because they are produced from different sources, the table is indicative of the numbers of troops in each geographic territory.

Table 9.1 Total manpower commitments of the British Army, 1808–15, including foreign and colonial troops but excluding artillery and wagon train (home and abroad) and militia (only served at home). Home includes Ireland and Channel Islands

	July 1808	*July 1809*	*July 1810*	*July 1811*	*July 1812*	*July 1813*	*July 1814*	*July 1815*
Northern Europe	300	39,237	288	391	404	1,128	14,180	54,406
Mediterranean	22,489	23,459	27,991	30,311	35,057	38,943	23,864	17,396
Peninsula	0	36,438	40,186	56,825	63,121	59,555	5,929	0
East Indies	22,734	24,729	26,819	28,815	26,667	29,679	29,701	26,587
North America and West Indies	29,426	29,059	30,123	27,706	30,307	38,524	43,803	30,418
Other overseas stations	8,924	7,989	8,147	6,059	7,307	6,335	6,085	5,611
Total overseas	83,873	160,911	133,554	150,107	162,863	174,164	123,562	134,418
At home stations	81,411	52,486	77,468	66,579	71,117	64,566	94,422	47,073
Total regular army	165,284	213,397	211,022	216,686	233,980	238,730	217,984	181,491

Source: Bamford, Sickness, Suffering and the Sword, *41.*

January 1808 there were 23,226 artillery troops and officers of whom 5,267 (23 per cent) were overseas; by January 1810 these figures had risen to 25,245 and 6,902 (27 per cent) respectively. The militia did not serve overseas except in Ireland.

The ability to keep the British army comparatively well provisioned from the sea enabled Wellington to resist the reinforced French army when it was virtually starving, then ultimately to drive it out of the Peninsula following Napoleon's decision to withdraw a significant number of his troops from Spain to join the campaign against Russia. Transports provided further essential support to Wellington's army by removing prisoners of war, moving supplies and artillery siege trains around the coast in support of its operations, and serving as floating warehouses providing ready access to those supplies. This gave him a distinct advantage over the French, who by this stage had to move most of their materials and men overland, a much slower and more cumbersome process. The service was provided despite the numerous other demands placed on the Transport Board, including the dispatch of hundreds of transports to Canada and the coast of North America to support the military and naval forces stationed there. Chart 9.2 clearly demonstrates the changes in the disposition of the transport fleet as the war in the Peninsula climaxed in 1812–13 and that the flow of troops, supplies and provisions was diverted to Canada and to the ships off the North American Atlantic coast. The repatriation of troops and materials from the Peninsula commenced at the end of 1813. In the same year the British government began to ship arms to northern Europe and the Baltic to support Russian and Prussian troops and then arms and troops to Holland from 1813 to 1815, initially to support the allies and subsequently in response to Napoleon's progress through France and then Belgium.

Chart 9.2. The principal dispositions of the transport fleet, 1812–15

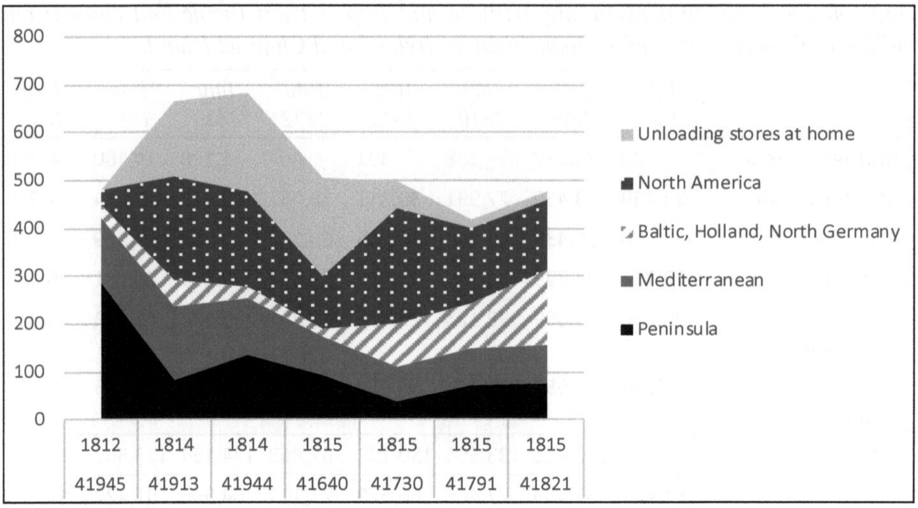

Source: TNA, WO, 1/820, 1 Apr. 1815, Abstract of all transport tonnage, also TNA, WO, 1/820, 1 May 1815; TNA, WO, 1/820, 1 Jun. 1815; TNA, WO, 1/820, 1 Jul. 1815.

By 1810 the war on the Peninsula had reached a critical stage. In the previous year Napoleon had dispatched massive reinforcements to Spain. This enabled his forces to defeat the Spanish in central Spain and opened up the way for a French offensive in the south, putting the defenders of Cadiz under terrific strain. Britain's long-standing offer of support for the town, which had previously been rebuffed, was finally welcomed. Between 11 and 17 February 1810, 2,500 Anglo-Portuguese troops were transported to Cadiz from Lisbon and Gibraltar and 3,000 Spanish troops from Ayamonte, a small coastal town at the mouth of the Rio Guadiana on the Spanish-Portuguese border, north of Cadiz. By July the British commander, Major-General Graham, had a force of 9,500 men at his disposal for the defence of the town. Subsequently the navy sent eleven ships of the line and two frigates to defend the port from attack from the sea and four bomb vessels to defend the harbour.

The French invasion of southern Spain put Wellington's army in Portugal at risk and he was forced to retreat behind the previously fortified lines of Torres Vedras, north of Lisbon, which he had established with great foresight in anticipation of such circumstances.[3] Despite the comprehensive planning of

[3] For a description of Wellington's three-line defensive plans, see Rory Muir, *Britain and the Defeat of Napoleon, 1807–1815* (New Haven, CT, 1996), 130–2: the Torres Vedras lines were 'a number of self-contained mutually supportive forts, enhanced by engineering works such as the flooding of rivers and the creation of cliffs and escarpments. The lines ran the whole twenty-nine miles from the Tagus to the sea.' The lines were mainly manned by the Portuguese army. The British army was placed behind to defend any breaches. 'In addition Wellington had fortified the Heights of Almada on the southern

his defences Wellington was far from certain of success and was determined that if he had to withdraw he would do so 'without disgrace and without loss, if possible'.[4] So he devoted considerable time and energy to planning an effective and speedy evacuation.

Planning for Evacuation

The recently formed Perceval administration was determined to continue to support the war in the Peninsula. However, the new Secretary of State for War, Lord Liverpool, shared the concern that Wellington's forces represented a major element of Britain's army and should they be overwhelmed by the French then the impact would be devastating.[5] He recognised that more transports were required to facilitate the removal of the whole force. In January 1810 the troop transports in the Tagus could not convey more than 22,000 men (at 1½ tons per man this equated to 33,000 tons) – not enough even for the British army let alone for the Portuguese contingent that would also need to be evacuated.[6] Liverpool pressed the Transport Board to advise him on progress towards assembling in the Tagus the 40,000 tons that Castlereagh had requested in the previous September.[7] Considerable progress had been made. Wellington acknowledged that 33,204 tons were already there. The 40,000 tons would be complete once the 7,000 tons then being prepared at Deptford and Portsmouth arrived.[8] The state of preparation of these transports is demonstrated in the statement of progress prepared by the Transport Office shown in table 9.3. Some of the transports were to sail direct for the Tagus once the relevant troops had embarked and the stores had been loaded; others would take troops, stores and ordnance to Gibraltar before making their way to the Tagus. However, even 40,000 tons would have been sufficient for only 27,000 men (at 1½ tons per man) and certainly no more than 40,000 (at 1 ton per man), which was probably the maximum that could be carried). Additional transports would still be required for horses, ordnance and stores.

A few days earlier Wellington, fearing the imminent arrival of even more French reinforcements, had urged Liverpool to increase the troop ship tonnage to 45,000 and he reiterated previous requests for six hospital ships and more horse ships. Even this tonnage would not have been sufficient to remove the British force of about 30,000 rank and file, which became almost 40,000 men when the officers and the sick and wounded were included. In addition there were artillery troops and officers, wagon-train drivers, engineers and staff including the commissariat. Wellington's dilemma was that he could not hire local ships to assist with the evacuation because that would forewarn the population of

bank of the Tagus together with a short line at St Julians which were intended as a last line of resort to cover the embarkation of the Anglo-Portuguese army.'

4 Muir, *Britain and the Defeat of Napoleon*, 129.

5 Duffy, 'Festering the Spanish Ulcer', 20; Hall, *British Strategy*, 190.

6 Gurwood, *Dispatches* III, 698, Wellington to Liverpool, 24 Jun. 1810.

7 TNA, ADM, 108/21/210, TB to Lieutenant Colonel Bunbury, 10 Feb. 1810.

8 TNA, ADM, 108/21/210, TB to Liverpool, 10 Feb. 1810.

Table 9.3. Statement of the progress of particular services under orders, 24 February 1810 (Transport Office)

Service	*Arrangements*	*Ships*	*Tons*
1st Battalion 9th Regiment, 787 men to embark from Ramsgate for the Tagus	In the Downs On the way to the Downs At Deptford fitting	1 1 1	537 324 310
3rd Battalion 1st Regiment, 700 men to embark at Portsmouth, 27 and 28 February 1810	At Portsmouth ready	5	1,278
		Ships	*Horses*
13th Light Dragoons, 800 horses from Portsmouth to Lisbon	At Portsmouth with troops embarked On the way to Portsmouth	22 1	758 38
Detachment of Royal Artillery drivers from Falmouth for Lisbon, 77 men and 102 horses	Sailed from Portsmouth for Falmouth 17 February	3	100
Detachment of wagon train from Portsmouth to Tagus with 100 horses	In the Downs On the way to the Downs At Deptford fitting	1 1 1	38 38 36
3 companies of Rifle Corps, 300 men, and 1 company of the staff, 60 men, at 1½ tons per man	Tonnage ordered to be appropriated		
		Ships	*Tons*
Stores for Lisbon 2,000,000 pounds of hay	Ready to load at Deptford	5	1,182
Stores for Gibraltar Freight wanted for 6,000 quarters of oats to Cadiz and 2,000 to Gibraltar	Tonnage to be accordingly provided		
Stores for Lisbon 1,085 tons of Ordnance stores from Woolwich Clothing for sundry Regiments	At Deptford loading At Deptford loading	6 1	1,149 166

Service	*Arrangements*	*Ships*	*Tons*
Troops for Gibraltar – 2nd Battalion 44th Regiment to embark from Guernsey for Gibraltar and the empty ships to proceed to the Tagus	At Portsmouth ready to proceed to Guernsey to embark the troops	4	1,060
Troops for Gibraltar – 1st Battalion 52nd Regiment to embark from Portsmouth to Gibraltar and the empty ships to proceed to the Tagus	This regiment embarked on 20, 21 and 22 February	5	1,393
Artillery and ordnance stores for Gibraltar		4	1,095
5 companies of the Royal Regiment of Artillery ready to embark from Woolwich		2	401
275 tons of ordnance stores for Gibraltar			
	Total – Troops ships	14	3,771 tons
	Total – Supply ships	18	3,993 tons
	Total – Both	32	7,764 tons
	Total – Horse ships	29	1,008 stalls

Source: TNA, WO, 1/806, Transport Office, 24 Feb. 1810.

his intentions and further exacerbate the tensions that already existed between the army and the populace. Consequently he asked Liverpool to request the Admiralty to send out more warships, anticipating that they would be required not only to complete the evacuation but to defend it should the French decide to interfere by bringing their fleet around from the Mediterranean.[9] The Admiralty subsequently allocated fourteen warships for this purpose.[10] Though he was hoping to remove about 30,000 Portuguese infantry Wellington did not believe that he had capacity for a single Spanish trooper. However, a few months later, in July 1810, he was coming to the conclusion that he would have to remove

[9] Gurwood, *Dispatches* III, 743–55, Wellington to Liverpool, 21 and 28 Feb. 1810.

[10] TNA, ADM, 1/4213, Liverpool to Admiralty, 23 Mar. 1810. Hall, *Wellington's Navy*, 96–9, four ships of the line and 9 converted warships carrying 4,000 infantry were subsequently dispatched to Lisbon.

10,000–12,000 Spanish troops serving under the Marques de la Romana. These were sound, battle-hardened troops who would be useful for service elsewhere if ever they had to leave Portugal.[11]

Lieutenant Fleetwood, Transport Agent at Lisbon, was charged with the task of preparing an embarkation plan but experienced some difficulty because the numbers of troops and the transports' availability were constantly changing. His task was made more complicated by the requirement to accommodate huge volumes of personal baggage on the store ships.[12] When the plan was completed in May 1810 it demonstrated that 12 ordnance transports had been allocated for the generals and their staff, 13 transports had been reserved for other senior officers and staff, and 128 for 36,518 rank and file. It established precisely to which transport each regiment was allocated. There was only accommodation for 1,186 horses, far fewer than were probably required. In addition 17 ships, including 12 hospital ships, had been allocated for up to 2,000 sick and wounded. Wellington and his staff would be accommodated on ships of the line. Fleetwood had managed to hold 25 troop and horse ships and 17 store ships in reserve for additional demand; these were sufficient for 270 officers, 2,960 rank and file and 542 horses.[13] The plans were comprehensive. However, unless there was sufficient time for orderly embarkation it is highly unlikely that they would have been observed. In an emergency withdrawal it was likely, as at Corunna, that troops would have been embarked on the first available ships as they arrived at the embarkation points. The plans did not anticipate how the hundreds of transports would manoeuvre into position and disperse in restricted waters.

In the meantime, Wellington was struggling to disguise the preparations to withdraw in a hurry and the shortage of cash (specie) exacerbated his dilemma. The local population, alerted to the possibility of withdrawal, was becoming increasingly reluctant to supply the army with essential provisions except for cash. 'If the army is not supplied with sufficient money the supplies must be taken from the country by force, without payment,' Wellington explained, and this would create disorder and great outrage in the local population.[14] This was also quite contrary to the modus operandi of the British army which generally paid for goods and services supplied by the local population, unlike the French army which confiscated such supplies. Wellington was also short of cash to

[11] Gurwood, *Dispatches* IV, 190, Wellington to Lieutenant-General Graham, 27 Jul. 1810.

[12] Wellington complained about the volume of baggage that accompanied the army: 'The baggage of the British army is always an embarrassment. The fact is that Regiments come on service with everything belonging to every individual in the corps, instead of there being a depot for the reception of baggage and other encumbrances that will never be wanted in the field. If it is lost there is no end of complaints and it is paid for out of the public purse.' Gurwood, *Dispatches* IV, 16, Wellington to Vice Admiral Berkeley, 22 Feb. 1810.

[13] TNA, ADM, 1/342/100, General Embarkation Scheme and Appropriation of all Transports in the Tagus and on the coast of Portugal, 9 May 1810.

[14] Gurwood, *Dispatches* VI, 788, Wellington to the Earl of Liverpool, 20 Mar. 1810. This was also quite contrary to the established conduct of the British army overseas.

pay wages, telling Liverpool, 'You cannot conceive how much want of money distresses us.'[15] Normally food and forage would be purchased locally in the vicinity of the army, but the shortage of specie meant that he had to continually move supplies around the coast from one magazine to another to keep the army supplied. This further increased the demand for transports but was a recurrent problem. Later, in 1813, he had to report again that not having enough specie to purchase provisions locally was obliging him to engage transports to move stores from Lisbon to the nearest port to the army and its allies, not only ordnance stores and ammunition but also provisions, forage and corn.[16]

When he received Fleetwood's plan Wellington advised Liverpool that 'There is sufficient transport tonnage and men of war in the river [Tagus] to embark [the British army] with its ordnance and stores and 2,000 horses' but there were not enough local vessels to take off the Spanish and Portuguese army. However, Vice Admiral Berkeley, the naval commander-in-chief in Portugal, was very concerned about the state of repair of many of the transports that were on station. These had been in the river for some considerable period. He considered that they were unfit to carry troops so he ordered a survey which led him to the conclusion that 'The state of the transports is too dreadful to be known publicly, and might almost occasion a mutiny in the army, if they knew to what rotten and dangerous ships they were confined.'[17]

Retaining transports on station was proving a challenge. There were constant demands for alternative uses. Wellington's dilemma was whether to release them for temporary assignments or to retain them for the original purpose. In August 1810 Berkeley was obliged to send twenty empty transports to England and Ireland to collect corn in lieu of subsidy for the Portuguese government. In October 3,000 tons of horse ships were required to sail to Oporto to collect cattle; in November 3,000 tons were required to bring corn from Algiers; in late December and early January 1811 there were demands for transports to transfer troops to Cadiz from Sicily and to move 6,000 stand of arms from Cadiz to the Tagus, and thirty horse ships were required to bring cattle from Oporto. In addition, there were a large number of prisoners of war to be ferried to England, but Wellington resisted the use of his precious transports for this purpose. This suited the Transport Board which was struggling to house the increasing number of prisoners already in England. For this reason the Admiralty issued 'positive directions that you do not send another French prisoner to England, on any consideration' to Berkeley. In January 1811, faced with all these conflicting demands and concerned about his position, Wellington declared that 'The plot is thickening upon us here to such a degree that I really think that we ought not to send away any more transports.'[18]

15 Gurwood, *Dispatches* VI, 698, 723, 761, 789, 793, Wellington to the Earl of Liverpool on preparing to evacuate, Jan. to Apr. 1810.

16 Gurwood, *Dispatches* VIII, 416, Wellington to Bathurst, 4 Jun. 1813.

17 Hall, *Wellington's Navy*, 98.

18 Gurwood, *Dispatches* IV, 512, 539, Wellington to Berkeley, 9 and 20 Jan. 1811.

However, this uncertainty was not to last long. No evacuation was necessary. Wellington's army was well fed and supplied thanks to the continuous cycle of convoys of transports with further provisions and materials. Defiantly, it held the lines at Torres Vedras. In contrast the French army was in real distress: it had not been paid for months and was badly clothed and seriously short of horses, carriages and equipment. The French troops were desperately short of food and survived only by plundering the region. Morale and discipline were low.[19] Inevitably, in March 1811 the French began to retreat from their positions in Portugal back into Spain. Feeling more secure with his army's situation, Wellington 'considered it very improbable that the British army can be under the necessity of embarking'. He consequently gave permission for the large fleet of transports to be reduced to a much more modest size, retaining the converted warships, coppered transports with a capacity for 3,000 troops, horse transports for 300 cavalry, a number of store ships with regimental baggage and all the hospital ships. In addition Wellington retained all the ordnance store ships to avoid having to unload them and 2,500 tons of ordnance transports to keep the battering train afloat to make it much easier to move from location to location.[20] Berkeley arranged for 148 troop, horse and store transports to be prepared for immediate return to England.[21] In September the Board was reporting a considerable increase in tonnage at home due to further arrivals from the Tagus; it sought guidance from the secretary of state about releasing excess tonnage.[22]

Further transports might have been released from Portugal but by July 1811 Berkeley was experiencing considerable difficulties storing ashore the vast quantities of provisions that were arriving daily in the Tagus. The relocation of the army behind the Torres Vedras lines had meant that the fully stocked stores from the Douro and from Montego had been transferred to Lisbon to avoid their being taken by the enemy. Berkeley was having to use transports as floating warehouses. This shortfall of storage accommodation was intensified by with the arrival of a major shipment of flour from America. Storage in transports was a considerable cost to the state so Berkeley took the unusual measure of requesting the suspension of shipments from England.[23]

[19] Gurwood, *Dispatches* IV, 389, Wellington to Liverpool, 3 Nov. 1810.

[20] Gurwood, *Dispatches* IV, 680–81, Wellington to Liverpool, and to Vice Admiral Berkeley, 21 and 20 Mar. 1811.

[21] TNA, ADM, 108/22/82, TB to Liverpool, 20 Apr. 1811.

[22] TNA, ADM, 108/22/166, TB to Liverpool, 25 Sept. 1811.

[23] Gurwood, *Dispatches* VIII, 76, Vice Admiral Berkeley to Wellington, 4 Jul. 1811. Regarding shipments of flour from America see Knight, *Britain against Napoleon*, 425–6 and Muir, *Britain and the Defeat of Napoleon*, 237. Even during the war with Britain, America exported flour for the use of the British army in Spain until 1814 when the trade links with northern Europe were re-established and the war on the Peninsula was running down. The exports, which were shipped in American boats licensed by the British government, were; in 1810, 220,000 barrels; in 1811, 830,000 barrels; in 1812, 940,000 barrels; in 1813, 970,000 barrels. Then in 1814 they reduced to 5,000 barrels.

Sometime later Wellington was invited to respond to a report on the cost of the Peninsular war. He accused the government of having 'taken an erroneous view of their expenditure and their situation in this country'. A significant element of the cost attributed to the war was in respect of the transports held to support the evacuation. Wellington argued that if he had been sent more men he would have been far more confident of victory and would not have needed such a large number of transports on standby for a swift evacuation: 'If I had 10,000 men more in 1810 then I would not have kept the transports.' In his view the saving on the cost of transports would have been much more than the incremental cost of the additional troops, which he believed had been held unnecessarily at Cadiz, in Gibraltar, in Sicily and at home. As might be expected, his view was that 'the largest army possible should have been sent to Portugal and no more than 2,000 or 2,500 should have been stationed in Cadiz … the expense at Cadiz was entirely thrown away'. He recommended that the government increase the force on the Peninsula by withdrawing troops from other parts of the empire, thus saving the 'whole expense' of the transports.[24] However, in this statement he was taking a very narrow view of the transport service.

Supporting the Peninsular army continued to demand a significant number of transports. In the five months from June 1812 transports shipped out 12,319 infantry, 3,005 cavalry and 1,588 artillery rank and file (a total of 16,912 troops), together with 4,593 horses. This would have required about 200 transports of almost 60,000 tons.[25] During the same period 3,000 troops were removed from Cadiz to join Wellington's army, and troops and materials were also being shipped to and from other stations in the Mediterranean, Sicily, Malta and Gibraltar as well as to British depots in North America and the West Indies.

By the end of 1812 the British position on the Peninsula appeared much stronger. Ciudad Rodrigo, Badajoz and Salamanca had been taken, 20,000 prisoners captured, the siege of Cadiz had been lifted and French arsenals in Madrid, Seville and on the lines before Cadiz had been taken or destroyed.[26] Nevertheless the extreme demands on the transport service continued throughout 1813. Lack of availability inevitably caused delays. In February Wellington requested Bathurst to send more horses; he had 1,000 cavalry troops without mounts and he was still awaiting artillery horses that had been promised.[27] Bathurst, in turn, complained to the Transport Board about the progress of vessels intended to convey the artillery horses from Cork to Lisbon and also detachments of infantry embarked for the Mediterranean.[28] In its defence the Board complained that Wellington

[24] Gurwood, *Dispatches* VIII, 388, Wellington to the Earl of Liverpool, 23 Mar. 1811, refers to HoCPP, 1812–13 (86) X1.215, 'An Account of the sums of money paid for service of the Army in the Peninsula and paid to Spain from 25 December 1811 to 30 March 1814' showed the cost to be: British Army £27,976,000; Spanish forces £1,791,000; subsidy £2,000,000. Total £31,767,000.

[25] Castlereagh, *Correspondence* VII, 248.

[26] Gurwood, *Dispatches* VI, 172, Wellington to Liverpool, 23 Nov. 1812.

[27] Gurwood, *Dispatches* VI, 293, Wellington to Bathurst, 10 Feb. 1813.

[28] TNA, WO, 6/159, War Department to TB, 11 Feb. 1813.

himself was contributing to the delay by preventing the return of horse transports from the Tagus.[29] Bathurst agreed to 'recommend' to Wellington that he use local and small transports, which were continually being sent out, locally and return some of the horse ships that were in the Tagus.[30]

At that time there were 259 transports on service with the army of the Peninsula engaged in a multiplicity of tasks. On the east coast 119 transports were supporting the army in amphibious operations and by transhipping ordnance stores, ammunition, provisions and corn to the advancing troops where a shortage of cash prevented supplies being obtained locally; four transports were at Cartagena and on passage from Gibraltar; four more were on the coast of Catalonia; twelve were at Cadiz; fifteen were at Corunna; eight were conveying Spanish troops to the coast of Catalonia; thirty-three were removing stores for the Spanish and Portuguese armies from one part of the country to another, whilst others were at Lisbon loading flour for Gibraltar and Alicante, refitting or preparing to return to England.[31] Some of the transports at Corunna were loaded with biscuit and flour and others with heavy train of artillery and ammunition and musket ammunition. They were ordered around to Santander a week later to await the arrival of the army that was advancing towards the port.[32] The eight transports conveying Spanish troops to Catalonia had embarked 2,000 Spanish troops at Corunna to transfer them around the south coast to join Lieutenant-General Sir John Murray's Anglo-Spanish army at Valencia.

Murray had been ordered by Wellington to recapture Tarragona. Before it fell to the French in the summer of 1811 the city had provided a secure harbour between Barcelona and Valencia for receiving supplies for the British army based on the eastern Spanish coast.[33] Murray's army of 17,059 troops with 1,453 horses and 152 mules were embarked on 28 May 1813 at Valencia and sailed on 31st to anchor to the east of Tarragona.[34] During the evening of 2 June and the following day the army disembarked, but not before a force under Lieutenant-Colonel Provost of the 2nd Battalion of the 67th Regiment had been detached under a separate convoy to capture the fort of St Felipe which dominated the main road from Valencia to Barcelona.[35] Despite the capture of the fort, Murray felt obliged to raise the siege of Tarragona when he learned that French reinforcements, which he estimated to be in excess of 20,500 men, were approaching. On 12 June he embarked the army 'to avoid a general action against superior forces approaching from Barcelona which I must have fought under every disadvantage'. In his haste to embark he ordered that the guns in the most advanced batteries be abandoned; eighteen

[29] TNA, WO, 1/813/141, TB to Bathurst, 12 Feb. 1813.

[30] TNA, WO, 6/159, War Department to TB, 18 Feb.1813.

[31] Gurwood, *Dispatches* VI, 512, Wellington to Bathurst, 4 Jun. 1813; Duffy, 'Festering the Spanish Ulcer', 24.

[32] Gurwood, *Dispatches* VI, 521, Wellington to Lieutenant-Colonel Bourke, 10 Jun. 1813

[33] Hall, *Wellington's Navy*, 137, 187; Muir, *Britain and the Defeat of Napoleon*, 271.

[34] Gurwood, *Dispatches* VII, 482, Lieutenant-General Sir John Murray's letter of events, 9 Jun. 1813.

[35] See page 140.

pieces of artillery were left behind. Murray overruled the local naval commander, Rear Admiral Benjamin Hallowell, who believed that the cannon might have been saved by waiting until dark when they could have been recovered. Murray felt that 'This was a risk that I did not wish to run for so trifling an object and preferred losing them to the chance of the embarkation being opposed and of a much more serious loss.'[36] This decision led him to face Wellington's displeasure and a court martial. Wellington's view was that the abandoned artillery and stores would be 'very important trophies for the enemy, of which he will make good use'. He also believed that Murray's failure to capture Tarragona 'entirely cripples our operations on the eastern coast'.[37] At his subsequent court martial Murray was found not guilty of most of the charges before him, but he was found guilty of abandoning his guns without due cause.

In July 1813 Wellington, as was his custom, complained bitterly about the delays in obtaining equipment for the second siege of San Sebastian. On this occasion there appears to have been some justification. An embarrassed Bathurst was forced to admit to an 'unaccountable confusion between the Transport Board and the Ordnance Board' when he discovered that a convoy of transports with guns and ordnance was not yet ready to sail even though the guns had been at Portsmouth for two months. The Transport Board protested that it had only been advised of their arrival three weeks earlier. One well-positioned observer reported to Wellington that 'nothing connected with the executive government of the country requires reform more than the Ordnance Department. It is, as now constituted, the greatest clog about the State.' A more balanced view might be that in general the performance of the Ordnance Board in 1813 was impressive, for it supplied vast quantities of arms, ammunition and equipment to every corner of Europe.[38]

A few months later, in October 1813, Wellington complained to Bathurst of a shortage of transports in the Tagus. The Transport Board pointed out that this was because transports had been sent from there to the north coast of Spain with troops and supplies and had been retained mainly to be used as store ships. There they fell under the command of Captain Sir George Collier, the senior naval officer on the north coast of Spain, who was attached to Lord Keith's Channel fleet rather than under the command of Vice Admiral Berkeley on the Lisbon station. Berkeley was therefore unable to order them back to the Tagus. Given the challenges of raising additional tonnage at home, the Board suggested that some of the 55,000 tons of transports that were supporting the army on the northern coast of Spain should be unloaded and sent to the Tagus. However, Collier protested that he had no empty transports to spare, and besides, at that time of year ships had great difficulty beating westwards against the winds, so he could not commit transports to the Tagus. With some difficulty and inconvenience, including dismantling the fittings in troop ships to convert them to store ships, 10,000 tons were raised in England and dispatched to the Tagus.[39]

36 Gurwood, *Dispatches* VI, 567, Sir John Murray to Wellington, 14 Jun. 1813.
37 Gurwood, *Dispatches* VI, 569, Wellington to Sir John Murray, 1 Jul. 1813.
38 Muir, *Britain and the Defeat of Napoleon*, 277.
39 TNA, ADM, 108/24, TB to Liverpool, 6, 9, 16 Oct. and 3, 13 Nov. and 2 Dec. 1813.

1812–14: The War with America

In 1812 already troubled Anglo-American relations deteriorated. The desire for territorial expansion into Florida and Canada was masked by expressions of frustration about the economic restrictions being enforced by Britain and by the practice of impressment of seamen from US vessels. On 18 June 1812 America declared war on Britain, despite being totally unprepared for the economic consequences that followed. It was particularly vulnerable because 92 per cent of its federal income came from customs revenue charged on imports and exports.[40] Britain itself could ill afford to release troops from Europe to defend Canada at that time. In December 1811 there were 7,625 British troops in North America. However, the surrender of Guadeloupe in 1810, which had completed the capture of all French possessions in the West Indies, meant that some troops could be sent from various stations in the West Indies to Canada. At the end of 1812 three more battalions were sent out from England by way of Bermuda to proceed to Canada once the St Lawrence was open. By the summer of 1813 the army had grown to a force of 20,000.[41]

The main British response to the declaration of war was the implementation of a Royal Navy blockade, initially in 1812–13 of the middle and southern states. It was extended in April 1814 to the coast of New England. This was no easy accomplishment as the eastern seaboard of America stretches for over 2,000 miles, with numerous harbours, river mouths and inlets. However, the blockade did not have to be watertight. It was sufficient that it dislocated and disrupted enemy trade.[42]

Although the transport service was engaged in delivering ordnance and materials, its main function was to provide troop ships and store ships for use as floating barracks and warehouses attached to the blockading force. The marines and troops barracked in these transports exploited the superior agility of sea-based troops to confuse and confound the enemy, launching numerous amphibious assaults against its depots at weak points. The store ships accommodated the victualling supplies for the troops and the huge naval presence, which comprised 77 warships in 1812, rising to 105 early in the following year.[43] Following the defeat of Napoleon in Europe, 20,000 troops were transported from Europe, including several battalions that were sent directly from Bordeaux during the spring and summer of 1814.[44] From February 1814 there were regular sailings of transports bearing troops and arms to Bermuda, Halifax and Quebec, and these continued well into 1815.[45]

40 Lambert, *The Challenge*, 63. For a comprehensive review of the economic war against America see Arthur, *How Britain won the War of 1812*.

41 Hall, *British Strategy*, 197; Muir, *Britain and the Defeat of Napoleon*, 238–9.

42 Arthur, *How Britain won the War of 1812*, 3.

43 Hall, *British Strategy*, 197; Donald R. Hickey, *The War of 1812* (Chicago, 2012), 170.

44 Lambert, *The Challenge*, 313, citing J. Latimer, *1812: War with America*, 307. The units were the 1st Battalions of 4th and 44th Foot, the 85th Light Infantry and 21st Foot.

45 TNA, WO, 1/160–2.

By October 1814 there were 217 transports on the North American station, peaking at 239 in April 1815 before the troops were returned to Britain following the Treaty of Ghent. Consequently the disposition of the transport fleet changed dramatically: as the number of transports on the American station increased, there were significant reductions in the Peninsula as troops and stores were returned to Britain (see chart 9.2). Ultimately the blockade played an important role in bringing America to the negotiating table. It had disastrous consequences for the US government's revenues through considerably reduced customs duties on declining imports and exports at a time when the cost of the war was escalating.[46] One historian has written that 'By late 1814 the British had won the war; it only remained for a bankrupt enemy to recognise the fact.'[47]

Arms and Troops to Northern Europe

The American war was a distraction, draining resources from the government's main focus on Europe. Equally it made the Transport Board's job of procuring sufficient shipping increasingly difficult. In this respect 1813 proved to be a particularly challanging year. The strain on the availability of transports required to support the army in the Peninsula and the blockade of the American coast increased in January 1813 when, following the news of Napoleon's disastrous retreat from Moscow, Castlereagh recognised the possibility of forging a new coalition against France involving Russia, Prussia and Sweden. To support this, the British government began a series of substantial arms shipments to its allies. Initially fifty-four cannon together with arms, ammunition and stores sufficient for 23,000 men were shipped to the Baltic to be allocated to the Russian and the Prussian armies.[48] Between March and November arms and equipment with a value of almost £2 million were loaded onto transports and shipped in support of the troops of Russia, Prussia, Austria and Sweden.[49] Stralsund had been occupied by the Swedes at the end of March 1813; subsequently a significant volume of arms sufficient to establish a well-provisioned magazine and a small garrison of 3,000 infantry troops were shipped there. On 24 April 1813 the Board was instructed to raise 2,000 tons of shipping which was required at Woolwich to be fitted and laden with heavy ordnance ammunition and engineers' stores.[50] On 28 April the 2,000 tons demanded was increased to 10,000 tons.[51] These vessels were for the use of Russian troops and were to be transferred to Colberg or Heligoland 'or any other such place designated by the Russian Ambassador'. On 14 May 1813 the Board was advised that the 1st Division of the Battering Train was to be embarked. This required 2,200 tons of shipping but the Board only had five transports of 1,042 tons available. Transport agents resorted to trawling

[46] Hickey, *War of 1812*, 200; Arthur, *How Britain won the War of 1812*, 227–47.
[47] Lambert, *The Challenge*, 331.
[48] Muir, *Britain and the Defeat of Napoleon*, 248.
[49] Hall, *British Strategy*, 200.
[50] TNA, WO, 6/160, War Department to TB, 24 Apr. 1813.
[51] TNA, WO, 6/160, War Department to TB, 28 Apr. 1813.

the Thames to find the other 1,150 tons. Yet more tonnage was required for the 2nd Division. It finally reached the Nore on the way to the Baltic on 21 June.[52] In light of these heavy demands for transports, Bathurst suggested that it was not efficient to send more empty tonnage to Portugal for infantry nor the promised horse ships.[53] A few days later he also agreed that scarce transports should not be used to transport the 4th Dragoon Guards, recently returned from Portugal, from Portsmouth to Hull. They were ordered to take the slower land route.[54] Transports continued to be engaged in the supply of ordnance to the allied armies throughout 1813 and into 1814.

On 15 November 1813, following Napoleon's defeat at Leipzig, the French garrison left Amsterdam. Two days later the long-expected uprising in Holland occurred. Sir Thomas Graham was appointed to lead a force into Holland. He was allotted 8,675 troops, 2,000 of whom had just returned from Swedish Pomerania. They were turned around when they reached Yarmouth Roads and sent to Holland without landing. They were followed by 1,500 foot guards who arrived on 6 December 1813 preceded by a shipment of 20,000 muskets.[55] The officers of the force were generally young and inexperienced, as were the majority of the rank and file. The expedition was a failure. There were significant losses: 381 killed, 533 wounded and 2,077 officers and men taken prisoner. They were later released on parole. The failure of the attack on Bergen op Zoom was, according to Graham, due to the officers in command of the various columns not carrying out the orders received.[56] Although the expedition to Holland failed to meet its specific purpose (the two fortresses of Antwerp and Flushing remained in French hands until the end of the war), no blame was attached to Graham. In fact he was elevated as Lord Lynedoch and remained in Flanders in command of troops from several nations until he was replaced in July 1814 by the Prince of Orange. Most of the troops remained in Holland and participated in the Battle of Waterloo in the following year.

Such was the pressure of demand for merchant shipping to accommodate both the significant rise in trade, principally due to the re-establishment of trading relationships with the Baltic States, and increased government demand that towards the end of the year charter rates rose to 30s per ton.[57] This represented an increase of 10s per ton for sheathed ships and 5s for coppered ships. Even so, very few additional ships were offered for charter. The Board even suggested to Bathurst that it might need to employ neutral ships on freight.[58]

52 TNA, WO, 6/160 War Department to TB, 14 May 1813.
53 TNA, WO, 6/160 War Department to TB, 15 May 1813.
54 TNA, WO, 6/160, War Department to TB, 24 May 1813.
55 Muir, *Britain and the Defeat of Napoleon*, 307.
56 War Office Intelligence Division, *British Minor Expeditions: 1746 to 1814* (London, 1884), 81.
57 See page 59.
58 TNA, ADM, 108/24/53, TB to Bathurst, 3 Dec. 1813.

The Repatriation of the Army

Bordeaux had already surrendered to Wellington's army when the Prussian and Russian forces fought their way to Paris on 31 March 1814. Napoleon abdicated on 6 April. The leaderships of the major powers of Europe were then locked in political negotiations to protect their individual interests in post-Napoleon Europe. Britain was interested in shaping a Europe that was free of domination by one state and that protected the security of the lowlands of Holland and Belgium. The drawn out negotiations resulted in the first Treaty of Paris, which was signed on 30 May 1814.

In Britain, the political focus was on reducing the costs of war by repatriating and discharging troops, reducing the size of the naval fleet by placing ships into ordinary and breaking up those considered old and worn, discharging officers and seamen, and releasing a significant number of dockyard workers.[59] Transports were heavily engaged in bringing home from Bordeaux, Cherbourg, Calais and Boulogne troops who had fought through Spain and France. The infantry were to be removed from Bordeaux and northern Spain; the cavalry were to ride north through France to embark from the Channel ports. In early July 1814 there were 12,828 tons of infantry transports and two hospital ships embarking troops in the Gironde and a further 2,000 tons preparing to sail there from Portsmouth. The Board had received instructions that the infantry troops had to be shipped to Cork. In addition a considerable tonnage of supply ships had been ordered to Cadiz and to the north coast of Spain to bring back more stores.[60]

Though a considerable number of transports were still involved in the repatriation of the army, the Transport Board was determined to reduce the number of transports on charter. By September 1814 there were a large number of unemployed transports at home. As no considerable quantity was required for the next few months, the Board was instructed by Bathurst to retain 20,000 tons of regular coppered transports and discharge the remaining regular transports and all three-month ships.[61] As more transports were released the Board felt more confident that, should the need arise, they could be replaced from the market without difficulty, probably at a reduced rate of hire and in a better condition.

There were still 811 transports of 231,589 tons in service on 1 October 1814; of these 217 were on the US station and 154 in the Mediterranean. At home there was a severe backlog of 132 store ships waiting to unload in the Thames. During the following month 160 transports were discharged. However, the number awaiting unloading in Britain had grown to 183. This again put severe pressure on the unloading facilities in the River and caused major congestion. The backlog was reduced to 104 by the beginning of January 1815. Under normal circumstances these store ships would have been discharged but such was the demand that the

[59] NMM, ADM, BP/34B, Navy Board to the Admiralty, 9 May 1814.
[60] TNA, ADM, 108/24, TB to Bathurst, 6 Jul. 1814.
[61] TNA, WO, 6/6/161, Bathurst to TB, 17 Sept. 1814.

transports were either sent out to collect more stores and materials from France and Spain or were fitted as troop or horse ships for further service.

By the terms of the Treaty of Chaumont (29 June 1814) Britain had committed to keep 75,000 troops in Europe under the command of the Prince of Orange. This was only achievable by adding troops to Lord Lynedoch's force, retaining the German regiments and hiring 22,000 Prussian troops at a cost of £50,000 per month. To support this army there were regular shipments of troops, ordnance, artillery regiments and material to Ostend and Antwerp. On 28 December 1814 a total of fifty-one transports of 18,451 tons were being prepared for this service: six troop ships of 2,235 tons at Cork, nineteen troop ships of 8,057 tons at Portsmouth, seventeen cavalry horse ships of 5,686 tons at Portsmouth, one artillery and stores ship of 420 tons at Plymouth, eight ordnance ships of 2,114 tons at Portsmouth.

The most complex task was in the Mediterranean where there were still seventy-seven transports in early 1815. Four were at Gibraltar, eleven at Genoa, sixteen at Malta, twenty-seven at Sicily, nine in the Ionian Islands and ten on passage from England. Lieutenant-General Lord William Bentinck and Rear Admiral Sir Josiah Rowley had decided that only tonnage for 4,000 men, 200 horses and 1,500 tons for stores was required. On their instructions all other transports were ordered to Sicily, Malta and Port Mahon for the purpose of embarking military and naval stores and provisions for England. The removal of the naval establishment at Palermo, Sicily, was already well advanced and all the transports that had been employed in the Adriatic and Ionian seas had been ordered to Malta on their way to England. The other military depots on Sicily, Corfu and the other islands in the Ionian Sea had not been dismantled so no stores of any consequence were to be removed from those places at that time. Any new transports arriving in the Mediterranean were to be ordered to Lisbon. When these arrangements were implemented it was expected that the transports in the Mediterranean at the close of the year would not exceed 12,000 tons.[62]

Another Case of Inter-service Friction

In July 1814 Bathurst took the usual step of accusing the Board of 'gross negligence'. His ire was raised when he read an army report written by Colonel De Lancey, the Quartermaster-General at Boulogne, which suggested that transports sent to Boulogne and Calais were incapable of clearing the harbours at high tide. The Board was forced to defend itself so that the charge of neglect of duty did not attach to the commissioners. It pointed out that the reason transports had been unable to leave the harbours was not related to the draughts of the ships but because of strong adverse winds.[63] The transport agent there was Lieutenant James Browne. His comprehensive report of events demonstrates quite clearly the challenges that agents experienced when dealing with representatives of

62 TNA, WO, 1/818, Extract of letter from Captain Withers Agent at Messina, 24 Aug. 1814.
63 TNA, WO, 1 / 817, TB to Bathurst, 15 Jul 1814.

other departments. He explained that the wind, which has continued blowing directly into Boulogne harbour, had 'baffled all my efforts to get any of the large ships to sea'. The strong flood tide which ran directly across the narrow channel of the harbour made it difficult to get vessels to sea unless the wind was favourable. In fact wherever possible he was employing such small vessels as were available for the conveyance of cavalry from French ports. This meant that he had been obliged to remove partitions and stalls in these ships to cram in more horses for the relatively short voyage. Between 11 and 13 July 1814 nineteen ships were loaded with 353 horses and departed for Dover.

Clearly the relationship between Browne and De Lancey had broken down at an early stage.[64] Browne explained that on his arrival at Boulogne, and after observing the tidal impact on the depth of water, he had decided that the best mode of embarkation was to order the transports into port. When he discussed the matter with Colonel De Lancey the latter had insisted that the cavalry should not be embarked unless there was a strong probability of the transports proceeding to sea immediately so that the horses would be on board for the shortest time possible. Browne disagreed; he recommended that it would be better to embark the horses in anticipation of getting the loaded transports to sea as and when the opportunity arose, but De Lancey was adamant. So when the first of the cavalry arrived on 5 July, Browne could only embark them on board such vessels as he could undertake to get to sea immediately in fair winds or foul. In spite of a foul wind, he loaded the 13th Regiment and part of 16th onto the *Lord Wellington* which then had to be warped and towed out of the harbour. This involved considerable risk to life and considerable effort by a large number of seamen and troops who manhandled the warps to get the vessel out against wind and sea. Indeed all the transports with the 13th Regiment on board were put to sea under similar circumstances.

Unfortunately during the first two days after the arrival of the cavalry the wind continued to be adverse. Then, when there was a prospect that the wind might change, Browne again asked Colonel De Lancey to embark the horses so that the vessels could be sent out on the first favourable occasion. De Lancey refused. On the morning of 8 July the wind became favourable and the agent wrote to De Lancey requesting that 300 of the cavalry embark as soon as possible, but De Lancey had gone to Calais. The commanders of the 18th Hussars and 14th Dragoons had also been ordered to take to their troops to Calais, so there were no horses to load and the first fair wind was entirely lost. From 11 July when the cavalry returned to the 16th the wind again blew directly into the harbour. However, on the 16th a favourable breeze sprang up; 1,140 horses were embarked on that day and the following morning and all the transports proceeded to sea without difficulty. Browne was adamant that no delay whatever had occurred at Boulogne solely on account of the depth of water; during the three tides on which the three largest ships could not float out of the harbour, the wind was

[64] TNA, WO, 1/818, James Browne at Boulogne to TB, 18 Jul. 1814.

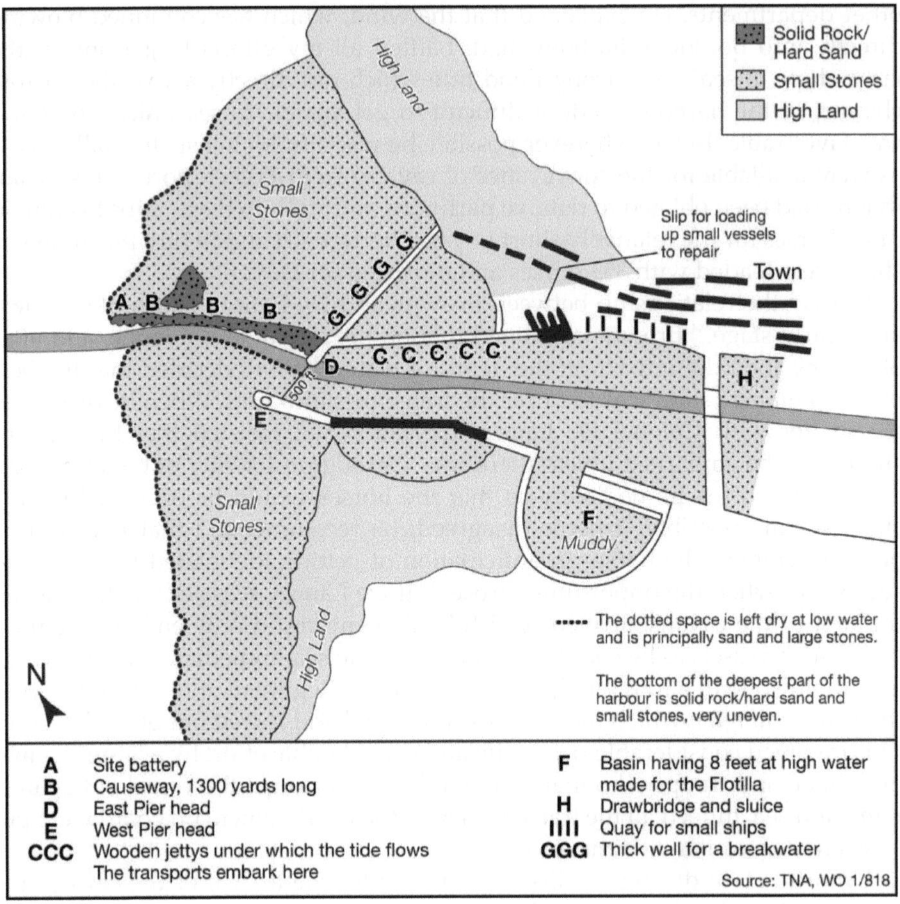

Map 7. Boulogne harbour in July 1814 by Transport Agent Lieutenant James Browne

contrary. Colonel De Lancey remained convinced that 'The transports should never have entered the harbour but should have anchored at the mouth where the horses might have been sent to them in boats.' Given that the transports could not anchor safely within 2 miles of the harbour, Browne claimed that this was the voice of inexperience. It was quite evident 'to every person well acquainted with the situation of the harbour and the state of the tides' that not only would loading so many horses whilst the ships were at anchor prove to be 'a most tedious business', it would have been exceptionally dangerous to tow boats laden with horses that distance across a strong tide. In addition the same winds which prevented the transports getting to sea would have made it very dangerous to attempt to carry horses in open boats. Browne illustrated his report to the Transport Board with a hand-drawn plan of Boulogne harbour in support of his stance (map 7).[65]

[65] TNA, WO, 1 / 818, James Browne at Boulogne to TB, 18 Jul 1814.

Browne and his team had embarked 4,800 horses. It was his opinion and that of 'others conversant on the subject' that less than half of these would have been embarked onto ships anchored in the roads in the equivalent time period, even if there had been sufficient small boats for the service. Browne and the other agents had worked tirelessly to load and dispatch a large number of troops and horses. In the period between 5 and 27 July, 4,923 horses had been landed in England from Boulogne and 3,348 from Calais, a total of 8,271.[66] There is no indication that Bathurst pressed this matter further.

100 Days to Waterloo

Napoleon escaped from Elba on 1 March 1815. This precipitated the so-called '100 days' leading up to the Battle of Waterloo on 18 June. Following his landing in the south of France, near Cannes, he gradually rebuilt his army as he progressed north towards Paris, persuading many of the troops sent to stop him to switch allegiance and join him. This forced the allies to concentrate on building up their armies in Holland and Belgium even as the peace talks progressed. Fortunately troops were still being removed from France, Spain and Portugal and some of those were sent to join the army in Holland.

Between April 1814 and March 1815 men and supplies were still being sent to North America. However, on 16 March 1815, following the news of Napoleon's escape, there was a sudden change of direction. Transports that had been allocated to bring back men from Bermuda were reassigned to convey troops and a dismounted squadron of the Light Dragoons from Cork to the Downs for onward conveyance to Ostend. A convoy of troop transports with Peninsular veterans of the 71st, 79th, 91st and 92nd Regiments, which was already on passage to Bermuda, was intercepted at sea and ordered to return to Deal from where it proceeded to Gravesend.[67] There the troops disembarked, immediately transferred to smaller vessels and sailed for Antwerp. Other troop ships were to be dispatched urgently to Halifax in order to bring directly to England the 29th, 62nd and 98th Regiments and a detachment of artillery that was urgently required to replace the troops removed from Ireland. In the meantime store ships being prepared for North America were redirected to Ostend and from that date the build-up of troops in Holland began to accelerate.

For the next few months there was a constant procession of transports back and forth across the southern North Sea between Ramsgate, Deal and Ostend. On 22 March Castlereagh announced that 8,000 men were being prepared and that 20,000 muskets were to be sent to the Dutch-Belgian forces.[68] On 25 March 1815 the Board advised Lieutenant Ross, Agent at Deal, to expect thirty-one small transports for the cavalry service to arrive at Ramsgate to ship a troop of the

66 TNA, WO, 1/818, TB, 27 Jul. 1814, Abstract of the number of Horses which have been landed from Boulogne and Calais.

67 Ian Fletcher, *'A Desperate Business': Wellington, The British Army and the Waterloo Campaign,* (Staplehurst, Kent, 2001), 12.

68 Muir, *Britain and the Defeat of Napoleon*, 353.

11th Light Dragoons that was due to arrive there the following day. Meanwhile Ross was busy dealing with five other troop ships of 1,664 tons that had arrived from Portsmouth with 1,399 men from the 51st and 23rd Regiments, 145 women and children, and 19 staff horses. There they were to be transhipped into smaller vessels as soon as these became available. More small ships were expected from Dover, although the 'boisterous' weather was causing some delays. Four other troop ships had already sailed for Ostend that day and three more were preparing to sail on the evening tide.[69]

Other ports in Kent and Essex were just as busy. On the same day the Board was advised that there were two troops of Royal Horse Artillery, a detachment of the Rocket Troop of Artillery and a detachment of Corps of Artillery drivers being held at Colchester, Canterbury, Chatham, Northfleet and Woolwich awaiting embarkation at Harwich, Ramsgate, and Sheerness – in all 155 officers, 1,204 men, 810 horses and 34 wagons.[70] On 29 March 1815 a further nineteen troop ships sailed from the Downs for Ostend.[71]

The troop ships were unloaded quickly and immediately ordered back to England. Some returned with wounded soldiers. However, victuallers took some time to unload, and store ships and ordnance ships were detained much longer. They were used as floating warehouses, but the large numbers created havoc in the harbours and were frequently diverted to ports where troops were not landing.

Amid this frenetic activity the Board was requested to send transports to bring home 3,000 men from Halifax and 22,000 from Quebec. Ten troop transports of 3,807 tons were appropriated for Halifax and seventy-eight of 28,526 tons for Quebec. It was planned that the shortfall of 700 tons for Halifax and 5,500 tons for Quebec would be supplied from transports already on the coast of America.[72] On 12 April 1815 the Board advised Bathurst that a further thirty troop, horse and ordnance transports had sailed for the Netherlands.[73] In total, during the month from 25 March to 25 April the British army in the Low Countries increased by 12,478 rank and file and by a further 9,181 during the following month. The remarkable build-up was achieved by stripping Britain and Ireland of their garrisons; this resulted in vigorous protests from the Irish government which was gravely concerned about stability and the risk of further internal rebellion.[74]

Wellington was appointed to take command of the allied army from the Prince of Orange and he arrived in Brussels in April 1815. One of his early requests to

[69] TNA, WO, 1/819, Lieutenant Ross, agent at Deal, to TB, 26 Mar. 1815.

[70] TNA, WO, 1/819, Detachment of Rocket troops and Royal Artillery held in readiness to embark.

[71] TNA, WO, 1/819, List of TNA, WO, 1/819, Lieutenant Ross, agent at Deal, to TB, 26 Mar. 1815; Detachment of Rocket troops and Royal Artillery held in readiness to embark; List of transports and small craft sailed from the Downs for Ostend, 29 Mar. 1815.

[72] TNA, WO, 1/819, Transport Office, 5 Apr. 1815.

[73] TNA, WO, 1/819, TB, 12 Apr. 1815, List of Transports and hired vessels that have sailed for the Netherlands.

[74] Muir, *Britain and the Defeat of Napoleon*, 353.

Bathurst was for more artillery. Only seventy-two pieces had arrived; he had requested 150 but had been advised that he would receive no more than eighty-four. Wellington still felt he needed 150. He was also short of musket balls, entrenching tools, carts for the engineers, pontoons and especially horses and drivers. He needed 4,800 horses and so far had 2,000. He believed that he might be able to procure the horses locally but the shortage of good drivers meant that he was having to use 1,400 valuable infantry troops to take care of the horses – troops he could ill afford. He requested that a detachment of Portuguese dismounted dragoons be sent to handle this task.[75]

Between 8 and 23 April 1815 eight regiments of cavalry – comprising 7,540 troops, 386 women, 135 children and 3,400 horses – were ferried in 160 transports from England to Ostend, with the loss of only one artillery horse.[76] By 23 April, there were twenty-nine transports at Ostend waiting to be unloaded, full of ordnance stores, guns, powder, shot and mortars as well as wagons and carts to transport them.[77] On 7 May 1815 another twenty-two transports sailed from Cork to Ostend with 759 men of the 32nd Regiment, 568 of the 92nd, 143 horses, a number of the 13th Light Dragoons and contingents of the 28th and 79th, troops and horses of the Wagon Train, medicines and ordnance stores.[78]

That same week a significant number of smaller horse ships, each carrying between eighteen and forty artillery horses and twelve to thirty drivers, sailed from Dover; two transports with men of the 28th Regiment sailed from the Downs; four victuallers sailed from Deptford; and finally eight transports with the battering train and a further troop ship sailed from Ramsgate. All were bound for Ostend.[79] Then on 15 May a further twenty-six horse ships with 800 men of the 1st and 13th Dragoons and their mounts sailed from Ramsgate to Ostend and seven troops ships with 639 men of the 42nd Regiment and 707 men of the 3rd Battalion of 1st Regiment sailed from Cork. In addition, two troop ships sailed from the Downs with part of the 28th Regiment and a victualler sailed from Deptford to Antwerp.[80]

This pace of activity did not falter throughout the second half of May. By the end of the month Wellington had 36,000 British and King's German Legion rank and file under arms compared with only 14,000 two months earlier. The build-up continued into June and beyond. There were regular sailings of troop ships,

[75] Gurwood, *Dispatches* XII, 320, Wellington to Bathurst, 21 Apr. 1815. To demonstrate that his request for field artillery was reasonable he explained that the Prussian army took into the field 80 batteries, manned by 10,000 artillerymen and the batteries were of eight guns each, so they would have about 600 pieces: 'They do not take this number for amusement.'

[76] TNA, WO, 1/820, Captain John Hill, Transport Agent at Ostend, 23 Apr. 1815, Disembarkation returns.

[77] TNA, WO, 1/820, List of Transports with Ordnance stores remaining at Ostend, 23 Apr. 1815.

[78] TNA, WO, 1/820, Transport Office, 7 May 1815, List of Transports reported to have sailed from Cork to Ostend.

[79] TNA, WO, 1/820, Transport Office, 12 May 1815.

[80] TNA, WO, 1/820, Transport Office, 15 May 1815.

horse ships with corps of artillery, ordnance transports and store ships. On 3 June there were thirty-four troop ships, five horse ships with cavalry and thirty-one store ships (seventy transports in all) preparing for immediate passage to Ostend. Transports arriving from Halifax with troops aboard were diverted to Ostend. By this time the transport agents had become very experienced in the speedy disembarkation of troops and the unloading of horses from transports onto the beach near Ostend. On 1 June, forty horses were unloaded in slings into the sea within 90 minutes, allowing the ship to return to England for more without delay, increasing the number of trips by individual transports, reducing the number of vessels required and thus minimising the cost to the state.[81]

The supply of men and ordnance continued unabated. However, after the war, in 1817, a review of the ordnance sent to support Wellington's army demonstrated a gross oversupply to the magazines, particularly of gunpowder. At Ostend there were 42,000 barrels of gunpowder whereas only 569 barrels were used at Waterloo. This was a potentially dangerous situation because there were no adequate means of storing it safely. The government had committed to procuring 80,000 barrels in 1815, exacerbating the oversupply situation in the ordnance depots in England. At a cost of £5 per barrel, it was an expensive investment. Of course no one could have predicted how long the army was going to be engaged against the enemy. It would have been a disaster if these magazines had been captured by the French.[82]

Napoleon was defeated at Waterloo on 18 June 1815 and abdicated for the second time on 22 June. On 7 July the allied armies re-entered Paris. The war may have been over but the transport service was still receiving requests to transport troops and provisions and supplies to Holland. On 4 July 1815 Bathurst was advised that fourteen troop ships had sailed on from Ramsgate to Ostend on the 1st. On board were detachments of the 1st and 2nd Regiments of the Life Guards and the Horse Guards.[83] On 4 July five transports with detachments of the Royal Wagon Train and staff horses sailed, also for Ostend, and three victuallers loaded with biscuits sailed from the Downs to Antwerp, as did five transports laden with the Battering Train and stores, one transport with forage and corn and one with storekeeper-general's stores.[84] Then on 8 and 9 July three troop ships with a garrison company of servants and orderlies for hospitals and detachments of various regiments, two empty troop ships to receive wounded troops and twelve store ships loaded with oats, Battery Train stores, biscuits and rum sailed from the Downs to Ostend, Antwerp and Rotterdam.[85] On the

[81] Knight, *Britain Against Napoleon*, 456.

[82] HoCPP, 1817 (275) IV.87, Third Report of the Select Committee on Finance: Ordnance. The reports illustrated that more gunpowder was used in sustained sieges than in battle. Even so only 1,800 barrels of gunpowder had been used at Copenhagen in 1807, and at the siege of San Sebastian 5,575 barrels had been used. Far more gunpowder had been destroyed, damaged or lost than had been expended in battle.

[83] TNA, WO, 1/820, TB to Bathurst, 4 Jul. 1815.

[84] TNA, WO, 1/821, TB to Bathurst, 6 Jul. 1815.

[85] TNA, WO, 1/821, Transport Office, 10 Jul. 1815.

evening of 11 July, seventeen troop ships with 3,573 rank and file from a number of foot regiments together with officers, women and children and one empty troop ship, which was to collect wounded troops, sailed from the Downs.[86] There were further sailings of fully laden troop ships and supply ships for Ostend between 22 and 25 July.

On 4 August thirty-three transports arrived at Portsmouth with 6,077 men from Quebec. One battalion, the 2nd of the 8th Regiment, was disembarked; the remainder were ordered to sail to Ostend. Likewise some troops arriving back from the Peninsula found themselves being shipped via Cork to Ostend. This transhipment of men who thought that they were returning home may have caused some distress. On 11 August a number of transports were delayed in sailing from the Downs by men of the 7th Foot who refused to be moved. Even 'mild and persuasive means' by agents and officers could not encourage the men to change their minds. A senior officer was dispatched from Canterbury, and after some further persuasion and the punishment of a few of the most disorderly, the troop ships were ready to sail, albeit a day late.[87]

Following the signing of the Treaty of Ghent on 24 December 1814, troops were arriving back from Canada. On 15 July 1815 a number of troop transports, with 11,320 troops and one transport with troops suffering from the highly contagious complaint of ophthalmia arrived in Portsmouth from Quebec, accompanied by four transports laden with oak and fir timber and one loaded with ordnance stores. At the same time troop ships and supply ships continued to arrive at Portsmouth from Lisbon.[88]

Under the terms of the Second Treaty of Paris, dated 20 November 1815, Wellington was appointed commander-in-chief of the 150,000-strong occupation force in France. Its role was to sustain peace in Europe by giving the restored Bourbon king, Louis XVIII, the time that he needed to rebuild the French army 'void of Bonapartist ambitions'.[89] Britain was willing to join this force because it continued to be sensitive to the security of the Netherlands and wished to deny France access to the port of Antwerp. Almost 30,000 British and Hanoverian troops served in France as part of this early multi-state peace-keeping force, which remained until November 1818. These forces required continued support from the transport service well after Waterloo.

The efficient and effective service provided by the Transport Board in the face of unprecedented competition from the mercantile sector was an integral feature of Britain's wartime recovery from the darker years of 1809 to 1811. During that period the limitations of the scale of feasible military expeditions had been exposed but the competence for mounting and sustaining significant operations overseas developed significantly, this despite the increasingly competitive

[86] TNA, WO, 1/821, Transport Office, 20 Jul. 1815.

[87] TNA, WO, 1/821, Lieutenant Ross, Agent at Deal, to TB, 11 Aug. 1815.

[88] TNA, WO, 1/821/367, TB to Bathurst, 13 Jul. 1815.

[89] Thomas Dwight Veve, *The Duke of Wellington and the British Army of Occupation in France, 1815–1818* (Westport, CT, 1992), 3.

demand for shipping from the mercantile sector. Throughout the wars mandatory requisitioning of shipping was avoided, allowing trade to prosper and thus generate national income to sustain the war and fund the all-important subsidies that were essential to buying the allegiance of Britain's allies in the fight for victory against Napoleon.

Conclusion

Despite its inglorious performance earlier in the wars, by 1815 the prestige of the British army was high because of its successes in the Peninsula and at Waterloo. The irony was that the prestige of the navy was not high at the end of the war. Despite the glorious fleet victories in the Revolutionary war, and of course at Trafalgar, its role in the later years of the Napoleonic war, though absolutely vital, was less visible. Nevertheless it was the British navy's virtual command of the oceans that made military success attainable by facilitating reasonably unfettered shipment of the army and also of the trade that sustained Britain's financial and fiscal stability and fed the nation when harvests were poor. However, in the end Napoleon was defeated not by British military and naval capacity alone. The proud nation had finite capacities in manpower which were insufficient to defeat the French single-handed or even to materially influence the foreign policy of its allies. The French were only quelled when the interests of the coalition of states of Britain, Russia, Prussia and Austria coalesced, significantly encouraged by heavy subsidies from Britain, which during the wars amounted to over £65 million. This demonstrates that British financial and fiscal stability, underpinned by naval hegemony, together with patient, often painful diplomacy were also important contributors to the defeat of Napoleon.[1] To this mix, should be added industrial inventiveness and manufacturing capacity, and the diligence of the government administrative backroom civil servants.[2]

Successful expeditions depended upon support from the local population and from allied armies. The repeated breakdown of coalitions tended to oblige the British government to be reactive to events. When it did react it tended to be over-ambitious, particularly in respect of attacks on the European mainland: den Helder in 1799, Hanover in 1805 and Walcheren in 1809 were organised relatively quickly on a scale that was too large for effective planning and co-ordination, but more importantly these expeditions failed because of the absence of effective support from local populations and the allies – from the Dutch (den Helder, 1799), from Prussia (Hanover, 1805) and from Austria (Walcheren, 1809). The West Indies expeditions of 1793–97 were also conducted on a large scale. They were initially judged unsuccessful due to the high rate of morbidity from yellow

[1] Harding, 'Expeditionary Armies and Naval Power', 73–4.

[2] Knight, *Britain Against Napoleon*, 467–8.

fever. Only more recently have these campaigns been assessed from a different perspective and considered successful in terms of economic warfare.[3]

However, there were a large number of successes, such as Egypt in 1801, Copenhagen in 1807, the Peninsula at Vimeiro in 1808 and thereafter, with the support of Spanish and Portuguese troops, at Talvera in 1809, Ciudad Rodrigo, Badajoz and Salamanca in 1812, Vitoria and San Sebastian in 1813 and Toulouse in 1814. Then finally victory was secured at Waterloo in 1815, in alliance with Generalfeldmarschall von Blücher's Prussian army. The British achieved the domination of islands in the Mediterranean, the East and West Indies and in Africa (the Cape of Good Hope in 1806 and Senegal in 1809).

Expeditions were sanctioned by the Cabinet, principally on the advice of the secretary of state for war. The first secretary was Henry Dundas (later Lord Melville). Although they were of quite different temperaments, both he and Viscount Castlereagh were the most influential holders of the post; both played a significant role in shaping the outcome of the wars. Lord Liverpool and Earl Bathurst, who followed Castlereagh, inherited and continued to support the Peninsular campaign but instituted few new campaigns, with the exception of Bathurst's involvement in Graham's campaign in Holland in 1814 and, of course, Waterloo. The other holders of the office, Lord Hobart (March 1801–May 1804), the 2nd Earl Camden (May 1804–July 1805) and William Windham (February 1806–March 1807), barely figure in the narration of expeditionary warfare; the latter was frankly incompetent.[4]

Military planning tended to be operational rather than strategic. It was invariably conducted by a small cadre who lacked comprehensive intelligence and who generally conducted limited consultation, thus leaving considerable unco-ordinated judgements and decisions to the operational departments whose expertise was not brought to bear until well into the proceedings. Subsequently ministerial interference in the detail and the constant fear of expanding the costs of the wars impacted on efficiency and effectiveness. Despite the inevitability of some delays due to the complexity of the preparations and the number of departments involved, the time taken to prepare various major expeditions from the point when instructions were first issued to the Transport Board to the date of sailing was between ten and sixteen weeks; this was not excessive, even by today's standards. The failure of campaigns should not be attributed to this phase of the operations. This particularly applies in the case of the Walcheren campaign, which was prepared in a shorter timeframe than all the others, in fact within ten weeks.

One factor which had a dominant impact on the outcome of military expeditions was the weather.[5] The reactive nature of the government meant that the timing of expeditions was usually beyond its control, so inevitably most expeditionary forces sailed too late in the year, creating delays, added

3 Duffy, *Soldiers, Sugar and Seapower*, 368 - 93.

4 Knight, *Britain Against Napoleon*, 461.

5 Hall, *British Strategy*, 47. Out of the navy's total loss of 317 ships in 1803–15, 223 were either wrecked or foundered, the great majority on account of hostile natural elements.

complications and sometimes disastrous consequences. Ministers, particularly Castlereagh, were forced to take great risks by sending the army overseas in adverse weather conditions. Seasonality influenced the availability of shipping. The most advantageous time to procure merchant shipping was early in the season, in March or April, before vessels were committed to foreign trade. With the exception of the 1807 Copenhagen expedition and Sir John Moore's expedition to the Baltic in 1808, the Board was invariably instructed to procure shipping for major expeditions well into the sailing season when ship-owners had already made commitments for their vessels.

Conjunct operations were familiar elements of British warfare. Molyneux had published a treatise on the subject in 1759, but the concept was mainly restricted to military operations in which the only naval involvement was to convey troops and materiel, ensure that supply lines remained open and then if necessary support a retreat, which Molyneux identified as the most challenging aspect of the role.[6] The modern concept of conjunct operations involves different services operating in a combined manner under a joint command structure. Combined command was not a recognised concept in the early nineteenth century. It was not until the Peninsular war, having witnessed the positive impact of Cochrane's amphibious assaults on coastal targets on the eastern coast of Spain, that the Admiralty actively instructed its senior commanders to participate wherever possible to support army commanders to achieve the common objective. Yet well before then, naval officers such as Horatio Nelson, Home Popham, Sidney Smith and Lord Hood had operated in conjunction with the army without specific instructions. Probably the best example of a conjunct operation during the wars was the 1807 attack on Copenhagen when the naval bombardment supported the land attack and forced the surrender. Then during the Peninsular war and the American war the navy participated heavily in amphibious warfare.

Between 1808 and 1814 the army strength of regular rank and file was over 200,000, peaking at 239,431 around December 1813; of these, well over 50 per cent (rising to 76 per cent in December 1813 when there were 180,991) were abroad. This manpower materially increased when officers and staff, artillery and wagon train divisions together with other support staff were included, adding up to 30 per cent to the force. The volume of shipping required to move these troops, horses and materiel and to replenish their numbers and supplies was enormous. Hundreds, nay thousands, of merchant ships were called into government service to fulfil this capacity and the process was managed by the Transport Board.

After its inception in 1794 the Board, hitherto regarded as a relatively insignificant branch of the naval service, was at the centre of all major military expeditions. The Board and the ships that it chartered were far more important than has previously been perceived. Its role was mainly directed towards supporting the military activities of the army. The government spent over £42 million pounds chartering merchant transports that were used as troop ships,

[6] Thomas More Molyneux, *Conjunct Expeditions, or, Expeditions that have been carried on jointly by the fleet and armyh: with a comment on littoral war* (London, 1759).

horse ships and supply ships; by comparison naval victualling was a minor role. By 1815, Pitt's decision, made in 1794, to establish an independent Transport Board under the auspices of the Treasury had been fully justified. Without doubt the previous administrative arrangements would have been totally inadequate and would not have been able to cope with the demands that were placed on the Transport Board by the size and complexity of the expeditions, the volume of shipping involved and the increasing competition from trade. The consequences would have been disastrous.

However, the politicians failed to get to grips with the issue of using warships as troop ships. The availability and effectiveness of troop shipping could have been improved enormously if commissioned naval ships had been used in much greater numbers. The benefits would have been significant; fewer ships would have been required, allowing regiments and equipment to be landed together rather than piecemeal, often in several locations. The fighting forces would have been more effective in a much shorter time. It was also argued that such ships would have had better defensive capability and would not have required the same level of convoy support and that fewer would have been lost due to maritime misadventures. The involvement of naval ships in the conveyance of troops was the subject of contentious debate throughout the wars. The army was strongly in favour of using naval vessels. The principal naval reluctance was the unresolved question of the command structure whilst at sea. Did the naval captain have the ability to direct the conduct of the troops whilst they were on board? The navy's preferred argument against the use of more naval vessels as troop ships was that neither ships nor, more importantly, crews could be spared. This argument withers under scrutiny because even during the Revolutionary war, when the balance of naval power with the enemy was more evenly matched, the 1801 Egypt landing involved forty-six naval troop ships. In the Napoleonic war naval ships were used to supplement transport capacity. In early 1809 naval ships were used to evacuate troops from Corunna and a few months later 17,000 troops were carried to the Scheldt in naval troop ships. Even greater use of naval troop ships would probably have improved the effectiveness of troop movements in the larger expeditions, would undoubtedly have eased the challenge of bringing forward merchant ships and would almost certainly have speeded up preparations. However, the use of vast quantities of merchant shipping was essential, even though this conflicted with the demands of trade, particularly after 1805.

Although the number and tonnage of ships on the British shipping register increased dramatically – from 16,329 ships of 1,564,520 tons in 1793 to 24,862 ships of 2,681,276 tons in 1815 – so too did the volume of international trade. Imports increased by 83 per cent and exports by 172 per cent over the same period. Government demand for transports put a great strain on the merchant ship fleet because the vast majority of British ships were either too small or otherwise unsuitable for government transports. The strain was particularly acute when demands for transports peaked in 1795, 1805, 1808, 1809 and 1814. In 1795 over 700 ships of almost 200,000 tons were taken up; after 1805 over 1,000 vessels of more than 250,000 tons were taken up by the Transport Board.

In the Revolutionary war, the government demand for transports materially subsidised the British shipping and shipbuilding industry, which would otherwise have suffered dramatic losses due to the closure of markets on the continental mainland such as France, Holland, Germany and Spain. In the Napoleonic war, when the demand for shipping for trade boomed despite Napoleon's Continental System, Castlereagh launched a number of large offensive operations in Europe which required unprecedented tonnages of merchant shipping to support them. This demand was considerably higher in terms of the available shipping than has previously been imagined: its impact on trade was substantial, and it caused higher freight rates, commodity shortages and higher commodity prices. It made the Transport Board's task exceptionally challenging and accounts for the difficulty that was experienced in bringing forward ships. The Board's success in procuring that level of shipping at such crucial times illustrates its vital role in the government's wartime administration.

The Board needed ships of over 220 tons to serve as troop ships. Over 73 per cent of regular charters were ships of over 220 tons but only 15 per cent of the ships registered were over 220 tons and many of those were overseas, out of reach of the Board. In reality only 9–10 per cent of the registered British fleet was of suitable size, manned and operational, and these vessels were also in great demand to support international and domestic trade. International trade was a vitally important contributor to the funding of the wars. Domestic trade, particularly the supply of coal to London and other major cities, was equally essential. Nevertheless, against this stiff competition and at critical times the Board secured 30–39 per cent of the pool of suitable ships. This is considerably higher than has hitherto been estimated. This significant accomplishment was achieved through the Board's familiarity with the state of the shipping market, its cultivation of a large cadre of ships' brokers and its skilful manipulation of the charter price, increasing the rate only when the necessity arose, taking the opportunity to reduce rates when the market permitted and resisting rate increases if it felt that the disadvantages outweighed the benefits

The cost of the war was under continuous parliamentary scrutiny; there was a constant desire for economy and efficiency. Inevitably, however, one normally came at the cost of the other. Clearly the preparation of major campaigns could have been expedited if transports had been retained rather than discharged. However, this would have been at a considerable cost. Lord Castlereagh was quick to identify this dilemma. Within two months of being appointed secretary of state for war in 1805 he proposed a 'disposable' army with transports on standby to enable troops to be moved quickly to any theatre of war. These reserves were dissipated by his successor, William Windham, only to be revived by Castlereagh in 1807. In 1812 when Earl Bathurst was secretary of state for war he considered implementing this proposal once again. However, from 1807 the demand for transports was so high that there were generally insufficient ships available for immediate action, let alone to support a standby fleet. Nevertheless, particularly after Corunna, there was a standby fleet under another guise when it was recognised that there was a moral obligation as well as a military necessity to leave sufficient transports on station to facilitate emergency evacuations.

Significantly the Walcheren expedition would not have sailed without the recall of many of those ships. Castlereagh and later Liverpool both subscribed to the philosophy of an emergency evacuation fleet, despite the cost of some £50,000 per month, when they insisted on retaining 40,000 tons of shipping in the Tagus to support any urgent withdrawal of the army from the Peninsula.

The cost of maintaining cavalry or horse ships in service was even greater. It cost £10 per horse per month to retain such transports. Horse ships were procured in large numbers for major expeditions. The cost, but more importantly the capacity requirements, invariably meant that forces sent overseas did not have as many cavalry or draft horses for hauling heavy guns and supply carts as the commanders would have wished. Colliers were the preferred vessels for horse ships but the impact of withdrawing large numbers of such ships from their routine task of ferrying coal from Northumberland and the Newcastle area in the north of England to London and other ports was an immediate shortage and a resultant increase in the cost of coal in those places.

Extraordinary levels of co-operation, co-ordination and communication between the many parties involved – including the Transport Board, the Victualling Board, the Ordnance Board, the War Office, the Commissariat and the Admiralty, and in the private sector numerous ships' brokers and provisions and material suppliers – were essential to the successful launch of an expedition. When these broke down there were inevitably delays. There were many other factors that also caused delays: the shortage of docking facilities for repairs and wharf facilities for loading and unloading ships; the requirements for installation of cabins for officers and stalls for horses; the shortages of beds and bedding; the late arrival of ships and troops; and the shortages of supplies and water. Castlereagh was well aware of the impact of such delays when he wrote, 'The interval between the wish to obtain transports and the power of obtaining them is much greater than seems generally to be imagined.'[7] Amassing large numbers of ships in one place demanded the competent application of well-developed practices and procedures conducted by the transport agents.

The result of the Board's establishment as a subsidiary board of the Treasury rather than the Admiralty or the secretary of state for war was that the chairman had three masters. A significant element in the Transport Board's success was Sir Rupert George's enduring presence. Although he was the Board's chairman for over twenty years, very little is known about the man himself and he remains an enigma. His tenure spanned seven administrations and the war years from 1795 to 1815. He survived the changes of administration and the organisational purges that had a significant impact on the other important departments, making his one of the longest tenures of any serving civil servant during that period. During this time the Board acquired greater responsibilities for prisoners of war and the care of sick and wounded seamen. This unique record points to Sir Rupert's qualities as an administrator and communicator. Under him the transport service was a highly centralised organisation; the commissioners retained very tight control

[7] HoCPP, Hansard, 7 Aug. 1807.

of the costs of the service and the disposition of the transports, though they relied heavily on various very experienced, long-serving transport agents who played a central role in bringing forward and preparing transports and keeping the Board informed of the progress and the movements of vessels.

Given the complexity of logistics management in the age of sail, the reader might wonder just how anything was achieved. It is testament to the efforts of all the parties involved that so many major expeditions were prepared reasonably successfully in timescales that even today would be considered acceptable. Without doubt, Sir Rupert George and his colleagues at the Transport Board performed a vital but hitherto unrecognised service, at a time that was critical for Britain, by skilfully harnessing the operations of the market to obtain the ships that played a major role in all the military adventures that ultimately contributed to Napoleon's downfall.

Appendices

Appendix 1. *An account of the number of ships, with their tonnage, which belonged to the several ports of the British Empire on the 30 September between 1799 and 1820.*

Year	*1799*		*1800*		*1801*		*1802*		*1803*	
	Ships	*Tons*	*Ships*	*Tons*	*Ships*	*Tons*	*Ships*	*Tons*	*Ships*	*Tons*
England	11,487	1,337,181	12,208	1,466,632	12,767	1,541,425	14,790	1,642,224	13,936	1,709,302
Jersey	61	4,611	53	4,244	52	3,922	55	4,367	83	7,264
Guernsey	78	6,199	77	6,403	88	7,622	93	8,610	137	11,988
Isle of Man	227	5,146	238	5,463	237	5,560	234	5,516	372	8,799
Plantations in America and West Indies	2,996	201,743	2,161	157,364	3,285	251,928	3,361	226,893	3,225	235,164
Scotland	2,031	148,110	2,155	161,807	2,279	173,564	2,349	183,935	2,662	209,222
Ireland	999	49,825	1,003	54,262	1,004	54,241	1,030	56,510	1,030	56,510
Total	**17,879**	**1,725,815**	**17,895**	**1,856,175**	**19,712**	**2,038,262**	**20,568**	**2,128,055**	**21,445**	**2,238,249**
Year	*1804*		*1805*		*1806*		*1807*		*1808*	
	Ships	*Tons*	*Ships*	*Tons*	*Ships*	*Tons*	*Ships*	*Tons*	*Ships*	*Tons*
England	detail not located		15,327	1,799,210	14,877	1,786,692	15,087	1,797,102	15,327	1,833,971
Jersey			74	6,769	68	6,485	77	6,891	66	5,439
Guernsey			111	9,759	108	10,193	106	9,927	110	10,850
Isle of Man			404	9,650	398	9,568	390	9,373	381	9,237
Plantations in America and West Indies	2,904	191,509	3,024	190,953	2,867	183,860	2,917	184,794	3,066	194,423
Scotland			2,581	210,295	2,788	211,431	2,615	216,553	2,592	211,950
Ireland			1,067	56,806	1,076	55,545	1,098	56,981	1,104	58,959
Total	**21,774**	**2,268,570**	**22,588**	**2,283,442**	**22,182**	**2,263,774**	**22,290**	**2,281,621**	**22,646**	**2,324,829**

Year	1809		1810		1811		1812		1813	
	Ships	*Tons*	*Ships*	*Tons*	*Ships*	*Tons*	*Ships*	*Tons*	*Ships*	*Tons*
England	15,687	1,875,224	16,048	1,918,089	16,164	1,942,406	16,295	1,951,234	16,602	2,029,637
Jersey	58	5,451	57	5,454	59	6,003	54	5,369	64	6,379
Guernsey	112	10,506	104	9,947	94	9,485	76	8,312	84	9,755
Isle of Man	372	8,989	366	8,785	398	9,585	393	9,439	342	8,513
Plantations in America and West Indies	3,188	201,243	3,450	215,383	3,628	227,452	3,470	216,068	2,689	165,591
Scotland	2,534	206,075	2,552	209,736	2,630	220,688	2,708	231,273	2,713	234,383
Ireland	1,119	60,979	1,126	58,650	1,133	59,155	1,111	57,104	1,146	60,226
Total	**23,070**	**2,368,467**	**23,703**	**2,426,044**	**24,106**	**2,474,774**	**24,107**	**2,478,799**	**23,640**	**2,514,484**
Year	*1814*		*1815*		*1816*		*1820*			
	Ships	*Tons*	*Ships*	*Tons*	*Ships*	*Tons*	*Ships*	*Tons*		
England	17,102	2,088,204	17,346	2,139,301	17,442	2,152,968	21,473	2,412,804	Including Scotland and Ireland.	
Jersey	62	6,794	69	7,519	77	7,992	}			
Guernsey	65	6,928	61	6,662	65	7,237	} 496	26,225		
Isle of Man	355	8,897	367	9,300	369	9,335	}			
Plantations in America and West Indies	2,868	202,795	2,991	203,445	3,775	279,643	3,405	209,564		
Scotland	2,783	241,578	2,863	254,926	2,958	263,536		See above		
Ireland	1,183	61,769	1,165	60,123	1,178	63,229		See above		
Total	**24,418**	**2,616,965**	**24,862**	**2,681,276**	**25,864**	**2,783,940**	**25,374**	**2,648,593**		

Source: Consolidated from various editions of Cobbett, The Parliamentary History of England *and HoCPP, Annual Trade and Navigation Accounts, TNA, CUST, 36/5*

Appendix 2. *Abstract of ships built in the several ports of England from 1786 to 5 January 1814*

	1786	1787	1788	1789	1790	1791	1792	1793	1794	1795	1796	1797	1798	1799
Barnstaple	27	22	5	3	5	3	4	3	5	6	9	6	8	7
Beaumaris	22	42	26	13	24	18	14	13	11	11	26	18	22	25
Bristol	10	23	18	10	4	17	10	6	11	3	4	7	4	11
Cardiff	25	23	21	13	15	4	12	9	0	1	1	0	0	0
Cowes	30	24	30	14	15	18	22	11	11	7	11	14	15	9
Dartmouth	13	26	24	20	18	18	14	11	12	19	15	14	21	23
Dover	7	20	17	17	15	29	38	22	13	10	16	30	32	34
Exeter	12	46	12	10	3	10	17	8	8	6	9	10	11	11
Hull	23	40	38	40	33	45	32	35	26	33	29	31	34	46
Liverpool	16	47	38	22	29	28	32	15	16	10	20	14	12	26
London	34	41	16	22	26	30	31	13	30	23	44	42	43	57
Newcastle	36	33	30	29	27	27	22	40	27	29	27	31	43	38
Plymouth	23	21	18	18	21	12	21	11	17	7	10	23	16	12
Southampton	20	30	18	8	4	8	10	5	7	6	7	6	6	8
Swansea	39	23	11	10	4	6	4	8	4	6	4	8	5	6
Whitby	15	23	26	15	25	21	24	21	20	21	19	22	20	21
Whitehaven	14	25	17	18	15	13	20	20	13	14	16	20	23	16
Yarmouth	43	31	22	28	25	26	32	33	33	35	49	30	41	31
Other 55 centres	339	289	260	192	192	174	157	191	172	199	194	195	211	230
Totals	748	829	647	502	500	507	516	475	436	446	510	521	567	611
Tonnage 000s	66	81	60	50	53	48	59	54	53	53	74	62	69	84
73 locations														

Source: TNA, CUST, 35/5/12

	1800	1801	1802	1803	1804	1805	1806	1807	1808	1809	1810	1811	1812	1813	Total
arnstaple	4	2	4	5	9	8	3	4	4	8	6	4	6	4	184
eaumaris	18	20	28	40	38	33	23	21	11	8	14	20	20	22	601
ristol	5	5	8	10	3	7	5	7	7	5	3	3	2	10	218
ardiff	1	0	0	4	3	3	1	0	0	0	0	1	3	1	141
owes	31	27	29	30	19	23	15	14	9	13	19	13	7	9	489
artmouth	21	27	36	44	17	21	26	22	11	21	18	21	25	20	578
over	26	16	25	15	15	14	23	20	17	4	3	10	4	3	495
xeter	16	11	21	21	10	9	7	10	8	7	8	10	9	6	326
ull	73	87	73	58	39	45	43	39	45	44	49	52	47	57	1236
iverpool	21	22	26	19	14	22	19	14	17	11	7	14	27	20	578
ondon	63	59	50	64	31	29	24	26	21	20	31	33	31	10	944
ewcastle	47	49	52	50	34	22	24	23	14	20	22	21	23	41	881
ymouth	14	23	27	32	11	15	17	16	11	5	7	6	15	6	435
outhampton	13	15	14	16	6	9	7	10	5	8	9	9	7	6	277
wansea	2	4	5	8	6	5	3	2	7	7	2	4	6	4	203
hitby	28	24	27	35	27	19	19	17	13	7	8	11	12	20	560
hitehaven	21	26	19	22	24	18	11	10	17	19	10	22	34	42	539
armouth	44	57	50	42	50	39	34	32	20	23	30	28	19	18	945
ther 55 centres	273	258	296	350	236	225	203	305	184	187	223	236	211	228	6410
otals	721	732	790	865	592	566	507	592	421	417	469	518	508	527	16040
onnage 000s	103	99	95	98	69	59	48	50	45	48	63	81	67	75	1863
3 locations															

***Appendix* 3.** *Return of the effective strength of the regular and militia forces on 25 June and 25 December 1808–16, specifying the numbers serving at home and abroad*

		Cavalry				*Infantry*						
	Date	*British*	*Foreign & Colonial*	*Total*	*Foot Guards*	*British*	*Foreign & Colonial*	*Total*	*Total Regulars*	*Militia*	*General Total*	*Sources*
Total												
1808	25-Jun	23,763	3,000	26,763	8,141	139,698	32,614	172,312	207,216	86,506	293,722	1810 (54) XIII.379
	25-Dec	24,419	2,972	27,391	8,080	142,047	33,096	175,143	210,614	81,577	292,191	ditto
1809	25-Jun	24,532	3,002	27,534	7,907	148,003	31,727	179,730	215,171	65,385	280,556	ditto
	25-Dec	24,540	3,446	27,986	7,594	146,574	34,005	180,579	216,159	71,467	287,626	ditto
1810	25-Jun	24,137	3,545	27,682	6,983	141,147	32,856	174,003	208,668	76,484	285,152	1810–11 (30) X.169
	25-Dec	23,797	3,594	27,391	6,877	141,162	35,729	176,891	211,159	84,300	295,459	ditto
1811	25-Jun	23,771	3,731	27,502	6,694	145,518	37,043	182,561	216,757	77,424	294,181	1812 (49) IX.189
	25-Dec	23,769	4,001	27,770	6,878	145,236	39,065	184,301	218,949	77,159	296,108	ditto
1812	25-Jun	24,032	4,253	28,285	7,558	150,583	43,240	193,823	229,666	71,803	301,469	1812–13 (34) XIII.1
	25-Dec	23,839	4,193	28,032	7,451	147,034	46,632	193,666	229,149	71,741	300,890	ditto
1813	25-Jun	24,919	4,164	29,083	7,480	149,199	47,554	196,753	233,316	67,989	301,305	1813–14 (117) XI.283
	25-Dec	26,611	4,471	31,082	8,852	152,463	47,034	199,497	239,431	63,059	302,490	ditto
1814	25-Jun	26,619	4,437	31,056	8,474	148,485	46,093	194,578	234,108	63,756	297,864	1814–15 (195) IX.321
	25-Dec	19,783	3,889	23,674	7,138	128,935	39,710	168,645	199,457	20,372	219,829	ditto
1815	25-Jun	20,241	3,419	23,660	7,114	124,177	35,622	159,799	190,573	11,588	202,141	1816 (99) XII.419

		Cavalry				*Infantry*						
	Date	*British*	*Foreign & Colonial*	*Total*	*Foot Guards*	*British*	*Foreign & Colonial*	*Total*	*Total Regulars*	*Militia*	*General Total*	*Sources*
	25-Dec	20,517	367	20,884	7,066	123,008	25,459	148,467	176,417	36,086	212,503	ditto
At Home												
1808	25-Jun	19,409	2,403	21,812	8,141	67,852	1,842	69,694	99,647	86,506	186,153	1810 (54) XIII.379
	25-Dec	15,970	2,378	18,348	5,638	50,682	611	51,293	75,279	81,577	156,856	ditto
1809	25-Jun	16,133	2,235	18,368	5,506	67,236	2,702	69,938	93,812	65,385	159,197	ditto
	25-Dec	15,704	2,857	18,561	5,308	65,258	3,167	68,425	92,294	71,467	163,761	ditto
1810	25-Jun	14,976	2,644	17,620	4,045	51,986	2,445	54,431	76,096	76,484	152,580	1810–11 (30) X.169
	25-Dec	14,908	2,540	17,448	3,768	49,949	2,787	52,736	73,952	84,300	158,252	ditto
1811	25-Jun	13,575	2,591	16,166	3,344	47,442	2,192	49,634	69,144	77,424	146,568	1812 (49) IX.189
	25-Dec	12,050	1,865	13,915	3,748	45,501	2,745	48,246	65,909	77,159	143,068	ditto
1812	25-Jun	12,146	2,138	14,284	4,115	48,109	3,496	51,605	70,004	71,803	141,807	1812–13 (34) XIII.1
	25-Dec	12,135	2,340	14,475	2,841	42,329	2,373	44,702	62,018	71,741	133,759	ditto
1813	25-Jun	12,051	1,622	13,673	3,364	44,633	2,132	46,763	63,800	67,989	131,789	1813–14 (117) XI.283
	25-Dec	12,554	1,037	13,591	2,928	39,491	2,430	41,921	58,440	63,059	121,499	ditto
1814	25-Jun	11,460	1,268	12,728	2,679	43,232	1,628	44,860	60,267	63,756	124,023	1814–15 (195) IX.321
	25-Dec	13,582	195	13,777	4,543	45,628	553	46,181	64,501	20,372	84,873	ditto
1815	25-Jun			7,974	2,992			35,339	46,305	11,568	57,873	1816 (99) XII.419

		Cavalry				*Infantry*						
	Date	*British*	*Foreign & Colonial*	*Total*	*Foot Guards*	*British*	*Foreign & Colonial*	*Total*	*Total Regulars*	*Militia*	*General Total*	*Sources*
	25-Dec			8,030	2,855			42,234	53,119	36,086	89,205	ditto
Abroad												
1808	25-Jun	4,354	597	4,951	0	71,846	30,772	102,618	107,569	0	107,569	1810 (54) XIII.379
	25-Dec	8,449	594	9,043	2,442	91,365	32,485	123,850	135,335	0	135,335	ditto
1809	25-Jun	8,399	767	9,166	2,401	80,767	29,025	109,792	121,359	0	121,359	ditto
	25-Dec	8,836	589	9,425	2,286	81,316	30,838	112,154	123,865	0	123,865	ditto
1810	25-Jun	9,161	901	10,062	2,938	89,161	30,411	119,572	132,572	0	132,572	1810–11 (30) X.169
	25-Dec	8,889	1,054	9,943	3,109	91,213	32,942	124,155	137,207	0	137,207	ditto
1811	25-Jun	10,196	1,140	11,336	3,350	98,076	34,851	132,927	147,613	0	147,613	1812 (49) IX.189
	25-Dec	11,719	2,136	13,855	3,130	99,735	36,320	136,055	153,040	0	153,040	ditto
1812	25-Jun	11,886	2,115	14,001	3,443	102,474	39,744	142,218	159,662	0	159,662	1812–13 (34) XIII.1
	25-Dec	11,704	1,853	13,557	4,610	104,705	44,259	148,964	167,131	0	167,131	ditto
1813	25-Jun	12,868	2,542	15,410	4,116	104,566	45,422	149,990	169,516	0	169,516	1813–14 (117) XI.283
	25-Dec	14,057	3,434	17,491	5,924	112,972	44,604	157,576	180,991	0	180,991	ditto
1814	25-Jun	15,159	3,169	18,328	5,795	105,253	44,465	149,718	173,841	0	173,841	1814–15 (195) IX.321
	25-Dec	6,201	3,694	9,897	2,595	83,307	39,157	122,464	134,956	0	134,956	ditto
1815	25-Jun			15,686	4,122			124,460	144,268	0	144,268	1816 (99) XII.419
	25-Dec			12,854	4,211			106,233	123,298	0	123,298	ditto

Bibliography

Manuscript Sources

British Library

BL, c.194.a456	*Regulations to be observed by Troops embarked in Transports for service abroad particularly those destined for the West Indies* (War Office, 10 Oct. 1795).

Huntington Library

HL, STG Collection	STG 136 (21). Correspondence of James Bowen, Commissioner, to Transport Board (Courtesy of Dr R. Knight)

Mulgrave Papers, Mulgrave Castle

VII, 19/29	Letter from Spencer Perceval, Chancellor of the Exchequer, to Lord Mulgrave, First Lord of the Admiralty, 18–19 Nov. 1808.

The National Archives

TNA, ADM, 1/1546	Admiralty in letters from Captains.
TNA, ADM, 1/3730–65	Admiralty in letters from the Transport Office, 1793–1815.
TNA, ADM, 1/3992–6	Admiralty correspondence from Lloyds Coffee House.
TNA, ADM, 2/1097–102	Convoy orders 1794–1809.
TNA, ADM, 7/374–400	Protections issued.
TNA, ADM, 8	List books.
TNA, ADM, 49	Convoy signals and instructions to ships on convoy.
TNA, ADM, 50/36	Admiral's Journals, VA Lord Keith Mar. 1800–May 1802.
TNA, ADM, 51/3521	*Venus* transport log book, Jan. 1802.
TNA, ADM, 51/4376	Transport agents' log books.
TNA, ADM, 95/108	List of chartered merchantmen.
TNA, ADM, 106/3096	Articles of General Instructions to Agents, 1803.
TNA, ADM, 106/3110	Board for Revision 9th Report. The Transport Board.
TNA, ADM, 106/3416	Regulations for transport agents.
TNA, ADM, 108/19–24	Transport Board Out letters to secretaries of state.
TNA, ADM, 108/30	Rupert George's letter book.
TNA, ADM, 108/31–83	Transport Board minutes.
TNA, ADM, 108/168	Register of transports.
TNA, ADM, 108/174	Transports burnt at Aix Roads.

TNA, BT, 6/185	Tables of trade, 1776–1801.
TNA, BT, 6/194–213	Register of licences to neutral vessels.
TNA, C, 108/114	Chancery files regarding ship-owner actions.
TNA, CUST, 36/5 and 12	Account book, ships register, ships built, ships in overseas trade, ships in coast-wise trade, numbers in crews, 1814–28.
TNA, FO, 83 and 95	Foreign Office correspondence with the Admiralty.
TNA, HO, 28/63	Home Office correspondence regarding transports.
TNA, HO, 42/29	Sir Jerome Fitzpatrick's comments on the embarkation of cavalry horses.
TNA, HRS/J/1	Captain Cochrane's instructions for landing troops at Aboukir Bay in 1801.
TNA, MFQ, 1/133/1	Egypt, Aboukir Bay, Plan of the coast showing the invading force in boats.
TNA, T, 1/959	Treasury Papers, statement of establishment at the Transport Office, 1795 and 1805.
TNA, WO, 1/798–822	War Department In letters from the Transport Board.
TNA, WO, 6/20	Instructions to Expedition Commanders-in-Chief.
TNA, WO, 6/154–61	War Department Out letters to the Transport Board.
TNA, WO, 37/10	Signals for landing troops, flags for transport agents.

National Maritime Museum

HNL	The Henley Collection
ADM, B/207	Admiralty to the Transport Board
ADM, BP/49b	Navy Office Transport Department memorandum, 26 Nov. 1829.
ADM, BPP, 35c	Transport Board to the Admiralty.

The Sim Comfort Collection

Henry Dundas Letter Book

Parliamentary Papers (available through parlipapers.proquest.com)

Relating to Copenhagen, 1807

Account of Transports employed against Copenhagen, 2 Feb. 1808 (10) IX.53.

Account of the Stores taken at Copenhagen, 1808 (236) IX.95.

Papers presented to the House of Commons relating to the Expedition to Copenhagen, 1808(9) IX.113.

Relating to the Expeditions to Spain and Portugal, 1808–09

Copies of instructions to the Commanders of HM's forces in Spain and Portugal, 1 Mar. 1809, (66) XI.1.

Copy of a letter from Major-General Brodrick to Viscount Castlereagh, 28 Aug. 1808 and Extract of a letter from Sir John Moore to Viscount Castlereagh, 13 Jan. 1809, (112) XI.277.

Papers relating to the Staff of the Army employed in Portugal and Spain 1808, 24 Mar. 1809, (86) XI.209.

Papers relating to supplies embarked for services in Portugal and Spain, Royal Artillery Embarkation, 24 Mar. 1809, (94) XI.225.

Papers relating to supplies embarked for services in Portugal and Spain, Storekeeper-General's Office, 24 Mar. 1809, (82, 83, 84, 85) XI.193.
Proceedings upon the Inquiry relative to the Armistice and Convention in Portugal 1808, 1809, (17) XII.1.

Relating to the Scheldt Expedition, 1809
An Account of the total Extraordinary Expense, so far as the same can be made up, of the late Expedition to the Scheldt, 12 Mar. 1810, (110) VII.337.
An Account of the transports reported to the Quarter-Master General or Commander-in-Chief at Walcheren for the reception of the sick and convalescents, between 1 September and 10 October 1809, 1810 (139) VII.267.
Minutes of Evidence, taken before the Committee of the Whole House, appointed to consider of the policy and conduct of the late expedition to the Scheldt, Session 23, Jan–Jun 1810 (12) VIII.1.

Relating to the Strength of the Army
Adjutant-General's Office: Return of the Effective Strength of the Regular and Militia Forces, distinguishing Foreign and Colonial Corps, Cavalry, Foot Guards and Infantry and specifying the number of each serving at Home and Abroad: 1806 (173), 17 May 1806; 1807 (133), 13 Aug. 1807; 1808 (316), 21 Jun. 1808; 1810 (54), 17 Feb. 1810; 1810–11 (30), X.169; 1812 (49), 11 Feb. 1812 ; 1812–13 (34) X.III.1; 1813–14 (117), 5 Apr. 1814 ; 1814–15 (195) IX. 321, 14 Mar. 1815; 1816 (99) XII.419; 1817 (206), 15 Apr. 1817.

Other Parliamentary Papers
The Eighteenth Report of the Commissioners of Military Enquiry, Office of the Commissariat, 20 Mar. 1812, (119), IV.247. Select Committee of the House of Lords on Policy and Operation of the Navigation Laws, 1847 (232), XI. 8.
Minutes of Evidence taken before the Select Committee of the House of Commons on petitions relating to East India built shipping, 1813–14 (115) VIII.1.
Papers presented to the House of Commons relating to the state and condition of the squadron Employed off Rochefort, under the command of Rear Admiral Sir Richard J. Strachan, Bart, 15 Mar. 1808, (103) IX.293.
Papers relating to the Licence Trade, 1812 (83) IX.345.
Report of the Commissioners Appointed to Examine, Take and State The Public Accounts of the Kingdom. Twelfth Report Relative to Passing the Accounts of the Treasurer of Ordnance, in the Office of the Auditor of the Impress, Session 18 May 1784–20 Aug. 1784: 11 Jun. 1784.
Reports of the Select Committee of the House of Commons to Inquire into the Operation and Policy of the Navigation Laws, 1847 (232) X.1.
Return of Deaths, Discharges and Desertions which have occurred in the Regular Army, 25 Dec. 1810 to 24 Dec. 1813, 1812 (52) IX.95 and 1813–14 (73) XI.227.
An Account of the Official Values of Imports from and Exports to The United States of America from 1801 to 1810, 1812 (63), X.1, 3.

Hansard Reports
Hansard, House of Commons: 31 Jul., 7, 13 Aug. 1807; 19, 25, Jan.. 1, 7, 24 Feb. and 9 May 1809; 21, 26 Mar. 1810; 4, 15 Mar. and 21 Jun. 1811.
Hansard, House of Lords: 9, 26 May 1809; 21 May 1810; 14 Jun. 1811.

House of Commons Sessional Papers of the Eighteenth Century: Reports of Commissions and Committees, in chronological order

Report of the Commissioners appointed to examine, take and state The Public Accounts of the Kingdom. Twelfth Report relative to passing the accounts of the Treasurer of Ordnance, in the Office of the Auditor of the Impress, (1784), 11–12, 43.

Eighteenth Report from the Select Committee on Finance (1798), 19 Jul. 1797, 1798 (11), 191, 206.

Thirty-First Report of the Select Committee on Finance on the Admiralty, Dockyards and Transports, 26 Jun. 1798, 1798 (113), 3, 74.

Further Proceedings on the 18th and 30th Report of the Select COmmittee on Finance 1715–1800 (114), 24.

Fifth Report of the Commissioners appointed to Inquire into the Fees, Gratuities, Perquisites and Emoluments which have lately been received in the Several Public Offices of the Navy, 1802–03 (111), 249.

Sixth Report of the Commissioners appointed to Inquire into the Fees, Gratuities, Perquisites and Emoluments which have lately been received in the Several Public Offices of the Dockyards, 1803–04 (11), 1.

Eighth Report of the Commissioners appointed to Inquire into the Fees, Gratuities, Perquisites and Emoluments which have lately been received in the Several Public Offices of the Victualling Office, 1803–04 (11),637.

Ninth Report of the Commissioners for Revising and Digesting the Civil Affairs of His Majesty's Navy, 73e of 1809, XXV, 354.

Thirteenth Report of the Commissioners For Revising and Digesting the Civil Affairs of His Majesty's Navy, 73e of 1809, XXV, 530.

Third Report of the Select Committee on Finance, Ordnance, 1817 (275), IV.87.

Acts of Parliament

1786 Registration Act, 26 Geo. III, c.60.

1794 Act for the Further Encouragement of British Mariners, 33 Geo. III, c.26.

1797 Manifest Act, 36 Geo. III, c.40.

1798 Convoy Act, 38 Geo. III, c.76.

1801 Merchant Shipping Act, 41 Geo. III, c.19.

1803 Convoy Act, 43 Geo. III, c.57.

Burney Collection of Newspapers (available through Athens)

Aberdeen Journal, 17 Mar. 1800, 20 Jun. 1810.

British Evening Post, 10 Oct. 1795.

Bury and Norwich Post, 20 Dec. 1809.

Courier and Evening Gazette, 10 Aug. 1799.

English Chronicle, 15 Apr. 1800.

Hampshire Telegraph, 10 Dec. 1804, 30 Mar. 1807, 23 Sept. 1811.

Hull Packet, 29 Sept. 1807.

Morning Chronicle, 27 Jul. 1808.

Morning Post and Gazetteer, 30 Apr. 1800.

New Lloyds Evening Post, 9 Aug. 1799, 24 Oct. 1800.

Trewman's Exeter Flying Post, 8 Jun. 1809.

True Briton, 16 May 1800.

Printed Primary Sources

Bonner-Smith, David (ed.), *Letters of Admiral of the Fleet the Earl of St Vincent*, vol. 2, Navy Records Society 61 (1927).

Bromley, J.S., *Manning of the Royal Navy; Selected Public Pamphlets, 1693–1873*, Navy Records Society 119 (1974).

Cobbett, William (ed.), *The Parliamentary History of England*, 36 vols (London 1806–20).

Fortescue, Sir John (ed.), *The Correspondence of King George the Third from 1760 to December 1783* (London, 1927).

Gurwood, John, *The Dispatches of Field Marshal the Duke of Wellington*, vols. 3–10 (reprinted Cambridge, 2010).

Hamilton, Sir R.V. (ed.), *Letters and Papers of Admiral of the Fleet Sir T. Byam Martin*, vol. 2 (London, 1898).

Hardy, Charles, *A Register of Ships Employed in the Service of the Hon the United East India Company from the Year 1760 to the Conclusion of Commercial Charter* (London, 1835).

Hattendorf, John, Knight, Roger, Pearsall, A.W.H., Rodger, N.A.M., and Till, Geoffrey (eds), *British Naval Documents 1204–1960*, Navy Records Society 131 (1993).

Lambert. S., *House of Commons Sessional Papers of the Eighteenth Century* (Wilmington, Delaware, 1975).

Lloyd, Christopher (ed.), *The Keith Papers*, 2 vols (London, 1950).

Molyneux, Thomas More, *Conjunct Expeditions, or, Expeditions that have been carried on jointly by the fleet and army: with a comment on littoral war* (London, 1759).

Rodger, N.A.M. (ed.), *Memoirs of a Seafaring Life: The Narrative of William Spavens* (Somerset, 2000).

Rose, George, *Observations respecting the Public Expenditure and Influence of the Crown* (London, 1810).

Ryan, A.N. (ed.), *The Saumarez Papers: The Baltic 1808–1812*, Navy Records Society 110 (1968).

Sinclair, Sir John, *History of the Public Revenues of the British Empire* (London, 1798).

Sinclair, Sir John, *Thoughts on the Naval Strength of the British Empire* (London, 1795).

Stockdale, J.J. (ed.), *Tracts of Naval Architecture: Substance of the speech of Viscount Lord Melville in the House of Peers, Monday 21 May 1810 on the subject of Troop Ships* (London, 1810).

Tomkinson, James (ed.), The *Diary of a Cavalry Officer in the Peninsular War and Waterloo Campaign, 1809–1815: Lieutenant-Colonel Tomkinson, 16th Light Dragoons* (London, 1971).

Tracy, Nicholas, *The Naval Chronicle: The Contemporary Record of the Royal Navy at War*, vols 4, 5 (London, 1999).

Vane, Charles William, Marquess of Londonderry, *Correspondence of Viscount Castlereagh*, vols I–VIII, X (London, 1851, reprinted in 2006).

War Office Intelligence Unit, *British Minor Expeditions: 1746 to 1814* (London, 1884).

Secondary Works

Abbott, Charles, *A Treatise of the law relating to merchant ships and seamen* (London, 1802).

Allen, J., *The Navigation Laws of Great Britain, Historically and Practically Considered* (London, 1848).

Arthur, Brian, *How Britain Won the War of 1812: The Royal Navy's Blockade of the United States, 1812–1815* (Woodbridge, 2011).

Aspinall, A. (ed.), *The Later Correspondence of George III* (Cambridge, 1966–67).

Bamford, Andrew, *A Bold and Ambitious Enterprise: The British Army in the Low Countries, 1813–1814* (Barnsley, 2013).

Bamford, Andrew, *Sickness, Suffering and the Sword: The British Regiment on Campaign, 1808–1815* (Norman, OK, 2013)

Bannerman, Gordon, *Merchants and the Military Establishments in Eighteenth-Century Britain* (London, 2008).

Barney, John, 'North Sea and Baltic Convoy 1793–1814: As Experienced by Merchant Masters Employed by Michael Henley & Son', *Mariner's Mirror* 95.4 (November 2009), 429–40.

Barrow, Tony, 'Corn, Carriers and Coastal Shipping: The Shipping and Trade of Berwick and the Borders, 1730–1830', *Journal of Transport History* 21 (2000), 6–27.

Bayley, George, *Tables Shewing the Progress of the Shipping Interest of the British Empire, United States and France* (London, 1844).

Batson, Charles, *The Convict Ships 1787–1868* (Glasgow, 1959).

Binney, J.E.D., *British Public Finance and Administration, 1774–92* (Oxford, 1958).

Blake, George, *Lloyds Register of Shipping 1760–1960* (London, 1958).

Bond, Gordon C., *The Grand Expedition: The British Invasion of Holland in 1809* (Athens, GA, 1979).

Brewer, John, *Sinews of Power* (London, 1989).

Bunbury, Henry, *Narratives of Some Passages in the Great War with France from 1799–1810* (London, 1854).

Burnham, Robert, 'British Bridging Operations in the Peninsula', in Rory Muir, Robert Burnham, Howie Muir, Ron McGuigan (eds), *Inside Wellington's Peninsular Army 1808–1814* (Barnsley, 2006).

Chalmers, George, *The State of the United Kingdom at the Peace of Paris, November 20, 1815* (London, 1816).

Cole, Gareth, *Arming the Royal Navy, 1793–1815: The Office of Ordnance and the State* (London, 2012).

Condon, M.E., 'The Establishment of the Transport Board – A sub-division of the Admiralty – 4 July 1794', *Mariner's Mirror* 58 (1972), 69–84.

Creswell, J., 'British Shipping at the End of the Eighteenth Century', *Mariner's Mirror* 25 (2 Apr. 1939), 197–207.

Crimmin, P.K., 'Admiralty Relations with the Treasury 1783–1806: The Preparation of Naval Estimates and the Beginnings of Treasury Control', *Mariner's Mirror* 53 (1967), 63–72.

Crouzet, François, 'The Continental System after Eighty Years', in Ronald Findlay (ed.), *Eli Hecksher, International Trade and Economic History* (Cambridge, MA, 2006), 323–46.

Crouzet, François, 'Great Britain's Response to the French Revolution and to Napoleon', in Crouzet (ed.), *Britain Ascendant Comparative Studies in Franco-British Economic History* (Cambridge, 1990), 267–85.

Crouzet, François, 'Wars Blockade and Economic Change in Europe 1792–1815', *Journal of Economic History* 24(1964), 567–88.

Crowhurst, Patrick, *The French War on Trade: Privateering 1793 –1815* (Aldershot, 1989).

Currie, Ann, 'Henleys of Wapping: A London Shipowning Family 1770–1830', *Maritime Monograph Report* 62 (1988).

Dancy, J. Ross, *The Myth of the Press Gang: Volunteers, Impressment and the Naval Manpower Problem in the Late Eighteenth Century* (Woodbridge, 2015).

Davey, James, 'The Repatriation of Spanish Troops from Denmark, 1808: The British Government, Logistics, and Maritime Supremacy', *Journal of Military History* 74 (July, 2010), 689–707.

Davey, James, *The Transformation of British Naval Strategy, Seapower and Supply in Northern Europe, 1808–1812* (Woodbridge, 2012).
Davies, Glyn, *A History of Money* (Cardiff, 2002).
Davis, Ralph, *The Industrial Revolution and British Overseas Trade* (Leicester, 1979).
Davis, Ralph, *The Rise of the English Shipping Industry in the 17th and 18th Centuries* (Newton Abbott, 1962).
Davis, Lance E., *Naval Blockades in Peace and War* (New York, 2006).
Doe, Helen, and Harding Richard (eds.), *Naval Leadership and Management 1650–1950: Essays in Honour of Michael Duffy* (Woodbridge, 2012).
Duffy, Michael, 'Festering the Spanish Ulcer: The Royal Navy and the Peninsula War 1806–1814', in Bruce A. Elleman and S.C.M. Paine (eds.), *Naval Power and Expeditionary Warfare: Peripheral Campaigns and New Theatres of Naval Warfare* (New York, 2012).
Duffy, Michael, *Parameters of British Naval Power 1650–1850* (Exeter, 1992).
Duffy, Michael, 'Samuel Hood, First Viscount Hood', in Peter Le Fevre and Richard Harding (eds.), *Precursors of Nelson: British Admirals of the Eighteenth Century* (London, 2000).
Duffy, Michael, 'Science and Labour: The Naval Contribution to Operations Ashore in the Great Wars with France, 1793 to 1815', in Peter Hore (ed.), *Seapower Ashore: 200 Years of Royal Navy Operations on Land* (London, 2001), 39–52.
Duffy, Michael, *Soldiers, Sugar and Seapower: The British Expeditions to the West Indies and the War against Revolutionary France* (Oxford, 1987).
Earle, Peter, *Sailors: English Merchant Seamen 1650–1775* (London, 2007).
Eaton, Rev. A.W.H., *The Cochran-Inglis Family of Halifax Nova Scotia* (Halifax, NS, 1899).
Elleman, Bruce A., and Paine, S.C.M. (eds), *Naval Power and Expeditionary Warfare: Peripheral Campaigns and New Theatres of Naval Warfare* (New York, 2012).
Fayle, Ernest, 'The Employment of British Shipping', in Northcote Parkinson (ed.), *Trade Winds: A Study of British Overseas Trade during the French Wars 1793–1815* (Leicester, 1948).
Fedorak, Charles John, 'The Royal Navy and British Amphibious Operations during the Revolutionary and Napoleonic Wars', *Military Affairs* 52.3 (Jul. 1988), 141–6.
Fisher, Lewis D. (ed.), *Research in Maritime History*, Vol.7: *The Market for Seamen in the Age of Sail* (London, 1994).
Flayhart, William Henry, *Counterpoint to Trafalgar: The Anglo-Russian Invasion of Naples, 1805–1806* (Columbia, 1992).
Fletcher, Ian, *'A Desperate Business': Wellington, The British Army and the Waterloo Campaign* (Staplehurst, Kent, 2001).
Furse, Lieut-Col. G. A., *Military Transport* (London, 1882).
Gardiner, Robert, *Frigates of the Napoleonic War* (London, 2000).
Gates, David, *The Spanish Ulcer: A History of the Peninsular War* (Cambridge, MA, 1986).
Gibson, John Fredrick, *Brocklebanks 1770–1950*, 2 vols (Liverpool, 1953).
Glover, Michael, *Wellington as Military Commander* (London, 1968).
Glover, Richard, *Britain at Bay, Defence against Bonaparte 1803 –1814* (New York, 1973).
Glover, Richard, *Peninsular Preparation: The Reform of the British Army 1795–1809* (Cambridge, 1963).
Hall, Christopher D., *British Strategy in the Napoleonic War 1803–1815* (Manchester, 1992).
Hall, Christopher D., *Wellington's Navy: Sea Power and the Peninsular War 1807–1814* (London, 2004).
Harding, Richard, 'Expeditionary Armies and Naval Power: The North German Campaign of 1805–6', in Helen Doe and Richard Harding (eds), *Naval Leadership and Management 1650–1950: Essays in Honour of Michael Duffy* (Woodbridge, 2012).

Harper, Lawrence A., *The English Navigation Laws: A Seventeenth-Century Experiment in Social Engineering* (New York, 1939).

Haythornthwaite, Philip J., *The Armies of Wellington* (London, 1998).

Haythornthwaite, Philip J., *Napoleon's Military Machine* (Staplehurst, Kent, 1995).

Hickey, Donald R., *The War of 1812* (Chicago, 2012).

Hill, Richard, *The Prizes of War* (Stroud, 1998).

Hinde, Wendy, *George Canning* (London, 1973).

Hope, Ronald, *A New History of British Shipping* (London, 1990).

Hore, Peter (ed.), *Seapower Ashore: 200 Years of Royal Navy Operations on Land* (London, 2001)

Imlah, A.H., *Economic Elements in Pax Britannica: Studies in British Foreign Trade in the Nineteenth Century* (Cambridge, MA, 1958), 27–42.

Imlah, A.H., 'Real Values of British Foreign Trade 1798–1853', *Journal of Economic History* 8 (1948), 87–102.

James, W.M., *The Naval History of Great Britain*, vol. 2 (reprinted London, 2001).

Keys, Dick and Smith, Ken, *Black Diamonds by Sea, North East Sailing Colliers 1708–1880* (Newcastle, 1998).

Kilmarx, Robert (ed.), *America's Maritime Legacy: A History of the US Merchant Marine and Shipbuilding Industry since Colonial Times* (Colorado, 1979).

Knight, Roger, *Britain Against Napoleon: The Organization of Victory 1793–1815* (London, 2013).

Knight, Roger, 'British Strategy at Sea in the Great Wars with France', Oxford Naval Historians' Conference, April 2014.

Knight, Roger, 'The First Fleet – Its State and Preparation 1786–1787', in J. Hardy and A. Frost (eds), *Studies from Terra Australis to Australia* (Canberra, 1989), 121–36.

Knight, Roger 'The Introduction of Copper Sheathing into the Royal Navy, 1779–1786', *Mariner's Mirror* 59 (1973), 299–309.

Knight, Roger, 'Politics and Trust in Victualling the Navy 1793–1815', *Mariner's Mirror* 94 (2008), 133–49.

Knight, Roger, *The Pursuit of Victory: The Life and Achievements of Horatio Nelson* (London, 2005).

Knight, Roger, and Wilcox, Martin, *Sustaining the Fleet 1793–1815: War, the British Navy and the Contractor State* (Woodbridge, 2010).

Lambert, Andrew, *The Challenge: Britain Against America in the Naval War of 1812* (London, 2012).

Lavery, Brian, *Nelson's Navy: The Ships, Men and Organisation* (London, 1989).

Lindsey, W.S., *History of Merchant Shipping and Ancient Commerce*, 4 vols (London, 1833).

MacDonagh, O., *The Inspector General: Sir Jeremiah Fitzpatrick and the Politics of Social Reform 1783–1802* (London, 1981).

Mackesy, Piers, *British Victory in Egypt: The End of Napoleon's Conquests* (London, 1995).

Mackesy, Piers. 'Problems of an Amphibious Power: Britain against France 1793–1815', in Merril L. Bartlett (ed.), *Assault from the Sea: Essays on the History of Amphibious Warfare* (Annapolis, 1983), 60–8.

Mackesy, Piers, *The War in the Mediterranean 1803–1810* (Cambridge, MA, 1957).

Mahan, Captain A.T., *The Influence of Seapower upon the French Revolution and the Empire* (Boston, 1892)

Middleton, Charles, *The Administration of British Foreign Policy 1782 –1846* (Durham, NC, 1977).

Mitchell, B.R. and Deane, Phyllis, *British Historical Statistics* (Cambridge, 1962).

Morgan, E. Victor, *The Study of Prices and the Value of Money* (London, 1950).

Morriss, Roger, 'Colonization, Conquest and the Supply of Food and Transport: The Reorganization of Logistics Management 1780–1795', *War in History* 14 (2007), 310–24.
Morriss, Roger, *The Foundations of British Maritime Ascendancy: Resources, Logistics and the State, 1755– 1815* (Cambridge, 2011).
Morriss, Roger, 'High Exertions and Difficult Cases: The Work of the Transport Agents, 1795–1797', in Helen Doe and Richard Harding (eds), *Naval Leadership and Management 1650–1950: Essays in Honour of Michael Duffy* (Woodbridge, 2012), 95–107.
Morriss, Roger, *Naval Power and British Culture 1760–1850: Public Trust and Government Ideology* (Aldershot, 2004).
Moyse-Bartlett, H., A *History of the Merchant Navy* (London, 1937).
Muir, Rory, *Britain and the Defeat of Napoleon, 1807–1815* (New Haven, CT, 1996).
Muir, Rory, Burnham, Robert, Muir, Howie, McGuigan, Ron, *Inside Wellington's Peninsular Army* (Barnsley, 2006).
Munch-Petersen, Thomas, *Defying Napoleon: How Britain Bombarded Copenhagen and Seized the Danish Fleet in 1807* (Stroud, 2007).
O'Byrne, Robert, *The Victories of the British Army in the Peninsula and the South of France, from 1808 to 1814* (London, 1889).
Oman, Sir Charles, *A History of the Peninsular War*, 7 vols (Oxford, 1902–30).
Palmer, Sarah, *Politics, Shipping and the Repeal of the Navigation Laws* (Manchester, 1990).
Palmer, Sarah, 'Shipbuilding in Southeast England 1800 –1913', in Simon P. Ville (ed.), *Shipbuilding in the United Kingdom in the Nineteenth Century: A Regional Approach* (St John's, Newfoundland, 1993), 45–74.
Palmer, Sarah, and Williams, David M., 'British Sailors, 1775–1870', in Paul Van-Royan, Bruijn Jaap and Jan Lucassen (eds), *Those Emblems of Hell* (St John's, Newfoundland, 1997), 93–118.
Parkinson, Northcote (ed.), *The Trade Winds: A Study of British Overseas Trade during the French Wars 1793–1815* (Leicester, 1948).
Pimlott, John, *British Light Cavalry* (London, 1977).
Pocock, Tom, 'Smith and the Siege of Acre 1799: Sea Power Decisive on Land', in Peter Hore (ed.), *Seapower Ashore: 200 Years of Royal Navy Operations on Land* (London, 2001), 26–38.
Robinson, D.E., 'The Secret of British Power in the Age of Sail: Admiralty Reports on the Coasting Fleet' *American Neptune* 48 (1987), 5–21.
Rodger, N.A.M., *The Command of the Ocean: A Naval History of Britain 1649–1815* (London, 2004).
Rodger, N.A.M., *The Wooden World* (London, 1988).
Rogers, Hugh, *Troopships and their History* (London, 1963).
Sayer, Arthur, Rostow, W.W., Jacobson, J., and Schwartz, Anna, *The Growth and Fluctuation of the British Economy 1790–1850* (Oxford, 1953).
Schuyler, Robert Livingstone, *The Fall of the Old Colonial System: A Study in British Free Trade* (Oxford, 1945).
Shaumann, August Ludolf Friedrich, *On the Road with Wellington* (London, 1999).
Slaven, A., 'The Shipbuilding Industry', in Roy Church (ed.), *The Dynamics of Victorian Business: Problems and Perspectives to the 1870s* (London, 1980), 107–26.
Starkey, David, *British Privateering Enterprise in the Eighteenth Century* (Exeter, 1990).
Starkey, David (ed.), *Pirates and Privateers: New Perspectives on the War on Trade in the Eighteenth and Nineteenth Centuries* (Exeter, 1997).
Stephenson, Charles, *The Admiral's Secret Weapon* (Woodbridge, 2006).
Syrett, David, *Shipping and the American War 1775–83: A Study of British Transport Organization* (London, 1970).

Syrett, David, *Shipping and Military Power in the Seven Years War: The Sails of Victory* (Exeter, 2008).

Syrett, D., and DiNardo, R.L. (eds.), *Commissioned Sea Officers of the Royal Navy, 1660–1815*, Navy Records Society Occasional Publications No. 1 (Aldershot, 1994).

Talbott, John, *The Pen and Ink Sailor: Charles Middleton and the Kings Navy, 1778–1813* (London, 1998).

Thomas, Donald, *Cochrane: Britannia's Sea Wolf* (London, 1978).

Torrance, J., 'Social Class and Bureaucratic Innovations: The Commission for Examining the Public Accounts 1780 – 1787', *Past and Present* 10 (1978), 56–81.

Tracy, Nicholas, *Attack on Maritime Trade* (Toronto, 1991).

van Royan, Paul, Jaap, Bruijn, and Lucassen, Jan (eds.), *'Those Emblems of Hell'? European Sailors and the Maritime Labour Market 1570 –1870* (St Johns, Newfoundland, 1997).

Veve, Thomas Dwight, *The Duke of Wellington and the British Army of Occupation in France, 1815–1818* (Westport, CT, 1992).

Ville, Simon P., 'Defending Productivity Growth in the English Shipping Industry in the Eighteenth and Nineteenth Centuries' *Economic History Review* 2nd ser. 40 (1987), 597–602.

Ville, Simon P., *English Shipowning during the Industrial Revolution: Michael Henley & Son, London Shipowners, 1770–1830* (Manchester, 1987).

Ville, Simon P., 'The Incidence of the Loss of Merchant Vessels during the French Wars', *Mariner's Mirror* 68.4 (1982), 450.

Ville, Simon P. (ed.), *Shipbuilding in the United Kingdom in the Nineteenth Century: A Regional Approach* (Newfoundland, 1993).

Ville, Simon P., 'The Size and Profitability of English Collier Vessels in the Eighteenth Century: A Reappraisal', *Business History Review* 58 (1984), 102–30.

Ville, Simon P., 'Wages, Prices and Profitability in the Shipping Industry during the Napoleonic Wars', in J. Armstrong (ed.), *Coastal and Short Sea Shipping* (Aldershot, 1997).

Walton, Gary, 'The New Economic History and the Burden of the Navigation Acts', *Economic History Review* 24.4 (1971).

Weller, J.A.G., *Wellington in the Peninsula 1808 -1814* (London, 1962).

Winfield, R., *British Warships in the Age of Sail 1793–1817* (Chatham, 2005).

Woodman, Richard, *A History of the British Merchant Navy*, Vol. 2: *Britannia's Realm: In Support of the State: 1763–1815* (Stroud, 2009).

Woodman, Richard, *The Sea Warriors* (London, 2001).

Unpublished Theses

Condon, Mary Ellen, 'The Administration of the Transport Service during the War against Revolutionary France 1793–1802' (University of London Ph.D. thesis, 1968).

Sutcliffe, Robert K., 'Bringing Forward Shipping for Government Service: The Indispensable Role of the Transport Service, 1793 to 1815' (University of Greenwich Ph.D. thesis, 2013).

Watson, Paula K, 'The Commission for Victualling the Navy, the Commission for Sick and Wounded Seamen and Prisoners of War, the Commission for Transports 1702–1714' (University of London Ph.D. thesis, 1965).

Index

Printed by Libri Plureos GmbH in Hamburg,
Germany

9 781843 839491